Eliezer Gileadi

Physical Electrochemistry

Related Titles

Izutsu, K.

Electrochemistry in Nonaqueous Solutions

2nd Ed.

2009

ISBN: 978-3-527-32390-6

Bard, A. J., *et al.* (eds.)

Encyclopedia of Electrochemistry

11 Volume Set

2007

ISBN: 978-3-527-30250-5

Hamann, C. H., Hamnett, A., Vielstich, W.

Electrochemistry

2007

ISBN: 978-3-527-31069-2

From the Series: **Advances in Electrochemical Sciences and Engineering**

Vol. 12

Alkire, A. C., Kolb, D. M., Lipkowski, J., Ross, P. N. (eds)

Photoelectrochemical Materials and Energy Conversion Processes

2011

ISBN: 978-3-527-32859-8

Vol. 11

Alkire, A. C., Kolb, D. M., Lipkowski, J., Ross, P. N. (eds)

Chemically Modified Electrodes

2009

ISBN: 978-3-527-31420-1

Vol. 10

Alkire, A. C., Kolb, D. M., Lipkowski, J., Ross, P. N. (eds)

Electrochemical Surface Modification

Thin Films, Functionalization and Characterization

2008

ISBN: 978-3-527-31317-4

Vol. 9

Alkire, A. C., Kolb, D. M., Lipkowski, J., Ross, P. N. (eds)

Diffraction and Spectroscopic Methods in Electrochemistry

2006

ISBN: 978-3-527-31317-4

Paunovic, M., Schlesinger, M.

Fundamentals of Electrochemical Deposition

2006

ISBN: 978-0-471-71221-3

Ohno, H. (ed.)

Electrochemical Aspects of Ionic Liquids

2005

ISBN: 978-0-471-64851-2

Eliezer Gileadi

Physical Electrochemistry

Fundamentals, Techniques and Applications

WILEY-VCH Verlag GmbH & Co. KGaA

The Author

Prof. Dr. Eliezer Gileadi
School of Chemistry
Tel Aviv University
69978 Tel Aviv
Israel

Library of Congress Card No.: applied for

British Library Cataloguing-in-Publication Data
A catalogue record for this book is available from the British Library.

Bibliographic information published by the Deutsche Nationalbibliothek
The Deutsche Nationalbibliothek lists this publication in the Deutsche Nationalbibliografie; detailed bibliographic data are available on the Internet at http://dnb.d-nb.de.

© 2011 WILEY-VCH Verlag GmbH & Co. KGaA, Weinheim

Typesetting   Thomson Digital, Noida, India
Printing and Binding   Strauss GmbH, Mörlenbach
Cover Design   Formgeber, Eppelheim

Printed in the UK

Printed on acid-free paper

ISBN: 978-3-527-31970-1

Contents

Physical Electrochemistry: Fundamentals, Techniques and Applications. Eliezer Gileadi
Copyright © 2011 WILEY-VCH Verlag GmbH & Co. KGaA, Weinheim
ISBN: 978-3-527-31970-1

Preface

Interfacial electrochemistry is a multi-disciplinary subject. Its core is the study of the mechanism of electron transfer and the theory of the double layer capacitance at the metal/solution interface, but it is related to many aspects of fundamental chemistry, biology and engineering, as well as to applied science. There are many applications of electrochemistry in industry, such as metal winning of Al and Mg, the production of chlorine, electroforming, electromachining and electropolishing, electro-organic synthesis, biosensors, corrosion protection, electroplating, batteries and fuel cells.

In spite of its importance, interfacial electrochemistry is rarely, if ever, included in the undergraduate curriculum of chemistry or chemical engineering in universities around the world. There are several texts that can be used for a graduate course in the field, but none that could be classified as a textbook, upon which such a course could be based.

The purpose of the present book is to satisfy this need. The book starts by covering the basic subjects of interfacial electrochemistry. This is followed by a description of some of the most important techniques (such as cyclic voltammetry, the rotating disc electrode, electrochemical impedance spectroscopy, and the electrochemical quartz-crystal microbalance). Finally, there is a rather detailed discussion of electroplating (including alloy deposition), corrosion, and electrochemical energy conversion devices (batteries, fuel cells and super-capacitors).

Admittedly, the book contains more than could be taught in one semester. This is deliberate, in order to allow some choice for the teacher to concentrate on aspects of the field that fit best the needs of the particular class. However, even covering one half or two thirds of the material in this book should provide students with some understanding of interfacial electrochemistry, the techniques applied, and at least one of the technologies of particular interest for them, and facilitate the learning and understanding of some specific subject which may be needed later in his or her professional life.

This book is also recommended as a text suitable for self-learning, which could be used to introduce scientists and engineers, who have not had an opportunity to participate in a formal course on interfacial electrochemistry, to aspects of this field needed for their research and development.

Physical Electrochemistry: Fundamentals, Techniques and Applications. Eliezer Gileadi
Copyright © 2011 WILEY-VCH Verlag GmbH & Co. KGaA, Weinheim
ISBN: 978-3-527-31970-1

There are different criteria by which the quality of a textbook could be judged. From my own point of view, the success or failure of this book will be judged by its ability to enhance and spread the teaching of interfacial electrochemistry, and establish it as the basis for graduate courses offered widely in universities around the world.

Acknowledgements

The author wishes to thank Professor E- Kirowa-Eisner for uncountable scientific discussions, sincere criticism and advice over the past forty years of scientific cooperation. Thanks also to Professor Y. Ein-Eli and Professor D. Golodnitsky for advice and criticism, helping to improving Chapter 20. I thank Mrs. D. Tzur for preparing the figures and Mrs. J. MacDougall for proofreading the manuscript.

Dedication

Dedicated to my wife Dalia, for her love, continued support, encouragement and patience. Without her, my scientific career would not be what it is and this book would not have been written.

Abbreviations

AC	alternating current
ASV	anodic stripping voltammetry
CE	counter electrode
CV	cyclic voltammetry
CVD	chemical vapour deposition
DC	direct current
DME	dropping mercury electrode
DMFC	direct methanol fuel cell
DP	differential pulse (polarography)
EC	ethylene carbonate
EDS	electron dispersive spectroscopy
EIS	electrochemical impedance spectroscopy
EQCM	electrochemical quartz-crystal microbalance
ETE	electric-to-electric
FE	Faradaic efficiency
HER	hydrogen-evolution reaction
HTSO	high-temperature solid-oxide (fuel cell)
ICE	internal combustion engine
IHP	inner Helmholtz plane
LPSV	linear potential sweep voltammetry
LSV	linear sweep voltammetry
MEMS	micro-electrical mechanical systems
mpy	thousandth of an inch per year (corrosion rate)
NP	normal pulse (polarography)
OHP	outer Helmholtz plane
PAFC	phosphoric acid fuel cell
PC	propylene carbonate
PEM	polymer electrolyte membrane
PVD	physical vapour deposition
PZC	potential of zero charge (E_2)
RConeE	rotating cone electrode
RCylE	rotating cylinder electrode

Physical Electrochemistry: Fundamentals, Techniques and Applications. Eliezer Gileadi
Copyright © 2011 WILEY-VCH Verlag GmbH & Co. KGaA, Weinheim
ISBN: 978-3-527-31970-1

RDE	rotating disc electrode
rds	rate-determining step
RE	reference electrode
RHE	reversible hydrogen electrode
RRDE	rotating ring–disc electrode
RT	room temperature
SASV	subtractive anodic stripping voltammetry
SDME	static dropping mercury electrode
SEI	solid electrolyte interface
SHE	standard hydrogen electrode
SW	square-wave polarography
TP	throwing power
UPD	underpotential deposition
WE	working electrode
XPS	X-ray photoelectron spectroscopy

Symbols

a	activity	$mol.cm^{-3}$
A	affinity	$J\,mol^{-1}$
A	surface area	cm^2
b	Tafel slope	$V\,decad^{-1}$
c_b	bulk concentration	$mol\,cm^{-3}$
c_s	concentration at the surface (x=0)	$mol\,cm^{-3}$
C_{dl}	double-layer capacitance	$\mu F\,cm^{-2}$
C_ϕ	adsorption pseudo-capacitance	$\mu F\,cm^{-2}$
C_L	adsorption pseudo-capacitance (based on the Langmuir isotherm)	$\mu F\,cm^{-2}$
C_F	adsorption pseudo-capacitance (based on the Frumkin isotherm)	$\mu F\,cm^{-2}$
C_H (C_{M-S})	Helmhotz double-layer capacitance	$\mu F\,cm^{-2}$
$C_{G,C}$ (C_{2-S})	diffuse double-layer capacitance	$\mu F\,cm^{-2}$
d	Distance between the tip of the Luggin capillary and the working electrode	cm
D	diffusion coefficient	$cm^2\,s^{-1}$
E	potential	Volt
$\bar{E}$	rational potential $(E-E_z)$	Volt
$E_{1/2}$	half-wave potential (in polarography)	Volt
E_b	breakdown potential (of passivity)	Volt
E_{pp}	primary passivation potential	Volt
E_{rev}	reversible potential	Volt
E_{rp}	re-passivation potential	Volt
E_z	potential of zero charge	Volt
E_-/E_+	potential with respect to a reference electrode that is reversible with respect to the anion/cation in solution	Volt
ΔE_{UPD}	difference between the reversible and the peak potentials during UPD formation	Volt
$\Delta E_{1/2}$	the width of the peak at half height (UPD)	Volt

Physical Electrochemistry: Fundamentals, Techniques and Applications. Eliezer Gileadi
Copyright © 2011 WILEY-VCH Verlag GmbH & Co. KGaA, Weinheim
ISBN: 978-3-527-31970-1

E^0	standard potential	Volt
E_0^0	the standard potential for UPD formation	Volt
F	Faradays constant (96,485)	Coulomb
$\vec{F}$	electrostatic field	$V\,m^{-1}$
f	the Frumkin lateral interaction parameter	dimensionless
f_0	resonance frequency of a quartz-crystal microbalance	Hertz
Δf	change of the resonance frequency of a quartz-crystal microbalance	sec^{-1}
G	Gibbs energy	$J\,mol^{-1}$
ΔG	change of the Gibbs energy (in a reaction)	$J\,mol^{-1}$
$\Delta \bar{G}$	change of the electrochemical Gibbs energy	$J\,mol^{-1}$
ΔG_{ads}	Gibbs energy of adsorption	$J\,mol^{-1}$
$\Delta G^{0\#}$	standard Gibbs energy of activation	$J\,mol^{-1}$
$\Delta \bar{G}^{0\#}$	standard electrochemical Gibbs energy of activation	$J\,mol^{-1}$
ΔG_{hydr}	Gibbs energy of hydration (of an ion)	$J\,mol^{-1}$
$\Delta \bar{H}^{0\#}$	standard electrochemical enthalpy of activation	$J\,mol^{-1}$
h	Plank constant 6.6261×10^{-34}	$J\,s$
h	roughness parameter	μm
I	current	Amper
ImZ/ReZ	imaginary/real part of the double layer impedance	Ohm
j	current density	$A\,cm^{-2}$
j_{ac}	activation-controlled current density	$A\,cm^{-2}$
j_0	exchange current density	$A\,cm^{-2}$
j_L	limiting current density	$A\,cm^{-2}$
k	rate constant	$mol\,cm^3\,s^{-1}$
$k_{s,h}$	standard heterogeneous rate constant	$cm\,s^{-1}$
k_B	Boltzmann constant 1.38065×10^{-23}	$J\,degree^{-1}$
k_f/k_b	forward/backward rate constant (heterogeneous)	$cm\,s^{-1}$
K	equilibrium constant	dimensionless
M	molecular weight	$gram\,mol^{-1}$
N	rotation rate	rpm
Na	Avogadro number 6.022×10^{23}	mol^{-1}
n	number of electrons transferred per molecule	dimensionless
nF	charge transferred per mol	$Coulomb\,mol^{-1}$
P	pressure	Atm., Pascal
Q	charge	Coulomb
q_F	Faradaic charge	Coulomb
q_1	surface charge density for $\theta = 1$	$Coulomb\,cm^{-2}$
q_M	excess surface charge density	$Coulomb\,cm^{-2}$
R_{ct}	charge-transfer resistance	$Ohm\,cm^2$
ReZ/ImZ	real/imaginary part of the double layer impedance	$Ohm\,cm^2$

R_F	Faradaic resistance	$Ohm\,cm^2$
R_p	polarization resistance (in corrosion)	$Ohm\,cm^2$
R_S	uncompensated solution resistance	$Ohm\,cm^2$
Re	Reynolds number	dimensionless
r	rate of change of the Gibbs energy of adsorption with coverage	$J\,mol^{-1}$
TP	throwing power	dimensionless
t	time	sec.
U_{hydr}	hydration energy of ions	$J\,mol^{-1}$
v	velocity	$m\,s^{-1}$
v	heterogeneous reaction rate	$mol\,s^{-1}\,cm^{-2}$
v	rate of potential sweep	$V\,s^{-1}$
v_{equi}	exchange rate (at equilibrium)	$mol\,s^{-1}\,cm^{-2}$
W	Warburg impedance (diffusion)	$Ohm\,cm^2$
Wa	Wagner number (in electroplating)	dimensionless
z	charge number, valency	dimensionless
Z	impedance	Ohm
α_{an}/α_c	anodic/cathodic transfer coefficient	dimensionless
β	symmetry factor	dimensionless
Γ	surface excess, surface concentration	$mol\,cm^{-2}$
Γ	imaginary part of the response of the EQCM	$Hertz\,(s^{-1})$
Γ'	relative surface excess	$mol\,cm^{-2}$
Γ_{max}	maximum surface excess, (full coverage)	$mol\,cm^{-2}$
γ	activity coefficient	dimensionless
γ	surface tension/excess surface Gibbs energy	$N\,m^{-1}/J\,m^{-2}$
δ	Nernst diffusion-layer thickness	μm
δ	thickness of the Helmholtz double layer	nm
ε_0	permittivity of free space 8.8542×10^{-12}	$Coul^2\,N^{-1}\,m^{-2}$
ε	dielectric constant	dimensionless
η	viscosity	cP
η	overpotential	Volt
$\eta_{ac}/\eta_{conc}/\eta_R$	activation/concentration/resistance overpotential	Volt
θ	fractional surface coverage (Γ/Γ_{max})	dimensionless
Θ	ratio of surface concentrations $[c_{Ox}(0;t)/c_{Red}(0;t)]$	dimensionless
κ	specific conductivity	$S\,cm^{-1}$
λ	solvent reorganization energy	$J\,mol^{-1}$
λ	dimensionless rate constant ($\lambda \equiv k_h\sqrt{t/D}$)	dimensionless
μ	chemical potential	$J\,mol^{-1}$
$\bar{\mu}$	electrochemical potential	$J\,mol^{-1}$
$\vec{\mu}$	dipole moment	Debye
μ_q	shear modulus (of a quartz crystal)	G Pascal
σ	lateral interaction parameter (c.f. electrosorption)	dimensionless

τ_c	time constant for the parallel combination of a capacitor and a resistor ($C_{dl} \times R_F$)	sec
τ_d	time constant for a diffusion controlled process	sec
υ	kinematic viscosity (η/ρ)	$cm^2 s^{-1}$
Π	two-dimensional surface pressure	$N\,m^{-1}$
p_1/p_2	reaction order at constant potential/constant overpotential	dimensionless
ρ	density	$g\,cm^{-3}$
ρ	specific resistivity of electrolyte	Ohm cm
ϕ	inner potential of a phase	Volt
$^M\Delta^S\phi$	metal-solution potential difference	Volt
χ	dimensionless distance $\left(\chi = x/\sqrt{4Dt}\right)$	dimensionless
κ	reciprocal Debye length	cm^{-1}
ω	angular velocity ($\omega = 2\pi f$)	sec^{-1}
∇^2	the Laplace operator (del-square)	dimensionless

1
Introduction

1.1
General Considerations

1.1.1
The Current–Potential Relationship

From a phenomenological point of view, the study of electrode kinetics involves the determination of the dependence of current on potential. It is therefore appropriate to start this book with a general qualitative description of such a relationship, as shown in Figure 1.1.

In the simplest case, E is the potential applied between two electrodes in solution and j is the current density. Curve a in Figure 1.1 represents the dependence of the purely activation-controlled current density on potential. Curve b is the actual current density measured, taking into account the effects of mass transport, represented by the limiting current density j_L. These concepts are explained in the following sections. In real experiments the potential E is always measured with respect to a suitable reference electrode and, instead of the current, one refers to the current density on the electrode being studied, but at this point we need not concern ourselves with these refinements.

It is immediately obvious from Figure 1.1 that Ohm's law does not apply, not even as a rough approximation. This observation is not as trivial as it may seem, when we recall that in the study of conductivity of electrolytic solutions, Ohm's law is strictly obeyed over a very large range of potentials and frequencies. The difference is that Figure 1.1 pertains to measurements conducted under direct current (DC) conditions, whereas ionic conductivity is measured, as a rule, with an alternating current or potential. The implication is that the impedance of the metal/solution interface is partially capacitive – a subject to be dealt with in considerable detail below.

Physical Electrochemistry: Fundamentals, Techniques and Applications. Eliezer Gileadi
Copyright © 2011 WILEY-VCH Verlag GmbH & Co. KGaA, Weinheim
ISBN: 978-3-527-31970-1

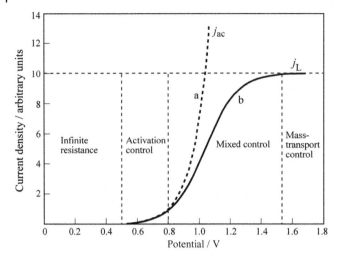

Figure 1.1 Schematic j/E plot for the electrolysis of a 1.0 M solution of KI in H_2SO_4, employing two Pt electrodes. The minimum potential for DC current flow is 0.56 V.

1.1.2
The Resistance of the Interface Can Be Infinite

Looking at Figure 1.1 carefully, one observes that up to a certain potential the current is zero. This is not a matter of limited sensitivity of the measuring instrument. The current is *exactly* zero (ignoring minor background currents that may be caused by impurities), corresponding effectively to an infinite resistance of the interface. The reason for this behavior is to be found in the realm of thermodynamics. Thus, when a current flows through a cell, electrical energy is used to produce chemicals. The reactions taking place are

$$\text{At the cathode} \qquad 2H_3O^+ + 2\,e_M^- \rightarrow H_2 + 2H_2O \qquad (1.1)$$

$$\text{At the anode} \qquad 2I^- \rightarrow I_2 + 2\,e_M^- \qquad (1.2)$$

$$\text{Overall reaction} \qquad 2H_3O^+ + 2I^- \rightarrow H_2 + I_2 + 2H_2O \qquad (1.3)$$

The reversible potential under standard conditions for this reaction is $E_{rev} = -0.56$ V, so this is the lowest potential that would allow passage of a current.

This is "up the Gibbs energy ladder", (hence the potential is given a negative sign). Evidently this reaction proceeds spontaneously in the opposite direction (forming 2HI), and, therefore, electrical energy must be supplied to make the reaction happen, according to the well-known relation:

$$\Delta G = -nFE \qquad (1.4)$$

The negative sign in this equation shows that when the Gibbs energy decreases, the potential is positive – the cell acts as a source of electrical energy, and vice versa.

Had we removed KI from the solution, the reaction taking place would be the electrolysis of water, forming molecular oxygen and hydrogen. The standard Gibbs energy for water electrolysis at room temperature is $+237.16 \, kJ \, mol^{-1}$, corresponding to $-1.229 \, V$. The positive value of ΔG^0 and the corresponding negative value of the standard potential E^0 show that this reaction will not occur spontaneously. A positive potential of at least $1.229 \, V$ has to be applied for it to occur. In this case the resistance of the cell shown in Figure 1.1 would be infinite up to this potential.

Replacing the platinum electrodes with copper, and adding some $CuSO_4$, changes the situation radically. Passing a current between the electrodes causes no net chemical change (copper is dissolved off the anode and deposited on the cathode). In this case current is observed as soon as a potential, small as it may be, is applied between the electrodes.

1.1.3
The Transition from Electronic to Ionic Conduction

If one were to describe the essence of electrode kinetics in one short phrase, it would be: the transition from electronic to ionic conduction, and the phenomena associated with, and controlling it. Conduction in the solution is ionic, whereas in the electrodes and the connecting wires it is electronic. The transition from one mode of conduction to the other requires charge transfer across the interfaces. This is a kinetic process. Its rate is controlled by the catalytic properties of the surface and adsorption on it, the concentration and the nature of the reacting species and all other parameters that control the rate of heterogeneous chemical reactions. In addition, the potential plays an important role. This is not surprising, since charge transfer is involved, which may be accelerated by applying a potential difference of the right polarity across the interface.

The current would continue to rise exponentially with potential, along line a in Figure 1.1, were it not for mass-transport limitation, represented by the horizontal part of line b. In the initial rising part of the curve, the reaction is said to be "charge-transfer controlled" or "activation controlled". The detailed dependence of current on potential in this region is discussed later.

1.1.4
Mass-Transport Limitation

The rate of charge transfer can be greatly increased by increasing the potential, but charge can be transferred across the interface only over a very short distance (of the order of 0.5 nm). Another process is required to bring the reacting species close enough to the surface, and to remove the species formed at the surface into the bulk of the solution. This process is called *mass transport*.

Mass transport and charge transfer are two consecutive processes. It is therefore always the slower of the two that determines the overall rate observed experimentally. When the potential applied is low, barely above the minimum value needed to pass a current, charge transfer is slow and one can ignore mass-transport limitation. The bottleneck is in transferring the charge across the interface to the electroactive species, not in getting the species to the surface. At high potentials, charge transfer becomes the faster process and ceases to influence the overall rate. Increasing the potential further will increase the rate of charge transfer, but this will have no effect on the observed current, which will be limited by mass transport. The result is a current density that is independent of potential, which is referred to as *the limiting current density*, j_L.

For the observed current density j, one can write the simple equation:

$$\frac{1}{j} = \frac{1}{j_{ac}} + \frac{1}{j_L}$$ (1.5)

Clearly, the smaller of the two currents is dominant. The mass-transport-limited current density can be written as

$$j_L = \frac{nFDc_b}{\delta}$$ (1.6)

in which nF is the charge transferred per mol ($C\,mol^{-1}$), D is the diffusion coefficient ($cm^2\,s^{-1}$), c_b is the bulk concentration ($mol\,cm^{-3}$), and δ is the Nernst-diffusion-layer thickness (cm). Calculated in these units, the current density is obtained in $A\,cm^{-2}$.

The corresponding equation for the activation controlled current density is

$$j_{ac} = nFkc_b$$ (1.7)

Where the rate constant k is a function of potential. From a comparison of the two equations it is seen that the ratio D/δ in Eq. (1.6) has the same role as the rate constant k in Eq. (1.7), except that it is independent of potential. This ratio may be regarded as the specific rate for diffusion.

Now, the essence of mass transport is the quantity δ. In certain favorable cases it has been calculated theoretically, in others it can only be determined experimentally. Sometimes it is a function of time, while under different circumstances it is essentially constant during an experiment. Stirring the solution and transporting it towards, past, or through the electrode all decrease the value of δ, and hence increase j_L. Moving the electrode (rotation, vibration) has a similar effect. In quiescent solutions δ increases linearly with $t^{1/2}$, hence j_L can be increased by taking measurements at short times.

In typical electrochemical measurements, the Nernst-diffusion-layer thickness attains values in the range 10^{-3}–$10^{-1}\,cm$. Since in aqueous solutions at room temperature the diffusion coefficient is of the order of $10^{-5}\,cm^2\,s^{-1}$, this yields limiting current densities in the range 0.01–$1.0\,mA\,cm^{-2}$, when $n = 1$ and the concentration of the electroactive species in solution is $1.0\,mM$.

The two most important things to notice in Eq. (1.6) are (i) that the limiting current density is independent of potential, and (ii) that it depends linearly on the bulk concentration. A less obvious, but equally important, consequence of this equation is that j_L is independent of the kinetics of the reaction (i.e., of the nature of the surface and its catalytic activity). These characteristics make it an ideal tool for probing the concentration of species in solution. This is why most electroanalytical methods depend, in one way or another, on measurements of the mass-transport-limited current density.

1.1.5
The Capacitance at the Metal/Solution Interface

When a metal is dipped in solution, a discontinuity is created. This affects both phases to some degree, so that their properties near the contact are somewhat different from their bulk properties. The exact position of the interface on the atomic scale is hard to define. "Where does the metal end?" we may ask. Is it the plane going through the center of the outermost layer of atoms, is it one atomic radius farther out, or is it even farther out, where the charge-density function of the free electrons in the metal has decayed to essentially zero? Fortunately, we do not need to know the position of this plane for most purposes, when we discuss the properties of the interface.

One distinct property of the metal/solution interface is a capacitance, called *the double-layer capacitance*, C_{dl}. It is a result of the charge separation between the two phases in contact. The double-layer capacitance observed depends on the structure of a very thin region near the interface, extending about 1–10 nm, called *the double layer*. If the surface is rough, the double layer will follow its curvature down to atomic dimensions, and the capacitance measured under suitably chosen conditions is proportional to the *real* surface area of the electrode.

The double-layer capacitance is rather large, of the order of $10–30\,\mu F\,cm^{-2}$. This presents a serious limitation on our ability to study fast electrode reactions. Thus, a $10\,\mu F$ capacitor coupled with a $10\,\Omega$ resistor yield a time constant of $R \times C_{dl} = \tau_C = 0.1\,ms$. It is possible to take measurements at shorter times by applying special techniques, but even so, the lower limit at present seems to be about $0.05\,\mu s$, six orders of magnitude slower than that currently achievable in the gas phase.

The double-layer capacitance depends on the potential, the composition of the solution, the solvent and the metal. It has been the subject of numerous investigations, some of which are discussed later.

1.2
Polarizable and Nonpolarizable Interfaces

1.2.1
Phenomenology

When a small current or potential is applied, the response is in many cases linear. The effective resistance can, however, vary over a wide range. When this resistance is high, we have a polarizable interface, meaning that a small current generates a high potential across it (i.e., the interface is polarized to a large extent).

When the effective resistance is low, the interface is said to be nonpolarizable. In this case a significant current can be passed with only minimal change in the potential across the interface. A nonpolarizable electrode is, in effect, a reversible electrode. The reversible potential is determined by the electrochemical reactions taking place and the composition of the solution, through the Nernst equation. For a copper electrode in a solution containing $CuSO_4$ this is

$$E_{rev} = E^0 + (2.3RT/nF)\log(a_{Cu^{2+}}) \tag{1.8}$$

in which $E^0 = +0.34\,V$ is the standard potential for the Cu^{2+}/Cu couple, on the standard hydrogen electrode (SHE) scale, and $a_{Cu^{2+}}$ is the activity of cupric ions in solution. In aqueous solutions that are not very concentrated, $(c_b \leq 1.0\,M)$ the error introduced by replacing the activity by the concentration is rather small, and often considered negligible.

A good reference electrode is always a reversible electrode. The inverse is not necessarily true. Not every reversible electrode is suitable as a reference electrode. For example, the correct thermodynamic reversible potential of a metal/metal-ion electrode may be hard to reproduce, because of impurities in the metal or complexing agents in the solution, even when the interface is highly nonpolarizable.

Polarizable interfaces behave differently. Their potential is not fixed by the solution composition, and it can be changed at will over a certain range, depending on the metal and the composition of the solution in contact with it. For such a system the potential may be viewed as an additional degree of freedom in the thermodynamic sense, as used in the Gibbs phase rule. To be sure, a so-called nonpolarizable interface can be polarized by passing a significant current through it. This, however, alters the concentration of both the reactant and the product *at the electrode surface* (without changing significantly their bulk concentrations). The potential developed across the interface will be in agreement with the Nernst equation, as long as the concentrations used are the surface concentrations, which depend on the current passing across the interface.

1.2.2
The Equivalent Circuit Representation

We have already seen that the metal/solution interface has some capacitance associated with it, as well as a (nonohmic) resistance. Also, the solution has a finite resistance that must be taken into account. Thus, a cell with two electrodes can be represented by the equivalent circuit shown in Figure 1.2.

Usually one considers only the part of the circuit inside the dashed line, since the experiment is set up in such a way that only one of the electrodes is studied at a time.

The equivalent circuit shown in Figure 1.2 represents a gross oversimplification, and interfaces rarely behave exactly like it. It does, nevertheless, help us gain some insight concerning the properties of the interface.

The combination of the double-layer capacitance C_{dl} and the Faradaic resistance R_F (also referred to as the charge-transfer resistance, R_{ct}) represents the interface. How do we know that C_{dl} and R_F must be put in a parallel rather than in a series combination? Simply because we can observe a steady direct current flowing when

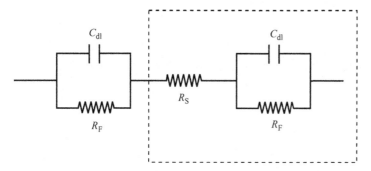

Figure 1.2 Equivalent circuit for a two-electrode cell. A single interface is usually represented by the elements inside the dashed rectangle. C_{dl}, R_F and R_S represent the double-layer capacitance, the Faradaic resistance and the solution resistance, respectively.

the potential is high enough (above the minimum prescribed by thermodynamics). Also when the resistance is effectively infinite under DC conditions, we can still have an AC signal going through.

The equivalent circuit just described also makes it clear why conductivity measurements are routinely done by applying a small AC signal. If the appropriate frequency is chosen, the capacitive impedance associated with C_{dl} can be made negligible compared to the Faradaic resistance R_F, which is thus effectively shorted, leaving the solution resistance R_S as the only measured quantity.

As pointed out earlier, the equivalent circuit shown in Figure 1.2 is meant to represent the simplest situation only. It does not take into account factors such as mass transport, heterogeneity of the surface and the occurrence of reaction intermediates absorbed on it. Some of these factors are discussed later. Even in the simplest cases, in which this circuit does represent the response of the interface to an electrical perturbation reasonably well, one should bear in mind that both C_{dl} and R_F depend on potential and, in fact, R_F depends on potential exponentially over a wide range, as will be discussed later.

The difference between polarizable and nonpolarizable interfaces can be easily understood in terms of this equivalent circuit. A high value of R_F is associated with a polarizable interface, whereas a low value of R_F represents a non-polarizable interface.

2
The Potentials of Phases

2.1
The Driving Force

2.1.1
Definition of the Electrochemical Potential

Knowledge of the driving force is of the utmost importance for the understanding of any system. It determines the direction in which a chemical reaction can proceed spontaneously, as well as its position of equilibrium, at which the driving force is zero along all coordinates.

The driving force in chemistry is the gradient of the chemical potential, μ.

$$\text{driving force} = -\text{grad}\,\mu \tag{2.1}$$

Let us turn our attention now to processes involving charged species, in particular charge-transfer processes. We recall that the chemical potential relates to the activity of the species:

$$\mu_i = \mu_i^0 + RT \ln a_i \tag{2.2}$$

The activity is related to the concentration via the activity coefficient γ_i, which is itself a function of concentration:

$$a_i = \gamma_i\, c_i \tag{2.3}$$

It would seem that μ_i, as given in Eq. (2.2), does not account for the effect of the electrical field or its gradient, unless we include that implicitly in the activity coefficient.

It has been found more expedient to define a new thermodynamic function, *the electrochemical potential*, $\bar{\mu}$, which includes a specific term to account for the effect of potential on a charged species:

$$\bar{\mu}_i = \mu_i + z_i F \phi \tag{2.4}$$

Physical Electrochemistry: Fundamentals, Techniques and Applications. Eliezer Gileadi
Copyright © 2011 WILEY-VCH Verlag GmbH & Co. KGaA, Weinheim
ISBN: 978-3-527-31970-1

where ϕ is the *inner potential* of a phase and z_i is the charge number on the *i*th species. When charged species are involved, the driving force is the gradient of electrochemical potential along some coordinate:

$$\text{driving force} = -\text{grad } \bar{\mu}_i \qquad (2.5)$$

It follows that the chemical potential is a special case of the electrochemical potential, applicable for an uncharged species.

2.1.2
Separability of the Chemical and the Electrical Terms

It was noted above that Eq. (2.4) is an attempt to separate chemical interactions, represented by μ_i, from electrical interactions, represented by the product of charge (per mol) $z_i F$ and the potential in the phase ϕ^α at a point x, y, z. But can such a separation be made?

One recalls that the potential $\phi^\alpha_{x,y,z}$ at some point (x, y, z) in space is defined as the energy required to bring a unit positive test charge from infinity to this point. This is fine as long as the charge is moved in free space or inside a homogeneous phase. But what happens when we try to determine the potential inside a phase, with respect to a point at infinity in free space, or the difference in potential between points in two different phases? As the "test charge" crosses the boundary of a phase, it interacts with the molecules in that phase, and it is impossible to distinguish between the so-called "chemical" and "electrical" interactions in this region.

We must conclude from the above considerations that, while the electrochemical potential $\bar{\mu}$ is a measurable quantity, its components μ and ϕ cannot be separately measured.

The potential ϕ^α in Eq. (2.4) is called the inner potential in a phase identified by the superscript. For the same reason that it cannot be measured, one cannot measure the value of $^\alpha\Delta^\beta\phi$, the difference between the inner potentials between two different phases. The foregoing statement may seem odd, since we are accustomed to measuring potential differences, say, the potential difference (i.e., the voltage) between two terminals of a battery. However, to do this we connect the two terminals of a suitable voltmeter with copper wires to the terminals of the battery. We are therefore measuring, in effect, the potential difference between two *identical phases*, which, as we shall show below, is possible.

Now consider an attempt to measure the potential difference across the metal/solution interface $\phi^M - \phi^S \equiv {}^M\Delta^S\phi$. Assume, that the metal used is copper, connected with a copper wire to one of the terminals of a voltmeter. This terminal will then be at the potential ϕ^M. Now, to determine the potential of the solution phase ϕ^S we would have to use a copper wire connected to the other terminal of the meter, and dip it in the solution. This, however, creates a new metal/solution interface, and the voltmeter would show the sum of two metal–solution potential differences.

It is important to realize that this is not a technical limitation, which may be overcome as instrumentation is improved. In any "thought experiment" one may devise, an attempt to measure the value of $\Delta\phi$ at a single interface necessarily creates at least one more interface.

To end this section on a positive note, it should be pointed out that, while $^{M}\Delta^{S}\phi$ cannot be measured, changes in it, $\delta\left(^{M}\Delta^{S}\phi\right)$, (caused, for example, by passing a current) can be readily determined. Indeed, it is this quantity that is measured when j/E plots are shown in electrochemistry.

If the two terms on the right-hand side of Eq. (2.4) cannot be separately measured, what is the point of using this equation? It turns out that in some special cases, of great practical importance, this equation does lead to results that can be tested by experiment. The usefulness of the electrochemical potential, as defined by Eq. (2.4), is in distinguishing between "short-range" interactions, represented by μ_i, and "long-range" interactions, represented by $z_i F\phi$. The energy of interaction for the former typically decays with r^{-6} while that for the latter decays with r^{-1}. This behavior is shown schematically in Figure 2.1.

In Figure 2.1 an initial value of $200\,kJ\,mol^{-1}$ was chosen for the energy of the chemical bond and the electrostatic energy was taken as $20\,kJ\,mol^{-1}$ at the same distance. The lines for μ and $z_i F\phi$ intersect when the bond length is increased by 58%, and the electrostatic energy becomes 10 times larger than the chemical energy when the bond length is increased by a factor of 2.5. Thus, electrostatic interactions between charged species predominate everywhere, except very close to the boundary between two phases.

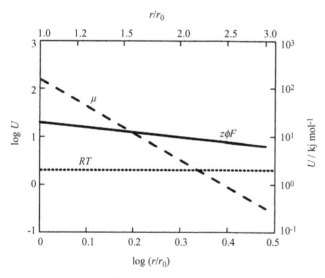

Figure 2.1 Variation of bonding energy with distance for short-range, covalent interactions and long-range, electrostatic interactions, marked by μ and $zF\phi$, respectively. The average thermal energy at 25 °C, RT, is given for comparison.

2.2
Two Cases of Special Interest

2.2.1
Equilibrium of a Species Between Two Phases in Contact

Consider Eq. (2.4) for the case of a species at equilibrium in two different phases – for instance, an electron in a copper wire and a nickel wire welded together. Since equilibrium is assumed, we can write

$$\bar{\mu}_e^{Cu} = \bar{\mu}_e^{Ni} \tag{2.6}$$

combining with Eq. (2.4) we have

$$\mu_e^{Cu} - F\phi^{Cu} = \mu_e^{Ni} - F\phi^{Ni} \tag{2.7}$$

$$^{Cu}\Delta^{Ni}\mu_e / F = {}^{Cu}\Delta^{Ni}\phi \tag{2.8}$$

In these and all following equations the subscript represents the species and the superscript shows the phase in which it is.

The physical significance of Eq. (2.8) is that, at the contact between two dissimilar metals, a certain potential difference will develop, generated by the difference in chemical potential of the electrons in the two metals. It might at first seem odd to have a potential drop inside a metal wire, (even if it is made of two metals welded together), unless a current is flowing. In the present case, however, one might consider $^M\Delta^S\phi$ as the potential difference needed to oppose the flow of electrons in the direction of decreasing chemical potential.

Consider a monovalent metal ion in solution, at equilibrium with the same ion in the crystal lattice. In this case

$$\mu_{M^+}^M + F\phi^M = \mu_{M^+}^S + F\phi^S \tag{2.9}$$

This leads to

$$^M\Delta^S\mu_{M^+} / F = -{}^M\Delta^S\phi \tag{2.10}$$

which is similar to Eq. (2.8). We recall that $^M\Delta^S\phi$ in these equations is not measurable, since any attempt to measure this quantity leads to the creation of at least one more interface, with its own potential difference. Equation (2.10) does, however, lead to some very interesting conclusions, as we shall presently see. To do this, let us substitute the expression for chemical potential from Eq. (2.2) into Eq. (2.10)

$$\mu_{M^+}^{0,M} + RT \ln a_{M^+}^M - \mu_{M^+}^{0,S} - RT \ln a_{M^+}^S = -F(^M\Delta^S\phi) \tag{2.11}$$

which can be written as

$$^M\Delta^S\phi = (^M\Delta^S\phi)^0 - (RT/F)\ln\left(a_{M^+}^M / a_{M^+}^S\right) \tag{2.12}$$

where we have defined $(^M\Delta^S\phi)^0$ as follows:

$$-(^M\Delta^S\mu_{M^+})^0/F = (^M\Delta^S\phi)^0 \tag{2.13}$$

Remembering that the activity of a species in a pure phase, $a_{M^+}^M$, is always defined as unity, we can rewrite Eq. (2.12) in a simplified form as follows:

$$\Delta\phi = \Delta\phi^0 + (RT/F)\ln(a_{M^+}^S) \tag{2.14}$$

which is a form of the Nernst equation, written for a single interface. Combining two such equations, corresponding to two half-cells, leads to the Nernst equation in its usual form (i.e., in terms of the measurable cell potential).

2.2.2
Two Identical Phases Not at Equilibrium

We turn our attention now to another special case of Eq. (2.4), in which the electrochemical potential of a species (an electron this time), $\bar{\mu}_e^M$, is considered in two identical phases that are not at equilibrium. In this case we have

$$\bar{\mu}_e^{M'} = \mu_e^{M'} - F\phi^{M'} \quad \text{and} \quad \bar{\mu}_e^{M''} = \mu_e^{M''} - F\phi^{M''} \tag{2.15}$$

and since

$$\mu_e^{M'} = \mu_e^{M''} \tag{2.16}$$

we obtain

$$\left(^{M'}\Delta^{M''}\bar{\mu}_e\right)/F = {}^{M'}\Delta^{M''}\phi \tag{2.17}$$

The quantity $^{M'}\Delta^{M''}\phi$ in Eq. (2.17) is the actual potential measured between two copper wires attached to the terminals of a battery. We note that potentials are measured, as a rule, with a device that has a very high input resistance, which prevents the establishment of equilibrium between the electrons in its two terminals.

The important physical understanding that can be gained from Eq. (2.17) is that the potential difference measured is nothing but the difference in the electrochemical potentials of the electrons in the two terminals of the measuring instrument (divided by Faraday's constant, F, for consistency of units).

It is important to understand clearly the difference between the two special cases of Eq. (2.4). First we discussed *equilibrium* between *dissimilar* phases. Then we discussed *non-equilibrium* between *identical* phases. The former led to:

$$\left(^{M'}\Delta^{M''}\bar{\mu}_e\right) = 0 \quad \text{hence} \quad \left(^{M'}\Delta^{M''}\mu_e\right)/F = \left(^{M'}\Delta^{M''}\phi\right) \tag{2.18}$$

where $\left(^{M'}\Delta^{M''}\phi\right)$ is not measurable. The latter led to:

$$\left(^{M'}\Delta^{M''}\mu_e\right) = 0 \quad \text{and hence to} \quad \left(^{M'}\Delta^{M''}\bar{\mu}_e\right)/F = \left(^{M'}\Delta^{M''}\phi\right) \tag{2.19}$$

where $\left(^{M'}\Delta^{M''}\phi\right)$ is the potential measured between the two (identical) terminals of a suitable voltmeter.

2.3
The Meaning of the Standard Hydrogen Electrode (SHE) Scale

Since it is impossible to determine experimentally the metal/solution potential difference at a single interface, it has been necessary to measure all such values against a commonly accepted reference. The reversible hydrogen electrode, operating under standard conditions (pH = 0; P_{H_2} = 1 atm) was chosen for this purpose, and its potential was assigned an arbitrary value of zero. The potential quoted for any redox couple on the SHE scale is then the actual potential measured in a cell made up of the desired redox couple in one half-cell (under standard conditions of concentration and pressure) and the SHE in the other half-cell. Taking the E^0 values of Cu/Cu^{2+} and Fe/Fe^{2+} as +0.340 V and −0.440 V vs. SHE, respectively, implies that these values will be measured vs. a SHE.

But what about sign convention? It has been internationally agreed that all potentials listed in the literature will refer to the reduction process. For the preceding two examples we can therefore write:

$$Cu^{2+} + 2e_M^- \rightarrow Cu \qquad E^0 = +0.340 \text{ V vs. SHE} \qquad (2.20)$$

$$H_2 \rightarrow 2H^+ + 2e_M^- \qquad E^0 = 0.000 \text{ V vs. SHE} \qquad (2.21)$$

leading to

$$Cu^{2+} + H_2 \rightarrow 2H^+ + Cu \qquad E^0 = +0.340 \text{ V vs. SHE} \qquad (2.22)$$

and similarly

$$Fe^{2+} + H_2 \rightarrow 2H^+ + Fe \qquad E^0 = -0.440 \text{ V vs. SHE} \qquad (2.23)$$

A positive value of E^0 indicates that the reaction proceeds spontaneously in the direction shown, since for $E^0 > 0$ one has $\Delta G^0 < 0$. We conclude from Eq. (2.20) and Eq. (2.23) that copper ions in solution can be reduced by molecular hydrogen while Fe^{2+} ions cannot. Looking at the same reactions proceeding in the opposite direction, we note that metallic copper cannot be dissolved in acid at pH = 0, while iron can.

If we were to combine copper with iron we would find:

$$Cu^{2+} + 2e_M^- \rightarrow Cu \qquad E^0 = +0.340 \text{ V} \qquad (2.24)$$

$$Fe \rightarrow Fe^{2+} + 2e_M^- \qquad E^0 = +0.440 \text{ V} \qquad (2.25)$$

$$\overline{\qquad\qquad\qquad\qquad\qquad\qquad\qquad\qquad\qquad\qquad\qquad\qquad}$$

$$Cu^{2+} + Fe \rightarrow Cu + Fe^{2+} \qquad E^0 = +0.780 \text{ V} \qquad (2.26)$$

showing that this reaction proceeds spontaneously in the direction of dissolution of metallic iron and precipitation of copper. This is, in fact, used as one of the industrial processes for copper recovery from ores. Note that Eq. (2.23) is written in the direction of reduction of Fe^{2+} ions, while Eq. (2.25) is written in the direction of oxidation of iron, hence the reversal of sign.

These observations are based on thermodynamic considerations alone. Thermodynamics can provide only the negative answers; it allows us to calculate and determine which reactions *will not* happen. We need kinetic information to determine what *will* happen at a rate that may be of practical interest, or at least at a rate that can be detected.

For the corrosion scientist it will be easy to remember that any metal for which E^0 is negative is liable to corrode in acid, while those having a positive value of E^0 will not. This rule of thumb should not be taken as being exact, since in situations of practical interest the system is rarely, if ever, under standard conditions. Pipelines rarely carry 1.0 M acid and metal structures are not, as a rule, in contact with a 1.0 M solution of their ions. For any specific system of known composition and pH, the reversible potential can readily be calculated from the Nernst equation, and the thermodynamic stability with respect to corrosion can be determined.

3
Fundamental Measurements in Electrochemistry

3.1
Measurement of Current and Potential

3.1.1
The Cell Voltage Is the Sum of Several Potential Differences

The measured potential is the sum of several potential differences. When a current is made to flow through the cell, these potential differences are affected to different degrees, and the change in cell potential resulting from an applied current density, j, reflects the sum of all these changes

$$\delta E = E_{(j)} - E_{(j=0)} = \delta(^{W}\!\Delta^{S}\phi) + j\,R_{S} + \delta(^{S}\!\Delta^{C}\phi) \tag{3.1}$$

where the superscripts W and C refer to the working and counter electrodes, respectively, and S stands for the solution phase. The voltage drop across the solution resistance, jR_{S}, can be calculated for certain simple and well-defined geometries. In most cases, however, it is measured and compensated for electronically. One is still left with δE representing the sum $\delta(^{W}\!\Delta^{S}\phi) + \delta(^{S}\!\Delta^{C}\phi)$, the changes of potential across the metal/solution interface at the working and the counter electrodes, respectively. Several methods to overcome this problem and to relate δE to $\delta(^{W}\!\Delta^{S}\phi)$ have been devised and are discussed below.

3.1.2
Use of a Non-Polarizable Counter Electrode

If we combine the working electrode with a highly non-polarizable counter electrode, the change of potential, $\delta(^{S}\!\Delta^{C}\phi)$ at the counter electrode will be negligible compared to that at the working electrode, $\delta(^{W}\!\Delta^{S}\phi)$, and practically all the change in potential observed will occur at the working electrode.

$$\delta E = \delta(^{W}\!\Delta^{S}\phi) + \delta(^{S}\!\Delta^{C}\phi) \cong \delta(^{W}\!\Delta^{S}\phi) \tag{3.2}$$

Physical Electrochemistry: Fundamentals, Techniques and Applications. Eliezer Gileadi
Copyright © 2011 WILEY-VCH Verlag GmbH & Co. KGaA, Weinheim
ISBN: 978-3-527-31970-1

This can be achieved either by using a highly reversible counter electrode or by making the counter electrode much larger than the working electrode. Since the same *total current* must flow through both electrodes, the *current density* at the counter electrode can be made much smaller than that of the working electrode. Now, the change in potential resulting from an applied current is determined by the current density, not the total current. Hence, $\delta(^{S}\Delta^{C}\phi)$ can be made negligible compared to $\delta(^{W}\Delta^{S}\phi)$, satisfying Eq. (3.2). This is so, even if the two electrodes are chemically identical and have the same inherent polarizability.

3.1.3
The Three-Electrode Measurement

A better method of measuring changes in the metal–solution potential difference at the working electrode (which we shall refer to from now on as "changes in the potential of the working electrode") is to use a three-electrode system, shown schematically in Figure 3.1.

A variable-current source is used to pass a current through the working and counter electrodes. Changes in the potential of the working electrode are measured versus a reference electrode, which carries practically no current. In this way the polarizing current flows through one circuit (which includes the working and the counter electrodes) while the resulting change in potential is measured in a different circuit (consisting of the working and reference electrodes), through which the current is essentially zero. Since no current flows through the reference electrode, its potential can be considered to be constant, irrespective of the current passed through the working (and counter) electrodes. Thus, the measured change in potential (between working and reference electrodes) is truly equal to the change in potential

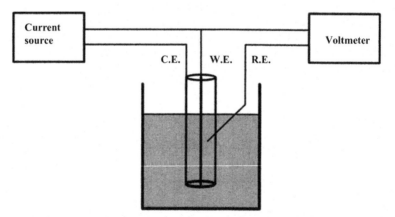

Figure 3.1 Schematic representation of a three-electrode circuit, showing the working electrode connected to a reference electrode through a high-input-resistance voltmeter and to the counter electrode through a low-input-resistance current source.

of the working electrode. As stated earlier, although $\Delta\phi$ cannot be measured, its variation $\delta(\Delta\phi)$ can be readily determined.

It should be noted that during measurement in a three-electrode cell, the potential of the counter electrode might change substantially. This, however, does not in any way influence the measured potential of the working electrode with respect to the reference.

The three-electrode arrangement can be used equally well if the potential between the working and reference electrodes is controlled and the current flowing through the working and counter electrodes is measured. Details of this mode of measurement are discussed below.

3.1.4
Residual $j R_S$ Potential Drop in a Three-Electrode Cell

Regarded superficially, it might appear that making a current–potential measurement in a three-electrode cell eliminates the need to consider any correction for the $j R_S$ potential drop in the solution, since there is practically no current flowing through the circuit used to determine the potential. Unfortunately, this is not quite true. The reference electrode (or the tip of the Luggin capillary leading to it) is situated somewhere between the working and counter electrodes. As a result, the potential it measures includes some part of the potential drop in solution between these electrodes. This is called the residual $j R_S$ potential drop. It is shown schematically in Figure 3.2, where a cell having parallel-plate geometry is chosen for simplicity.

Placing the reference electrode near the working electrode can decrease, but not totally eliminate, the residual potential drop caused by solution resistance. The methods used to determine this potential drop are discussed later. It is important to understand that the $j R_S$ potential drop *cannot* be determined by measuring the resistance between the terminals of the working and reference electrodes, since such a measurement includes resistive elements through which there is no flow of current during determination of the j/E relationship.

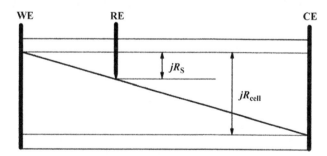

Figure 3.2 The residual $j R_S$ potential drop between the working and the reference electrodes. The total potential drop in the cell, between the working and the counter electrodes is also shown.

3.2
Cell Geometry and the Choice of the Reference Electrode

3.2.1
Types of Reference Electrodes

A good reference electrode consists of a reversible electrode having a reproducible and stable potential, connected to the "outside world" with a Pt wire. A typical commercial reference electrode (such as calomel) is a complete system, with electrode and electrolyte enclosed in a small compartment and connected to the rest of the cell through a porous plug. The latter is designed to allow passage of ions, yet keep the flow of solution to a minimum. A mercury electrode in contact with mercurous chloride (Hg_2Cl_2), commonly known as calomel, is placed in a saturated solution of KCl. Leaving some solid KCl in contact with the saturated solution ensures that its composition will be constant, even if some of the water has evaporated with time, as long as the temperature is maintained constant, leading to a stable reference potential. Compatibility with the various components in solution is, however, important. Thus, a calomel reference electrode should not be used in a solution of $HClO_4$ or $AgNO_3$ because $KClO_4$ or $AgCl$, respectively, could precipitate in the porous plug and isolate the reference electrode from the solution in the main compartment. Also, since chloride ions are strongly adsorbed on electrode surfaces and can hinder the formation of passive films in corrosion studies, a calomel reference electrode should not be used, unless the test solution itself contains chloride ions, such as in seawater. The calomel electrode is an example of a so-called *secondary reference electrode.* This type of reference electrode consists of a metal (M), an insoluble salt of the same metal (MA) and a solution containing a soluble salt having the same anion (A), for example Ag|AgCl|KCl; H_2O. (the vertical lines represent boundary between phases). A so-called primary reference electrode comprises a metal at equilibrium with a solution containing its soluble salt, for example Cu|CuSO$_4$; H_2O. Secondary reference electrodes are usually preferred over primary reference electrodes, because their potential is more stable and more reproducible.

Another class of reference electrodes, often called *indicator electrodes,* are electrodes in direct contact with the solution. The most common among these is the reversible hydrogen electrode, formed by bubbling hydrogen over a large-area platinized Pt electrode in the test solution. This electrode is reversible with respect to the hydronium ion, H_3O^+, serving, in effect, as a pH indicator electrode. Similarly, one could use a silver wire coated with AgCl in a chloride-containing test solution as a reversible Ag|AgCl|Cl$^-$ electrode, which responds to the concentration of Cl$^-$ ions in solution, following the Nernst equation.

The advantage of indicator electrodes is that they always measure the reversible potential with respect to the ion being studied, regardless of its concentration in solution. Their disadvantage is that they must be prepared for each experiment and often end up being less stable and less reliable than commercial reference electrodes. Also, being in intimate contact with all ingredients in the test solution, they can be contaminated, either by impurities or by components of the test solution, such as additives for plating and corrosion inhibitors, which might lead to changes in the reference potential.

3.2.2
Use of an Auxiliary Reference Electrode for the Study of Fast Transients

One of the advantages of making measurements in a three-electrode configuration is that the resistance of the reference electrode should not affect the measured potential, since the current passing through it is extremely low. This allows us to use reference electrodes well separated from the main electrolyte compartment, thus minimizing the danger of mutual contamination. Typical values of the resistance of the reference-electrode assembly may be about $10 \, k\Omega$. Combined with an input resistance of $10^{12} \, \Omega$ of the voltmeter, the error in the measured potential is negligible. As is often the case, in practice the situation is more complicated, and it is advantageous to lower the resistance of the reference electrode. For one thing, the electrode and its connecting wires may act as antennas, picking up stray electromagnetic signals. The resulting noise in the measurement of potential will then increase with increasing resistance of the reference electrode. The problem is aggravated in the study of transients. The inherent reason for this is that there is a trade-off between response time and input impedance in all measuring instruments. Whereas, for steady-state measurements, an input impedance of $10^{12} \, \Omega$ is commonplace, fast oscilloscope and transient recorders may have an input impedance as low as $10^6 \, \Omega$.

Using an indicator-type reference electrode could alleviate this problem. When this is not possible (because of chemical incompatibility), an "auxiliary reference electrode" can be used. This usually consists of a platinum wire placed near the working electrode. While the potential of such an electrode is not stable or well defined, it can be measured just before application of the transient, and it can be safely assumed to be constant during the transient. The transient is then applied, with the platinum wire (which has a very small resistance) acting momentarily as the reference electrode. The potential of the Pt wire is measured before and after application of the transient, against a proper reference electrode.

3.2.3
Calculating the Uncompensated Solution Resistance for a Few Simple Geometries

As a rule, the jR_S potential drop is measured and a suitable correction is made, either directly during measurement or in the analysis of the data. When the geometry of the cell is simple, it is possible to calculate this quantity. Such calculations are important because they can yield clear criteria for the design of cells and for positioning the reference electrode with respect to the working electrode, as will be shown below.

3.2.3.1 Planar Configuration
The planar configuration was shown in Fig. 3.2. The jR_S potential drop is given by

$$jR_S = j\frac{d}{\kappa} \tag{3.3}$$

in which j is the current density ($A cm^{-2}$), R_S is the uncompensated solution resistance (Ωcm^2), κ is the specific conductivity of the solution ($S cm^{-1}$), and d is the effective distance (cm) between the working electrode and the tip of the Luggin capillary connected to the reference electrode. The way to decrease jR_S is to increase the conductivity (e.g., by adding an inert supporting electrolyte), or by positioning the tip of the Luggin capillary close to the surface of the working electrode. The latter approach is not recommended, because it could result in local distortion of the current-density distribution, and the potential measured could deviate significantly from the metal–solution potential difference that exists over most of the electrode surface. We note that $j R_S$ is independent of the electrode area, because R_S is given in units of Ωcm^2 and j has units of $A cm^{-2}$.

3.2.3.2 Cylindrical Configuration

A cylindrical configuration is shown schematically in Figure 3.1. A thin-wire of radius r, used as the working electrode, is positioned at the center of a cylindrical counter electrode. The equation relating $j R_S$ to the distance d and the radius of the working electrode r is

$$j R_S = \frac{j r}{\kappa} \ln(1 + d/r) \tag{3.4}$$

It is interesting to consider two extreme cases of this equation. Close to the surface of the working electrode, where $d/r \ll 1$, the approximation

$$\ln(1 + d/r) \approx d/r \tag{3.5}$$

applies, yielding

$$j R_S = \frac{j d}{\kappa} \tag{3.3}$$

which brings us back to the equation for planar configuration. This should not be surprising, because any probe "looking" at a curved surface from a distance that is short compared to the radius of curvature, responds to it as though it were flat. On the other hand, for large distances from the electrode surface, when $d/r \gg 1$, one has

$$j R_S = \frac{j r}{\kappa} \ln \frac{d}{r} \tag{3.6}$$

There is little to be gained by decreasing the distance d, (unless it can be reduced to well below the radius of the electrode), since $j R_S$ changes logarithmically with the distance. On the other hand, we note that $j R_S$ decreases in this case with decreasing radius of the working electrode. There is, therefore, a clear advantage in using a very fine wire in this type of measurement.

3.2.3.3 Spherical Symmetry

Next we consider the case of an electrode in the shape of a sphere, located at the center of a spherical counter electrode. The potential drop across the solution resistance is expressed in this case by

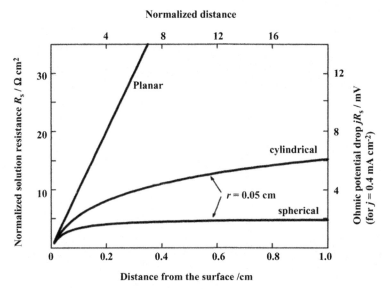

Figure 3.3 Uncompensated solution resistance and the corresponding potential drop, as a function of the distance from the electrode surface. The specific conductivity, κ; is 0.01 S cm^{-1}, $r = 0.05$ cm. The normalized distance is defined as d/r.

$$j R_S = \left(\frac{j\,d}{\kappa}\right) \left(\frac{r}{r+d}\right) \tag{3.7}$$

For very short distances, $d/r \ll 1$, Eq. (3.7) reverts to the equation for planar configuration, as for the cylindrical case. For large distances, however, the same equation yields

$$j R_S = \frac{j\,r}{\kappa} \tag{3.8}$$

What this equation shows is that, for spherical symmetry, most of the $j R_S$ potential drop occurs in the vicinity of the working electrode (within, say, five radii), beyond that it approaches a constant value, independent of distance. The variation of potential with distance is shown in Figure 3.3 for the three configurations just discussed.

It should be clear that the spherical configuration is the best in reducing the error that results from a residual $j R_S$ potential drop, and the planar configuration is the worst. In spite of this, the cylindrical configuration is often used in research, because it is only a little worse than the spherical, but much better than the planar configuration, and is easier to implement experimentally than the spherical configuration.

The normalized resistance is shown for three radii of a spherical electrode in Figure 3.4. It decreases linearly with decreasing radius, according to Eq. (3.8). Similar behavior applies for the cylindrical configurations. In both cases the error introduced by the potential drop associated with the uncompensated solution resistance can be reduced by reducing the radius of the electrode.

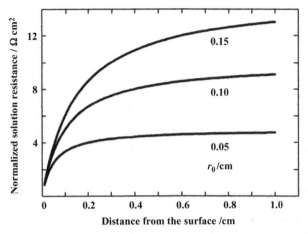

Figure 3.4 The normalized solution resistance at a spherical electrode, as a function of distance from the electrode surface, shown for different radii.

3.2.4
Positioning the Reference Electrode

In the design of an electrochemical cell, the position of the reference electrode has to fulfill two conflicting requirements. On the one hand, it should be far from the working electrode and well separated from the solution in the main cell compartment, in order to reduce the possibility of mutual contamination. On the other hand, it should be as close as possible to the working electrode, in order to reduce the residual jR_S potential drop. This set of requirements is partially met by the use of a Luggin capillary, shown in Figure 3.5.

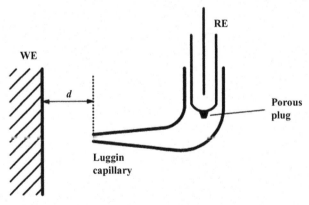

Figure 3.5 Schematic representation of a Luggin capillary. Working electrode (WE); reference electrode (RE).

The reference electrode compartment is separated from the rest of the solution by a porous plug and the tip of the capillary. Since the current to the reference electrode is essentially zero, the potential anywhere inside the capillary (and up to the reference electrode compartment) is the same as the potential at its outer rim, a distance d away from the working electrode.

But how close should the tip of the capillary be to the surface of the electrode? Should it be in the middle of the electrode or near the edge? The former position measures a more typical value of the potential, not influenced by edge effects, but the body of the Luggin capillary may "cast a shadow" on the working electrode, disturbing the uniformity of the current density on its surface. The latter position is influenced by edge effects but interferes less with the current flow. Many configurations have been suggested in the literature, and they all share the following drawbacks (albeit to varying degrees), which follow from the basic laws of electrostatics. Bringing the Luggin capillary close to the surface causes a nonuniformity of the current density in that area (usually it is a decrease in local current density, caused by the existence of a nonconducting body, the glass capillary, in the path of the flow of current). If the Luggin capillary is small, this anomaly may not affect the total current to a significant extent. However, the potential is measured near the tip of the capillary, where the deviation of the current density from its average value is a maximum.

The configuration shown in Figure 3.5 was proposed by Luggin more than a century ago, when electronic instrumentation was in a very primitive stage, so that the Luggin capillary may have been the best way to reduce the residual solution resistance. Fortunately, modern instruments allow accurate measurement and dynamic compensation for the residual $j R_S$ potential drop, obviating the need to use the Luggin capillary, except as a means for separating the solution in the reference electrode compartment from that in the working electrode compartment. If a Luggin capillary is used, it is therefore better to move it farther away from the working electrode, (a distance of about five times the outside radius of the Luggin capillary is usually enough), in order to minimize the inhomogeneities of the current density distribution. In the case of cylindrical and spherical electrodes this approach has the added advantage that the potential drop changes little with distance (at $d/r \geq 5$). Thus, the measured potential is not sensitive to the exact position of the Luggin capillary, and reproducibility is improved.

3.2.5
Edge Effects

Uniformity of the current density over the whole area of the electrode is important for the interpretation of current–potential data. We recall that the measured quantities are the total current and the potential at a certain point in solution, where the tip of the Luggin capillary is located. From this we can calculate the average current density, but not its local value. As for the potential, we have already noted that, unless the cell is properly designed, the potential measured may be grossly in error, if the reference electrode is located at a point where the local current density deviates significantly from its average value.

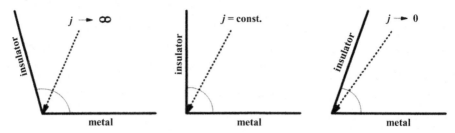

Figure 3.6 The angle φ between metal and insulator determines the current distribution near the edge of the electrode.

An interesting case to be discussed is the edge effect calculated for the point of contact between an electrode and the insulator in which it is cast. The equation describing the current density as a function of distance d from the point of contact is

$$j = K \, d^{(\pi/2 - \varphi)} \tag{3.9}$$

Three cases of interest are shown in Figure 3.6. It is easy to see that, according to Eq. (3.9), a uniform current density is expected only for a right angle ($\pi/2 = \varphi$). For obtuse angles the exponent in Eq. (3.9) is negative; hence current density at the point

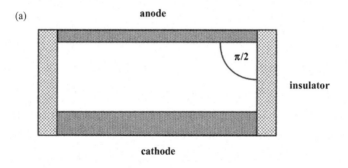

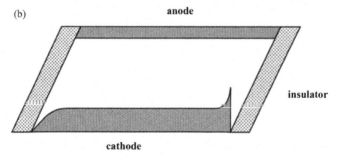

Figure 3.7 Plating thickness (indicative of current distribution) for different geometries. (a) Rectangular cell yielding uniform plating thickness. (b) Paralleloid-shaped geometry yielding near zero thickness at the acute angle and high thickness at the obtuse angle.

of contact between the electrode and the insulator should approach infinity. For acute angles the exponent is positive and the current density approaches zero at the point of contact. This type of behavior is indeed observed experimentally and is well known in the plating industry.

A complete evaluation of the uniformity of current distribution should take into account the finite rate of the charge-transfer reaction that is taking place at the interface. When this is done, the current density becomes more uniform, and it neither declines to zero, nor does it increase to infinity at the edge, as implied above. It turns out that both lower exchange-current density for the metal deposition process and lower applied current density enhance the uniformity of the current distribution.

Maintaining a uniform current distribution is of great importance in the electro-chemical industry. In plating it determines the uniformity of the thickness of the deposit; in electro-organic synthesis it affects the uniformity of the products. A nonuniform current distribution can lead to the formation of undesired side products and to waste of energy in all areas of the electrolytic industry.

In studies of electrode kinetics it should be borne in mind that the fundamental equations used are derived with the tacit assumption of uniformity of current distribution. Hence, the analysis of the current–potential relationship is valid only when this assumption applies to a very good approximation.

The variation of the thickness of the electroplated layer is shown schematically in Figure 3.7. The deviations from uniformity at the edges for obtuse and acute angles are shown, in contrast to the uniformity expected when the contact between the insulator and the conductor is at right angles.

4
Electrode Kinetics: Some Basic Concepts

4.1
Relating Electrode Kinetics to Chemical Kinetics

4.1.1
The Relation of Current Density to Reaction Rate

The current density is proportional to the rate of the heterogeneous reaction taking place at the interface. The relationship

$$j = nF\mathrm{v} \qquad (4.1)$$

follows directly from dimensional analysis, since its right-hand side yields

$$(\mathrm{C\,mol^{-1}}) \times (\mathrm{mol\,cm^{-2}\,s^{-1}}) = (\mathrm{C\,cm^{-2}\,s^{-1}}) = \mathrm{A\,cm^{-2}} \qquad (4.2)$$

Substituting the appropriate numbers into Eq. (4.1) shows that the electrochemical reaction rate can be measured with very high sensitivity, without causing significant changes in the concentration of reactants or products in solution. This follows from the high sensitivity in the measurement of current, hence also of charge. Thus, a current of $1\,\mu A$, which can readily be measured accurately, corresponds to a very low reaction rate of about $10^{-11}\,\mathrm{mol\,s^{-1}}$. Hence, one can ordinarily measure the rate of an electrode reaction for hundreds of seconds without causing a significant change in the concentrations of reactants or products. As a result, most electrochemical reactions can be studied under *quasi-zero-order* conditions, since the change in concentration can be kept negligible during measurement of the current.

Consider the following example, showing the high sensitivity that can be achieved by measurement of the current. The charge required to form a monolayer of hydrogen atoms adsorbed on platinum in the reaction

$$\mathrm{H_3O^+ + Pt + e_M^- \rightarrow Pt\text{-}H + H_2O} \qquad (4.3)$$

can be estimated as follows. The area taken up by a single platinum atom on the surface is of the order of $0.1\,\mathrm{nm^2}$ and $1\,\mathrm{cm^2} = 10^{14}\,\mathrm{nm^2}$. Hence there are about 10^{15} platinum atoms $\mathrm{cm^{-2}}$ of the metal surface. Considering that one electron is discharged per platinum atom in Eq. (4.3), the total charge required to form a monolayer of adsorbed hydrogen atoms is

$$1.6 \times 10^{-19} \times 10^{15} = 1.6 \times 10^{-4} \, C \, cm^{-2} = 0.16 \, mC \, cm^{-2} \qquad (4.4)$$

This is an order-of-magnitude calculation. The correct number, obtained experimentally for hydrogen atoms adsorbed on platinum, is $0.22 \, mC \, cm^{-2}$. Thus, it is necessary to pass a current of, say, $10 \, \mu A \, cm^{-2}$ for $22 \, s$ to form a monolayer. As far as the electrical measurement is concerned, one could easily measure a very small fraction of a monolayer. Now, $0.22 \, mC \, cm^{-2}$ equals about $2 \times 10^{-9} \, mol \, cm^{-2}$ (for $n = 1$). In the case of hydrogen atoms, this amounts to $2 \, ng$ of added weight. Thus, measurement of the current and time makes it possible to determine quantities of adsorbed hydrogen well below $1 \, ng \, cm^{-2}$, which is better than the sensitivity achieved by the electrochemical quartz-crystal microbalance, itself by far the most sensitive method of weighing small amounts of materials adsorbed from solution on a surface (cf. Chapter 17).

The same high sensitivity that makes measurements so convenient causes great difficulty in the electrolytic industry. It takes $(9.65 \times 10^4 \times n) \, C = (26.80 \times n) \, A \, h$ to generate $1 \, mol$ of a product formed by electrolysis. A cylinder of compressed hydrogen contains about $0.5 \, kg$ of the gas. It would take a water electrolyzer running at a total current of $1000 \, A$ about $13 \, h$ to produce this small amount of hydrogen! Incidentally, the same amount of electric charge would produce 35.5 times as much Cl_2 in the electrolysis of a solution of NaCl.

4.1.2
The Relation of Potential to Energy of Activation

In kinetics, the rate constant can be written in the form,

$$k = k_0 \exp\left(-\Delta G^{0\#}/RT\right) \qquad (4.5)$$

where $\Delta G^{0\#}$ is the standard Gibbs energy of activation and k_0 is a chemical rate constant. Now we have already shown that for a charge-transfer process it is advantageous to separate the electrochemical potential into so-called "chemical" and "electrical" terms.

$$\bar{\mu}_i = \mu_i + z_i F \phi \qquad (2.4)$$

In much the same way, we can express the change in standard electrochemical Gibbs energy of a reaction as:

$$\Delta \bar{G}^0 = \Delta G^0 \mp zF\Delta\phi \qquad (4.6)$$

For the standard electrochemical Gibbs energy of activation, we write

$$\Delta \bar{G}^{0\#} = \Delta G^{0\#} \pm \beta F \Delta\phi \qquad (4.7)$$

Although Eqs (4.6) and (4.7) look similar, the transition from one to the other is by no means trivial, and is the subject of detailed discussion below. Here we shall limit ourselves to a brief discussion of two points. First, the charge, z, on the particle, which appears in Eq. (4.6), has been dropped from Eq. (4.7), since it is tacitly assumed that electrode reactions occur by the transfer of one electron at a time[1]. Second, the parameter β, called *the symmetry factor*, has been introduced. By definition β can assume values between zero and unity,

$$0 \le \beta \le 1 \tag{4.8}$$

and can be viewed as representing the fraction of the total change in the standard electrochemical Gibbs energy of the reaction that is applied to its standard Gibbs energy of activation:

$$\beta \equiv \frac{(\delta\Delta\bar{G}^{0\#}/\delta\Delta\phi)}{(\delta\Delta\bar{G}^{0}/\delta\Delta\phi)} = \frac{\delta\Delta\bar{G}^{0\#}}{\delta\Delta\bar{G}^{0}} \tag{4.9}$$

The important consequence of Eq. (4.7) for our present discussion is that the standard electrochemical Gibbs energy of activation is a linear function of potential. The negative sign in Eq. (4.7) is applicable to anodic reactions, in which the rate is enhanced by increasing the potential in the positive direction, and the positive sign is applicable to cathodic reactions. In either case, the electrochemical Gibbs energy of activation decreases when the potential is changed "in the right direction", namely positive for an anodic process and negative for a cathodic process. We can thus relate the anodic current density to the exponent of the potential:

$$j = nF\text{v} = n\,F\,k_0\,c_b\exp\left(-\frac{\Delta G^{0\#}}{RT}\right)\exp\left(\frac{\beta\,\Delta\phi\,F}{RT}\right) \tag{4.10}$$

Consider how the rate of a chemical or an electrochemical reaction can be changed. The best means of accelerating a chemical reaction is by increasing the temperature. This approach does have its limitations, however, because of possible decomposition of reactants or products, the range of stability of reaction vessels, the cost of energy, and so on. Increasing the concentration of reactants is another means; employing a suitable catalyst (which decreases the value of $\Delta G^{0\#}$) is a very effective method, most often used in industrial processes.

For an electrochemical reaction all the above apply, but there is an additional, very powerful, means of controlling the rate of the reaction, namely by adjusting the potential. From Eq. (4.10) we can calculate the magnitude of this effect. To do this, we follow the common practice in electrochemistry and assume β = 0.5, although we shall see later that this may not always be the case and, in any event, one should not take the symmetry factor to be β = 0.500 (i.e., exactly one half).

For a change of potential of $\delta(\Delta\phi) = 1.0$ V. Eq. (4.10) yields a change of current density by a factor of:

$$\exp\left[\frac{0.5 \times 9.65 \times 10^4}{8.31 \times 298}\right] = 2.9 \times 10^8 \tag{4.11}$$

---

1) It has recently been shown that simultaneous transfer of two electrons is possible. Nevertheless, we consider here the transfer of one electron at a time, for simplicity.

The effect of temperature on the reaction rate depends on the enthalpy of activation. Taking a reasonable range of $(40-80)$ kJ mol^{-1} for this quantity, we find that the rate of a reaction at $100\,^{\circ}$C is larger than at $0\,^{\circ}$C by a factor of $1.3 \times 10^2 - 1.6 \times 10^4$. Thus, the effect on the reaction rate of raising the temperature by $100\,^{\circ}$C is many orders of magnitude less than the effect of changing the potential by 1 V.

4.1.3
Mass-Transport versus Charge-Transfer Limitation

We began this book with a schematic presentation (cf. Figure 1.1) of the current–potential relationship in an electrolytic cell, from the region where no current is flowing, in spite of the applied potential, to the region where the current rises exponentially with potential, following an equation such as Eq. (4.10) and through the limiting-current region, where the current has a constant value, determined only by the rate of mass transport to the electrode surface or away from it.

Mass-transport limitation is more often encountered in electrode kinetics than in any other field of chemical kinetics, because the activation-controlled charge-transfer rate can be accelerated (by applying a suitable potential) to the point where it is much higher than the rate of mass transport, and therefore no longer controls the observed current. From the laboratory-research point of view, mass transport is an added complication to be either avoided or corrected for quantitatively, in order to obtain the true kinetic parameters for the charge-transfer process. From the engineering point of view, mass transport is often the main factor determining the space–time yield of an electrochemical reactor, that is, the amount of product that can be generated in a given reactor (or a given plant) per unit time. The task of the applied scientist is then twofold: first to find better catalysts (or rather electrocatalysts), which accelerate the rate of charge transfer, and then to design cells with high rates of mass transport, to take full advantage of the improvement in catalytic activity.

Mass transport to the interface can occur through three independent mechanisms: *migration, convection and diffusion.* The driving force for migration is the electric field in solution. In setting up an experiment, one usually tries to eliminate this effect by adding a high concentration of supporting electrolyte (compared to the concentration of the electroactive species). The supporting electrolyte is chosen such that it will not take part in the charge-transfer process. It is added to reduce the electric field by increasing the conductivity in solution. In addition, most of the electricity is carried by the inert ions of the supporting electrolyte, so that the contribution of the electrical field to the rate of mass transport is reduced to a negligible level. In the following discussion, it is assumed that mass transport by migration has been essentially eliminated, unless we refer to it specifically. It should be borne in mind, however, that this is not always the case. Specifically, when films are formed on the surface of electrodes (such as anodic oxide films on aluminum and some other metals of the so-called solid electrolyte interphase), mass transport through the film may depend exclusively on migration. Also, in many industrial processes, a high concentration of the electroactive ions is used to allow high rates of production (e.g., Cl_2 from NaCl,

Cu-plating with an acid bath containing H_2SO_4 and $CuSO_4$), and migration is viewed favorably, as an additional mode of mass transport.

Convective mass transport is caused by the movement of the solution as a whole. The driving force in this case is external energy, usually in the form of mechanical energy of stirring the solution, rotating the electrode, or pumping the liquid at, past, or through the electrode. Most efficient convective mass transport is achieved when the flow at the electrode surface is turbulent; yet turbulent flow tends to be sensitive to minor changes in cell configuration, electrode area, the nature of the electrode surface and so on, often making it difficult to obtain reproducible results.

Mass transport by diffusion can be regarded as the last resort. When movement of the electroactive species is not promoted by the input of external energy, either electrical (migration) or mechanical (convection), diffusion takes over. The driving force in this case is the gradient in chemical potential caused by the gradient in concentration. It is a relatively slow process, with diffusion coefficients for small molecules in dilute aqueous solutions at room temperature, in the range of $2 \times 10^{-6} - 8 \times 10^{-5} \text{ cm}^2 \text{ s}^{-1}$.

4.1.4
The Thickness of the Nernst Diffusion Layer

We alluded to the Nernst diffusion layer thickness, δ, in the first chapter. This quantity is related to the mass-transport-limited current density through the equation

$$j_L = nFDc_b/\delta \tag{1.6}$$

For a diffusion-controlled process, δ is proportional to the square root of time

$$\delta = (\pi Dt)^{1/2} \tag{4.12}$$

and hence the limiting current density decreases gradually with time.

The concept of the diffusion-layer thickness can also be applied, however, when the main mode of mass transport is convection. In such cases one assumes that there is a thin layer of liquid at the surface that is stationary, while the rest of the solution is stirred. Concentration profiles near the interface, following the application of a potential step, are shown in Figure 4.1.

In Figure 4.1a the concentration profile is presented in dimensionless form, as the relative concentration, c/c_b, versus the dimensionless distance, $x/(4Dt)^{1/2}$. This is an elegant way of representing the results calculated from a function that depends on several parameters. Thus, this curve is independent of the initial concentration in solution, the diffusion coefficient of the reacting species, and time. A great deal of information is condensed into a single curve, which can then be used to calculate values of the concentration as a function of distance and time for any specified system.

But how far is one unit of dimensionless distance in real terms? That depends on the diffusion coefficient and on time. For $D = 1 \times 10^5 \text{ cm}^2 \text{ s}^{-1}$ it is 2.0 µm after 1 ms and 63 µm after 1 s. Figure 4.1b shows concentration profiles calculated for the same system, but plotted versus the distance in centimeters. Each curve corresponds to a

(a)

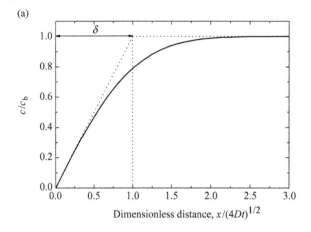

(b)

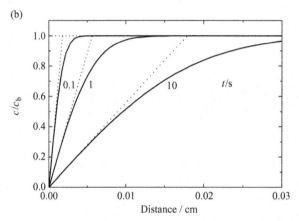

Figure 4.1 Calculated concentration profiles following a potential step. $D = 8 \times 10^{-6}$ cm^2 s^{-1}. (a) dimensionless coordinates, (b) distance in cm, with the time marked on each curve.

different time after application of the potential step, and the evolution of the concentration profile with time is presented. All the information contained in the curves in Figure 4.1b exists, of course, in the single curve shown in Figure 4.1a. This is both the advantage and the disadvantage of presenting the data in dimensionless form.

The current flowing across the interface is given by

$$j = -n \, F \, D (dc/dx)_{x=0} \tag{4.13}$$

It is determined by the gradient of concentration at the electrode surface, which decreases with time, as seen in Figure 4.1b.

In a stirred solution, a steady state is reached when the concentration inside the diffusion layer varies linearly with distance. Under such conditions we can express the current density by

$$j = n F D \frac{c_b - c_s}{\delta} \tag{4.14}$$

It is evident that it can reach its mass transport limited value, j_L, when $c_s = 0$, namely

$$j_L = n F D c_b / \delta \tag{1.6}$$

Combining the last two equations yields

$$j/j_L = 1 - c_s/c_b \tag{4.15}$$

or

$$c_s/c_b = 1 - j/j_L \tag{4.16}$$

Since the reactant is consumed at the interface, its surface concentration, c_s, is always smaller than the bulk concentration, c_b. The ratio between the two depends on the ratio j/j_L. For proper kinetic measurements, without the influence of mass transport, this ratio must be maintained below a chosen level, say 0.01 or 0.05, depending on the accuracy desired. This can be done by taking measurements at very low current densities, by increasing j_L, or both. Alternatively, the concentration at the surface can be allowed to deviate significantly from its bulk value, as long as this deviation is taken into account quantitatively, by solving the appropriate equations for mass transport.

For a diffusion-controlled process, δ is a rough estimate of the distance over which molecules can diffuse in a given time. For $\delta = 50\,\mu m$ and $D = 8 \times 10^{-6}\,cm^2\,s^{-1}$, Eq. (4.12) yields a value of $t = 1\,s$. During this time some molecules will be found further away from the surface than $50\,\mu m$. Thus, δ should be considered as "the characteristic length" for diffusion. This kind of simple calculation allows us to estimate how long it would take, for example, for chloride ions from a calomel reference electrode to penetrate the test solution compartment through a side arm of given length, or how fast ions will diffuse into the pores of an anodized-aluminum electrode, to be deposited there electrochemically.

4.2
Methods of Measurement

4.2.1
Potential Control versus Current Control

An experiment in electrode kinetics usually consists of determining the current–potential relationship under a given set of fixed conditions. The measurement may then be repeated under a set of gradually changing conditions, to obtain the j/E plots as a function of temperature, concentration, and so on.

In many cases, the experiment can be performed either by controlling the current externally and measuring the resulting changes in potential at the working electrode (i.e., between the working electrode and a suitable reference) or by controlling the potential and measuring the resulting current. The former is referred to as a

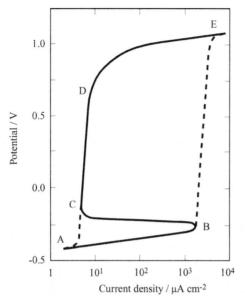

Figure 4.2 Schematic j/E plots for a system undergoing corrosion and passivation. Active dissolution region: A–B, passive region: C–D, transpassive region where pitting may occur: D–E.

galvanostatic measurement and the latter a *potentiostatic* measurement. These commonly used terms are misnomers to some extent, since they include cases in which either E or j is changed in the course of measurement, and the suffix "static" does not really apply. They serve, however, to identify the variable that is controlled externally and that which is measured.

In Figure 4.2 we show schematically the current–potential relationships for a system undergoing corrosion and passivation. If the experiment is conducted potentiostatically, the curve A–B–C–D–E is observed, in which both the active dissolution region (A–B) and the passive region (C–D) are seen. If measurement is conducted galvanostatically, all the information concerning the passive region is lost, and the potential follows the dashed line, from B to E. Clearly, a system exhibiting the type of behavior shown in Figure 4.2 should be studied potentiostatically.

On the other hand, if one is studying only the active region (A–B), or if passivity does not occur, galvanostatic measurements could yield equally accurate results and, in certain cases, may be experimentally advantageous, as we shall see.

Even though galvanostatic and potentiostatic measurements may yield the same results in many cases, it is important to understand the inherent difference between them and to determine the advantages and disadvantages of each.

Controlling the potential in measurements of electrode kinetics is, in many ways, "the natural way" to conduct a measurement. It is similar to any other measurement in chemical kinetics in that all the experimental parameters (concentration, temperature, the nature of the catalyst, and, in electrochemistry, also the potential) are fixed externally and the current, which is proportional to the reaction rate, is measured (cf. Eq. (4.10)). Just as in chemical kinetics, we might repeat the experiment changing the concentration of one of the components or the temperature. In electrochemistry we have the additional option of changing the potential.

The potential of an electrode with respect to a reference electrode represents its oxidizing or reducing power. In a mixture of, say, silver and copper ions (E^0 equals $+0.79\,V$ and $+0.34\,V$ vs. SHE, respectively), one could set the potential at a value that ensures that only silver ions can be reduced, producing a pure deposit of this metal, even if we leave the system operating for a very long time. Similarly, it is possible, in principle at least, to determine the course of an electro-organic synthesis, to produce one desired product or another, by prudent choice of the potential.

A galvanostatic measurement represents a different situation, unparalleled in chemical kinetics. Here the rate of the reaction (i.e., the current density) is controlled externally, and the potential is allowed to assume a value appropriate to that rate.

The choice between galvanostatic and potentiostatic measurements depends on the circumstances. From the instrumentation point of view, galvanostats are much simpler than potentiostats. This is not only a matter of cost, but also a matter of performance. Thus, where it is desired to measure very low currents (e.g., on single microelectrodes), a battery with a variable resistor may be all that is needed to set up a low-noise galvanostat. At the other extreme, when large currents must be passed – for instance, in an industrial pilot plant for electrosynthesis – power supplies delivering controllable currents in the range of hundreds of amperes are readily available, whereas potentiostats of comparable output are either nonexistent or extremely expensive. In large-scale industrial processes such as the production of chlorine or aluminum, the current is always the controlled parameter.

A potentiostat is inherently a more complex apparatus. It operates with a feedback loop, in which the potential between the *working* and the *reference* electrodes is compared to a value determined externally, the difference is amplified, and a current is passed between the *working* and *counter* electrodes, of such magnitude and sign as to reduce that difference of potential to a very small value, typically $1\,mV$ or less (cf. Figure 3.1).

Potentiostats did not become commercially available until the late 1950s. Earlier work was conducted either galvanostatically or potentiostatically, but with a two-electrode cell, in which one electrode served as both the counter and the reference electrode. Because of their complexity, potentiostats tend to have slower response times than galvanostats. It should be pointed out, however, that some of the limitations of potentiostats alluded to above are a matter of the past. With present-day (2010) electronic components, it is possible to build home-made potentiostats, or to purchase commercial units, that make use of all the inherent advantages of potentiostatic measurements with few instrumental limitations.

4.2.2
The Need to Measure Fast Transients

So far, we have concentrated our attention on activation-controlled processes. Since the rate of such processes increases exponentially with potential, it is usually possible to drive them at a rate sufficiently high, so that mass transport becomes the limiting factor. For a measurement to be truly activation controlled, the current density must be small compared to the mass-transport-limited current density. The latter is given by

$$j_L = nFDc_b/\delta \qquad (1.6)$$

in which δ is the thickness of the Nernst diffusion layer. The activation-controlled current density is given by an equation of the form

$$j_{ac} = nFv = n\,F\,k_0\,c_b\exp\left(-\frac{\Delta G^0}{RT}\right)\exp\left(\frac{\beta\Delta\phi F}{RT}\right) \qquad (4.10)$$

To maintain the ratio j_{ac}/j_L small (say, <0.05) over a wide range of potentials, we would like to find ways of increasing j_L without changing j_{ac}. The best way to achieve this is to decrease the diffusion-layer thickness, δ. This can be done by improving the efficiency of stirring, or by taking measurements in unstirred solutions at short times. As stated earlier, in quiescent solutions, in which diffusion is the only mode of mass transport, the diffusion-layer thickness is given by:

$$\delta = \sqrt{\pi Dt} \qquad (4.12)$$

Taking measurements at short times increases the value of j_L allowing us to study j_{ac} over a wider range of potentials.

Equation (4.12) is strictly applicable only in unstirred solutions, but as long as the value of δ calculated from it for a purely diffusion-controlled process is small compared to the thickness of the diffusion layer set up by stirring, the latter will have no effect on j_L.

Now that the importance of transient measurements in electrode kinetics has been established, it is of interest to discuss the effect of the residual jR_S potential drop in solution on the analysis of such transients.

Consider the equivalent circuit shown in Figure 4.3. It is important to understand that, although the potential that is applied in a potentiostatic measurement (or is being measured in a galvanostatic measurement) is between points A and C, the

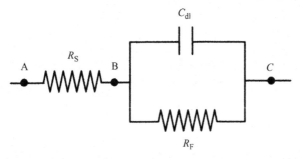

Figure 4.3 The equivalent circuit for an interface, showing the difference between the measured potential E_{AC} and the potential actually applied across the interface, E_{BC}, which drives the charge-transfer process.

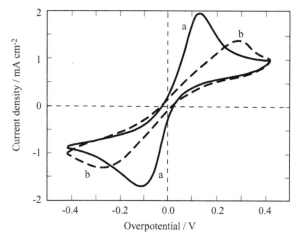

Figure 4.4 Variation of current density with time during linear potential sweep measurement in a poorly conducting electrolyte. (a) With dynamic $j R_S$ compensation, (b) without $j R_S$ compensation. 5 mM quinhydrone and 1 mM H_2SO_4, $v = 75$ mV s^{-1}.

potential driving the charge-transfer process is that between points B and C. The difference between these two potentials is

$$E_{AB} = j R_S \qquad (4.17)$$

Consider the effect of the uncompensated solution resistance on a cyclic voltammetry experiment, in which a cyclic linear sweep is applied between the working and the reference electrodes. A typical current–time response is shown in Figure 4.4.

The corresponding variations of the potential E_{AC} applied externally and the potential E_{BC} actually imposed on the interface are shown in Figure 4.5.

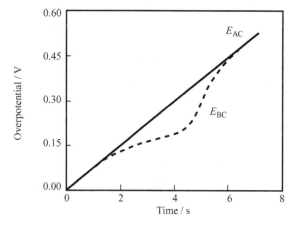

Figure 4.5 Schematic representation of the variation of potential with time during linear potential sweep. Solid line: the applied potential, E_{AC}, dashed line: the potential imposed across the interface, E_{BC}.

While the potentiostat delivers a constant rate of change of potential with time,

$$v = \frac{dE}{dt} = \text{constant} \tag{4.18}$$

the actual rate of change of potential imposed on the interface (dashed line in Figure 4.5) varies during the transient. This can cause a major shift in the position of the peak current density along the potential axis and modify the numerical value of the peak current, j_p, as seen in Figure 4.4. Analysis of the j/E plots obtained is based on the assumption that the potential applied across the interface varies linearly with time. If the potential drop across the uncompensated solution resistance is significant, it could lead to gross errors in the interpretation of the results.

It may be concluded that potentiostatic transients can be analyzed correctly only if the value of $j\,R_S$ is negligible throughout the transient or if it is has been compensated for in real time.

4.2.3
Polarography and the Dropping-Mercury Electrode (DME)

Polarography is one of the earliest electrochemical techniques, which had a major impact on the development of electroanalytical chemistry. Developed by Heyrovsky in the 1920s (and recognized by the Nobel Prize he received in 1959), it was implemented by connecting a fine capillary to a mercury reservoir and dipping the end of the capillary into the test solution. A mercury pool at the bottom of the cell served as both the counter and the reference electrode in a single-compartment, two-electrode set-up. In the classical configuration, the flow rate of mercury was determined by the hydrostatic pressure between the top of the mercury reservoir and the tip of the capillary, and by the dimensions of the capillary. Mercury drops were formed at the tip of the capillary, and were detached at equal intervals, when the weight of the drop just exceeded the surface tension holding it.

Mercury electrodes played a major role in the early development of electrochemistry, not only for electroanalytical purposes, but also for understanding the thermodynamics and kinetics of the metal/solution interface, the potential dependence of the double-layer capacitance and its relation to the dependence of the surface tension on potential. Electrosorption, the potential-dependent adsorption of ions and neutral molecules on electrode surfaces, was also studied initially on mercury electrodes. The unique features of mercury electrodes are that the metal can be purified by distillation, and the electrode is renewed periodically, each new drop presenting a highly reproducible, fresh, structureless and clean surface. Moreover, mercury is a very poor catalyst for hydrogen evolution. Consequently, it acts as an (almost) ideally polarizable interface over a fairly wide range of potentials, particularly in the cathodic direction, allowing the study of many cathodic electrode reactions for which the standard potential is negative with respect to the standard hydrogen electrode, with little interference from decomposition of the solvent, (i.e. of hydrogen evolution). When metal deposition is concerned, one finds the added advantage that many metals are soluble in mercury, forming an amalgam. The potential for formation of the amalgam can be significantly shifted in the positive direction compared to deposition on a solid electrode. For the alkali metals this shift can be as much as 1 V, allowing the study of the rate of deposition of such metals on mercury, but not on solid metals.

Mercury was used in large quantities in the chlor-alkali industry, for the production of chlorine and pure NaOH. However, its use in the industry was essentially stopped when its deadly impact on the environment was realized, and even in research its use has been limited as far as possible. In modern polarography, pressure is applied by an inert gas and the experiment is conducted using very small amounts of mercury.

For the classical dropping mercury electrode, the diffusion-limited current is given by the Ilkovic equation

$$I_d = 708 \, m^{2/3} t^{1/6} D^{1/2} c_b \tag{4.19}$$

where m is the flow rate of Hg (mg s^{-1}), t is the drop time(s), D is the diffusion coefficient (cm^2 s^{-1}) and c_b is the bulk concentration (mM). The current calculated from Eq. (4.19) with these units is given in μA. Under typical polarographic conditions, the value of I_d, calculated from the Ilkovic equation, is about 3 μA for $n = 1$, a drop time of 1 s, in a 1 mM solution. Note that this is a total current, I, not a current density, j, and the concentration is in units of mM, in agreement with the accepted notation in the polarographic literature. (In most other areas of electrode kinetics the current density and the concentration are given in units of [A cm^{-2}] and [mol cm^{-3}], respectively). The area of the drop, just before it falls, is about 0.01–0.04 cm^2, yielding current densities in the range 0.08–0.30 mA cm^{-2} under the above conditions.

The classical dropping-mercury electrode (DME) represents a rather complex system. As the volume of the drop grows, its surface moves towards the solution, while the diffusion layer is also growing. As a result, the thickness of the diffusion layer is less than that which would have been calculated for a stationary electrode. Therefore, the Ilkovic equation is an approximation, not an exact solution of the diffusion problem. The dependence of the diffusion-limited current on $t^{1/6}$ results from the combined effects of the increasing surface area (at a rate that is proportional to $t^{2/3}$) and the decreasing diffusion-limited current (that is proportional to $t^{-1/2}$).

For a reversible process, the potential depends on the current and is expressed by

$$E = E_{1/2} + \frac{2.3RT}{nF} \log \frac{I_d - I}{I} \tag{4.20}$$

where $E_{1/2}$ is the polarographic half-wave potential, given by

$$E_{1/2} = E^0 + \frac{2.3RT}{nF} \log \left[\left(\frac{\gamma_{Ox}}{\gamma_{Red}} \right) \left(\frac{D_{Red}}{D_{Ox}} \right)^{1/2} \right] \tag{4.21}$$

where γ_i stands for the activity coefficient of the species indicated by the subscript. Since the logarithmic term on the right-hand side of this equation is usually quite small, the measured value of $E_{1/2}$ is close to that of E^0, and it can serve to identify the species being reduced.

The current–potential relationship for irreversible polarographic waves is more complex than that for the reversible situation. Instead of Eq. (4.20), one can use the approximate expression

$$E = E_{1/2} + (0.916\,b)\log\frac{I_d - I}{I} \tag{4.22}$$

The half-wave potential in the above equation is given by

$$E_{1/2} = E^0 + b\log\left[1.32k_{s,h}(t/D)^{1/2}\right] \tag{4.23}$$

where $k_{s,h}$ is the heterogeneous rate constant at the standard potential, E^0 and b is the Tafel slope, to be discussed below in detail. The important point to note here is that the half-wave potential is a function of the heterogeneous rate constant and of the time of measurement: it is no longer determined by the standard potential alone.

In classical polarography (also referred to as DC polarography) the potential is changed linearly with time at a slow rate of a few mV s^{-1}, and the drops are allowed to fall freely. In modern instruments, the mercury drop is extruded in a few milliseconds to its final size, determined by the dimensions of the capillary and the gas pressure. The surface area is then maintained constant during measurement. This type of set-up is referred to as static-dropping-mercury electrode (SDME). In normal-pulse polarography (NP), the commonly used technique, the current is sampled once on each drop, and the potential is applied for a short time, just before the drop is knocked off. For a pulse width in the range 1–50 ms, the thickness of the diffuse double layer is 10 μm or less, compared to a radius of curvature of 250–350 μm for typical mercury drops. The thin diffusion layer can increase the limiting current (and hence the sensitivity) by more than an order of magnitude, compared to DC polarography. Moreover, the system can be treated as semi-infinite linear diffusion, yielding an accurate equation for the diffusion-limited current, since the complications attributed above to the expansion of the electrode surface during the measurement have been eliminated

$$I_d = n\,F\,A_p\,c_b\left(\frac{D}{\pi\,t_p}\right)^{1/2} \tag{4.24}$$

where A_p and t_p represent the area of the drop during the application of the pulse and the duration of the pulse, respectively.

Several other techniques related to polarography have been developed over the years, mainly to increase the sensitivity, but also to reduce the time of measurement and the amount of mercury used. The most common among them are differential-pulse (DP) and square-wave polarography (SW). These methods of employing the SDME for analytical applications have been widely treated in the literature and will not be discussed here.

4.2.4
Application of the Stationary Dropping-Mercury Electrode for Kinetic Studies

The SDME is also used to obtain kinetic information. Measurement at the foot of the polarographic wave, where the current is activation-controlled, can yield the most accurate I/E relationship achievable with any electrode, because of the very high reproducibility of the surface. Equations have also been developed that allow quantitative correction for mass-transport limitation up to higher currents ($I \leq 0.95\ I_d$) along the polarographic wave, extending the region where the activation-controlled current can be evaluated.

When the reaction being studied is relatively slow, the I/E relationship does not follow Eq. (4.20). This problem was treated by *Koutecky*, who was able to express the activation-controlled current I_{ac} in terms of a dimensionless parameter, χ, which is linearly related to the heterogeneous rate constant of the reaction:

$$\chi = (12/7)^{1/2}(t/D)^{1/2}k_h \tag{4.25}$$

Values of χ can be evaluated numerically as a function of I/I_d. From these, the activation-controlled current is obtained as a function of the measured current, hence also as a function of potential. In fact, the Tafel slope can be obtained directly from a plot of $\log\chi$ versus E. An example of the treatment of polarographic data following the theory developed by Koutecky is shown in Figure 4.6, for the case of the reduction of hydroxylamine (NH_2OH). These results were obtained by analysis of the polarographic data for values of j/j_d ranging from 0.001 up to 0.95, where the activation-controlled current is nineteen times larger than the diffusion-controlled current. This

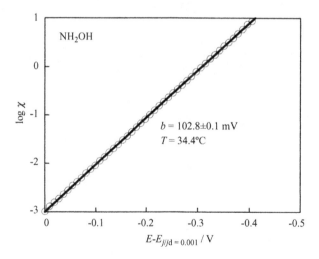

Figure 4.6 Semi-logarithmic plot of the dimensionless rate constant $\chi \equiv k_h(12/7)^{1/2}(t/D)^{1/2}$ versus E, for the reduction of 2 mM hydroxylamine in 0.2 M phosphate buffer, (pH = 6.8), 0.6 M KCl. Drop-time 1 s. Based on data from E. Kirowa-Eisner, M. Schwarz and E. Gileadi. *Electrochim. Acta*, **43** (1989) 1103.

may sound odd, because the total current cannot exceed the limiting current, but the activation controlled current can be larger than the mass-transport limited current, as seen in Eq. (1.5).

$$\frac{1}{j} = \frac{1}{j_{ac}} + \frac{1}{j_L} \tag{1.5}$$

4.3
Rotating Electrodes

4.3.1
The Rotating Disc Electrode (RDE)

There are many ways to increase the rate of mass transport by stirring. Moving the electrode in solution turns out to be more efficient, as a rule, than moving the solution by gas bubbling, using a magnetic stirrer, and so on.

One of the best methods of obtaining efficient mass transport in a highly reproducible manner is by the use of the rotating disc electrode. The RDE consists of a cylindrical metal rod embedded in a larger cylindrical plastic holder (usually Teflon, due to its great chemical stability and inertness). The electrode is cut and polished flush with its holder, so that only the bottom end of the metal cylinder is exposed to the solution.

The configuration of a RDE is shown schematically in Figure 4.7a, compared to a rotating ring–disc electrode (RRDE) shown in Figure 4.7b, which is discussed below (cf. Section 4.3.3). The most important feature of the RDE is that it acts as a so-called *uniformly-accessible surface*, which, in simple language means that the rate of mass transport to the surface is uniform. This property is by no means self evident, considering that the linear velocity of points on the surface of the rotating disc increases with their distance from the center of rotation. The other important property of the rotating disc is that flow of the solution at the surface is laminar up to rather high rotation rates.

In hydrodynamics, the transition from laminar to turbulent flow is characterized by a dimensionless parameter called the *Reynolds number, Re*. This number is the product of a characteristic velocity, v, and a characteristic length l, divided by the kinematic viscosity, ν. The latter is defined as the viscosity divided by the density, and has the dimensions $cm^2 s^{-1}$ (the same as the dimensions of the diffusion coefficient, D).

$$Re = \frac{vl}{\nu} \tag{4.26}$$

The characteristic velocity and length must be defined for each geometry separately, and in each case there is a critical value of the Reynolds number at which transition from laminar to turbulent flow takes place. For example, for flow

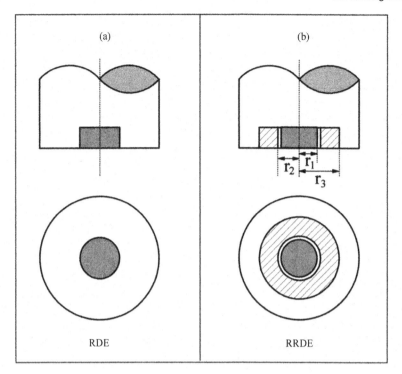

Figure 4.7 The rotatingdisk (a) and rotating ring–disk (b) electrodes. The electrodes and insulating material are marked by filled and clear areas, respectively.

in a tube $l = r$. The corresponding critical Reynolds number is about 2×10^3. In dilute aqueous solutions at room temperature, $v \approx 10^{-2}$ cm^2 s^{-1}. Thus, flow in a major pipeline is usually turbulent (for $r = 25$ cm, $v = 10$ cm s^{-1} one finds $Re = 2.5 \times 10^4$) whereas in small capillaries it is almost always laminar (for $r = 0.1$ cm, $v = 10$ cm s^{-1}, $Re = 1 \times 10^2$).

In the case of the rotating disc, the characteristic velocity is the linear velocity at the outer edge of the disc, given by $v = \omega r$, where ω is the angular velocity, expressed in radians per second. The characteristic length is taken as the radius of the disc and the critical Reynolds number is about 1×10^5. The condition for laminar flow is then:

$$Re = \omega r \frac{r}{v} = \frac{\omega r^2}{v} \leq 1 \times 10^5 \qquad (4.27)$$

For a typical radius of 0.2 cm, we find that laminar flow should be maintained up to $\omega \leq 2.5 \times 10^4$ rad s^{-1}, which corresponds to a rotation rate of about 2.5×10^5 rpm.

It should be noted that critical Reynolds numbers represent, as a rule, the upper limits for laminar flow at ideally smooth surfaces. If the surface is rough, turbulence may set in at a lower Reynolds number. Most RDEs are operated at a maximum rotation rate of $N \leq 10^4$ rpm ($\omega \approx 1 \times 10^3$ rad s^{-1}), well within the range of laminar flow. The lower limit for the rotation rate is determined by the requirement that the limiting current density resulting from rotation should be large compared to that which would exist in a stagnant solution due to natural convection. In practice, this corresponds to a lower limit of about 400 rpm, which can be extended to 100 rpm under carefully controlled experimental conditions.

One should be careful in applying the foregoing considerations to electrodes of widely different dimensions. For example, if one were to employ the RDE in an industrial cell using, say, an electrode area of 1000 cm^2, the critical Reynolds number would be reached at the rim of the electrode at a rotation rate of only 30 rpm, and turbulence could set in at even lower rotation rates. Such electrodes may still be of practical value in an industrial process, as long as one is aware that flow may become turbulent beyond a certain radius, and the surface will no longer be uniformly accessible.

At the other extreme, the assumptions made in solving the hydrodynamic equations under most circumstances imply that one would be ill advised to use an RDE with a radius of less than about 0.05 cm.

Since flow is laminar, it is possible to calculate the rate of mass transport. The corresponding equation, for the limiting current density, developed by Levich, is

$$j_L = 0.62 \, nFD^{2/3} \nu^{-1/6} \omega^{1/2} \, c_b \tag{4.28}$$

The current density calculated from this equation is in A cm^{-2} if the angular velocity is given in rad s^{-1} and the concentration is given in mol cm^{-3}. We can rewrite Eq. (4.28) in the more convenient form:

$$j_L = 0.20 \, nFD^{2/3} \nu^{-1/6} N^{1/2} \, c_b \tag{4.29}$$

where the angular velocity is replaced by the rotation rate, N, in units of rpm. Comparing the last two equations with the equation for mass-transport-limited current density,

$$j_L = n \, F \, Dc_b / \delta \tag{1.6}$$

we find, for the diffusion-layer thickness, δ, at the RDE, the expressions

$$\delta = 1.61 \, D^{1/3} \nu^{1/6} \omega^{-1/2} \tag{4.30}$$

with ω in rad s^{-1} and

$$\delta = 5.00 \, D^{1/3} \nu^{1/6} N^{-1/2} \tag{4.31}$$

for the rotation rate expressed in rpm.

For typical values of $D = 1 \times 10^{-5}$ cm^2 s^{-1} and $\nu = 1 \times 10^{-2}$ cm^2 s^{-1}, the diffusion-layer thickness turns out to be 5–50 μm for rotation rates of 10^2–10^4 rpm. This range should be compared to values of 50–150 μm obtained by stirring of the solution. For a diffusion-controlled process, a value of $\delta = 5$ μm is reached after about 8 ms (cf. Eq. 4.12), showing the advantage of even moderately fast transients (say, in the range of 0.1–1.0 ms) in enhancing the rate of mass transport. On the other hand, measurement with a RDE is a steady-state measurement, which is often advantageous.

Considering Eq. (4.29), we note that j_L is a linear function of concentration. Hence the RDE can be used as a tool in electroanalytical measurements. It has also been used extensively to determine the diffusion coefficients of different electroactive species in solution.

One may wonder why the movement of the RDE in a fixed plane has an effect on mass transport in a direction perpendicular to that plane. This comes about because the disc drags the solution nearest to it and imparts to it momentum in the tangential direction. As a result, solution is pushed out of the surface sideways (in a plane parallel to it) and is replaced by solution moving in from the bulk, in a direction perpendicular to the surface. The rotating surface acts, in effect, as a pump, pulling the liquid up toward it, as shown schematically in Figure 4.8.

So far, we have discussed the properties of the limiting current density at the RDE. What about activation or mixed control? For a purely activation-controlled process, the current should be independent of rotation rate, or should at least become independent of it beyond a certain rotation rate. Under conditions of mixed control, the activation-and mass-transport-controlled current densities combine to yield the total current density as the sum of reciprocals, namely

$$\frac{1}{j} = \frac{1}{j_{ac}} + \frac{1}{j_L} \tag{1.5}$$

Since j_L is proportional to $\omega^{1/2}$, this can be rewritten as:

$$\frac{1}{j} = \frac{1}{j_{ac}} + \frac{1}{B\omega^{1/2}} \tag{4.32}$$

Equation (4.32) is very useful for the study of electrode kinetics. It is clear that a plot of $1/j$ versus $1/\omega^{1/2}$ should yield a straight line having an intercept of $1/j_{ac}$. By

(a)

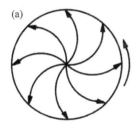

(b)

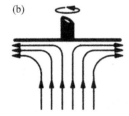

Figure 4.8 Motion of the liquid at the surface of an RDE. (a) Tangential motion in the plane of the electrode, (b) perpendicular motion towards the electrode.

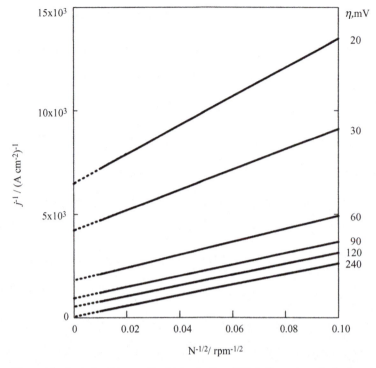

Figure 4.9 Analysis of the results obtained with a RDE, in the region of mixed control. The activation-controlled current density is obtained from the intercept of the lines at infinite rotation rate, where $(N^{-1/2} \rightarrow 0)$.

repeating the experiment at different potentials, one can obtain the dependence of j_{ac} on potential, as shown in Figure 4.9.

Extrapolation of $1/j$ to $N^{-1/2} = 0$ is equivalent to measuring at a point where the rotation rate, and thus the limiting current density, is extrapolated to infinity. At this point the current would evidently be controlled only by the rate of charge transfer.

The constant B, as defined here (Eq. (4.32)), appears to be independent of potential. This is indeed true in most cases, but a detailed analysis shows that it is a function of potential where the system is studied close to equilibrium. It is nevertheless reassuring to note that the value of j_{ac} obtained by extrapolating $1/j$ versus $N^{-1/2}$ in the foregoing manner is correct, irrespective of the dependence of the parameter B on potential.

As might be expected, several methods have been developed that make use of the basic idea of the RDE, but improve upon it or extend it for the purpose of studying electrode kinetics. Some of these are discussed later. Some caution should be exercised, however, in the analysis of such data, because of the nonuniformity of current distribution. Thus, although the RDE represents a *uniformly accessible surface*, implying that the mass-transport-limited current density is uniform, the same cannot be said for the activation-controlled current density. The configuration of a metal electrode and its insulating holder in the same plane, necessarily leads to large edge effects (cf. Eq. (3.9) and Figure 3.6), if the rate of mass transport is high. The resulting error can be lessened, but not entirely eliminated, by placing a ring around the electrode, separated from it electrically and held at the same potential.

The best way to avoid the problem of nonuniform current distribution on the electrode surface is by operating at lower current densities (which increases the value of R_F, as shown below, cf. Section 19.2) or by increasing the conductivity in solution, to decrease R_S, or both.

4.3.2
The Rotating Cone Electrode (RConeE)

The usefulness of the RDE stimulated the development of several other rotating configurations that are worth mentioning. Some of these have evolved in response to specific experimental needs, whereas others serve as possible extensions of the technique.

Often bubbles are generated at the electrode surface (e.g., when the reaction being studied is the hydrogen evolution reaction) or reach it from the bulk of the solution from the gas used to deaerate it. A bubble trapped at the surface of the RDE is driven toward its center by centrifugal force. This phenomenon can cause major errors in measurement because (i) part of the surface of the electrode is blocked and (ii) the hydrodynamic flow in the vicinity of the electrode may be greatly distorted.

A solution to this problem is to use a *rotating cone electrode* (RConeE). It turns out that RCone electrodes are similar to RDE in that they act as uniformly accessible surfaces, producing a uniform rate of mass transport to all parts of the electrode surface. For a cone having an opening angle of θ, the current density is given by

$$j_{L, \text{cone}} = j_{L, \text{disc}} \sqrt{\sin(\theta/2)} \tag{4.33}$$

When a bubble forms at the rotating cone, it is also driven to its center, namely to the tip of the cone. However, it is in a rather precarious state there, and if it is not dislodged spontaneously by some minor instability in rotation, it can be easily removed.

4.3.3
The Rotating Ring–Disc Electrode (RRDE)

The rotating ring–disc electrode (RRDE) is perhaps the most useful extension of the idea of the rotating disc. The ring current has the same form as the disc current, namely, it is also proportional to the bulk concentration of the electroactive species and to the square root of the rotation rate[2].

$$I_R = 0.62 \, nF\pi \left(r_3^3 - r_2^3\right)^{2/3} D^{2/3} v^{-1/6} \omega^{1/2} \, c_b \tag{4.34}$$

---

2) Note that for the rotating-ring and rotating-ring–disc electrodes the common practice is to use the total currents, I, rather than the current densities, j.

The great advantage of the RRDE is that it can be used to analyze short-lived intermediates in a steady-state measurement.

The RRDE is constructed in such a way that both electrodes are in the same plane, and in close proximity to each other. The dimensions of the electrode are defined by three radii: the radius of the disc (r_1), the inner radius of the ring (r_2) and its outer radius (r_3), as shown in Figure 4.7b.

Since both electrodes are part of the same flow regime (evidently the convective flow of the liquid cannot distinguish between the disc electrode, the insulator and the ring), the equations for the limiting current at both the disc and the ring electrodes can be solved simultaneously.

In order to operate an RRDE, one needs a four-electrode potentiostat, often referred to also as a "bi-potentiostat". Now, we began this book with simple two-electrode cells and proceeded to the three-electrode configuration. How does the fourth electrode fit into this scheme? The best way to understand this is to view both the ring and the disc as working electrodes, with common counter and reference electrodes. The important point is that the four-electrode potentiostat controls the potentials of the disc and of the ring with respect to the reference electrode *independently* and measures the current going through each of them separately.

We could, for example, hold the potential of the ring at a fixed value and vary the potential of the disc, allowing an analysis of the products formed on the disc by studying the ring current as a function of the potential applied to the disc, or vice versa.

The RRDE is calibrated by determining its collection efficiency N. This parameter represents the fraction of a species formed at the disc that reaches the ring and reacts there. To do this, consider a simple oxidation reaction, say, Fe^{2+} to Fe^{3+}. The disc potential is set at a positive value, where the oxidation process occurs at some rate I, while the ring potential is set at a negative value, at which the oxidized species is reduced at its limiting current. If there is initially no Fe^{3+} in solution, the ring current results only from Fe^{3+} ions produced at the disc and transported to the ring by convection (there is always some diffusion, but its rate is negligible compared to the rate of convection in an RRDE experiment). The ratio of ring current to disc current observed under these conditions is defined as the collection efficiency N.

$$N = I_R/I_D = f(r_1; r_2; r_3) \tag{4.35}$$

where I_R and I_D represent the *total* currents measured for the ring and the disc, *not* the corresponding current densities. The important thing to notice in this equation is that the collection efficiency is not a function of the rotation rate; rather, it depends exclusively on the dimensions of the electrodes. The function $f(r_1, r_2, r_3)$ is somewhat complicated, but its values have been tabulated, and it can be obtained, given the values of the above three radii. In practice, it is always preferable to evaluate N experimentally, since it depends strongly on the width of the gap, $(r_2 - r_1)$, which cannot be measured very accurately. Typical values of the collection efficiency for commercial RRDEs are in the range of 0.1–0.3. Decreasing the insulating gap between the disc and the ring increases N. Decreasing the width of the ring, $(r_3 - r_2)$, leads to a lower collection efficiency, but improves the accuracy of measurement of the lifetime of unstable intermediates formed at the disc.

If an unstable intermediate or product is formed at the disc, only a fraction of it will reach the ring, and the ratio I_R/I_D will be smaller than N. In this case, the extent to which the collection efficiency has decreased is a function of the rate of rotation. The dependence of the collection efficiency on the rotation rate can be used to evaluate the lifetime (or rate of decomposition) of the unstable intermediate. As in any kinetic measurement of this type, one attempts to design the system for the fastest possible transition time from disc to ring, to allow detection of short-lived intermediates. The gap in commercial RRDEs is of the order of 0.01 cm, but electrodes with substantially narrower gaps have been built. One might be tempted to use modern techniques of microelectronics to construct an RRDE with a very small gap, say, 0.1 µm. A closer examination of the hydrodynamics involved reveals, however, that this may not work and, in fact, there is little or no advantage in reducing the gap below about 5–10 µm.

The RRDE can also be very useful in other studies, where the species formed at the disc are stable. For example, the corrosion of a metal is always associated with a cathodic reaction, which in deaerated solutions is hydrogen evolution and in the presence of air it might be the reduction of molecular oxygen. The rate of corrosion of a metal (used as the disc) at open circuit can therefore be followed indirectly by measuring the amount of molecular hydrogen reaching the ring electrode or by the decline in the amount of molecular oxygen reaching it. The RRDE can also be used to study complex reactions in which two processes occur in parallel. For example, the electrodeposition of zinc is accompanied by hydrogen evolution. The partial current for hydrogen evolution at the disc can then be determined by measuring the amount of molecular hydrogen reaching the ring. Similarly, when a $Ni(OH)_2/NiOOH$ electrode (used in a Ni/Cd battery) is charged, oxygen evolution is known to start before the electrode has been fully charged. The partial current for charging can be determined by monitoring the amount of oxygen reaching the ring electrode, and calculating the corresponding current, corrected for the collection efficiency, from the total observed disc current.

4.3.4
Rotating Cylinder Electrode (RCylE)

Using a rotating cylinder electrode is a good way to achieve high rates of mass transport. This is very different, however, from the RDE in that the flow is turbulent rather than laminar. As a result, it is not possible to derive theoretical equations that relate the rate of mass transport to the various parameters in the reaction, and one must resort to empirical correlations. These tend to be critically dependent on dimensions and on the specific configuration of the cell, and so are less reproducible. The equation most often used for mass transport to a rotating inner cylinder of radius r is:

$$j_L = 0.079 \, n \, Fr^{0.4} D^{0.65} \nu^{-0.34} c_b \, \omega^{0.7} \tag{4.36}$$

It should be noted specifically that j_L is proportional to $\omega^{0.7}$ and not to $\omega^{0.5}$, as in the case of the RDE. Laminar flow around a rotating cylinder does not enhance the rate of mass transport to it, because laminar flow, by definition, implies that the liquid must move in circles parallel to the surface, with no components of movement perpendicular to it. On the other hand, turbulence sets in at very low rotation rates, of the order of 10–20 rpm for electrodes of sizes typically used in the laboratory.

4.4
The Physical Meaning of Reversibility

It may be appropriate to ask here why the potential at a reversible electrode should change at all with current density. This does not occur because "no system is really ideally polarizable", and one is observing a small polarization. Indeed the relationships shown in Eq. (4.20) holds strictly only when the interface is ideally nonpolarizable.

The physical meaning of reversibility is that the concentrations of reactants and products *at the interface* are equal at all times to their equilibrium values, corresponding to the applied potential. For the system to be maintained under reversible conditions, the rate of charge transfer must be high enough to allow the surface concentrations to maintain their equilibrium values corresponding to the Nernst equation during the measurement. This naturally depends not only on the heterogeneous rate constant but also on the experimental conditions, such as the timescale of the measurement and the rate of diffusion. In this context it may be argued that there is neither a reaction that is always reversible, nor one that is always irreversible – it all depends on the timescale of the experiment.

This is presented in Figure 4.10a, which shows polarograms calculated for different values of the standard heterogeneous rate constant $k_{s,h}$, and a constant time of measurement of the current. A similar plot is shown in Figure 4.10b, calculated for a fixed value of $k_{s,h} = 10^{-2.5}$ cm s^{-1} and variable times of measurement. Close examination of these figures shows that there is a gradual transition from reversible to irreversible behavior, determined primarily by the ratio between the kinetically controlled rate constant and the characteristic rate of diffusion, expressed by the ratio D/δ. Comparing the equations for j_{ac} and j_L:

$$j_{ac} = n F k_{s,h} c_b \tag{4.37}$$

$$j_L = n F (D/\delta) c_b \tag{1.6}$$

it can be seen that the characteristic dimensionless parameter for transition from irreversible to quasi-reversible and then to reversible behavior is given by

$$\frac{k_{s,h}}{D/\delta} = \frac{k_{s,h}}{D}(\pi D t)^{1/2} \tag{4.38}$$

When the above parameter is ≥ 10, the system can be considered to be essentially reversible. When it is ≤ 0.1 the system can be regarded as irreversible.

(a)

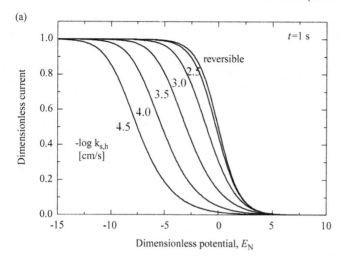

(b)

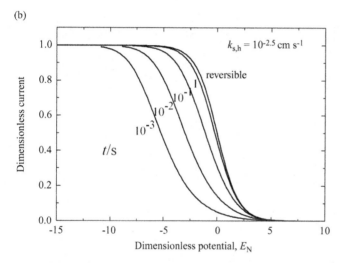

Figure 4.10 Polarographic waves calculated for (a) a fixed time of measurement but different values of the heterogeneous rate constant, and (b) for a fixed value of the rate constant but different times of measurement. $D = 8 \times 10^{-6}$ cm^2 s^{-1}, $\alpha = 0.5$, $n = 1$. $E_N \equiv E/(RT/F)$.

Similar considerations can be used to determine reversibility in stirred solutions, for example, when measurement is conducted on a rotating disc electrode, where the Nernst diffusion layer is independent of time.

$$\delta = 1.61\, D^{1/3} v^{1/6} \omega^{-1/2} \tag{4.30}$$

Substituting this value into Eq. (4.38) yields

$$\frac{k_{s,h}\delta}{D} = 1.61\, k_{s,h} D^{-2/3} v^{1/6} \omega^{-1/2} \tag{4.39}$$

This equation can be used to determine the values of $k_{s,h}$ for which the reaction will behave reversibly for any rotation rate. We use the requirement that $k_{s,h}\delta/D \leq 0.1$ for totally irreversible behavior and $k_{s,h}\delta/D \geq 10$ for reversible behavior. For a rotation rate of 2500 rpm, for example, this leads to irreversible behavior for $k_{s,h} \leq 3.4 \times 10^{-4}$ cm s^{-1} and reversible behavior for $k_{s,h} \geq 3.4 \times 10^{-2}$ cm s^{-1}, which are rather similar to the values shown in Figure 4.10 for polarographic measurements at 1 s.

A qualitative way of looking at reversibility is to consider the total overpotential as a sum of three terms

$$\eta = \eta_{ac} + \eta_{conc} + \eta_{jR_s} \tag{4.40}$$

Reversibility is then defined as the situation where the activation overpotential, η_{ac}, is negligible compared to the sum of the concentration overpotential, η_{conc}, and the overpotential associated with the solution resistance, η_{jR_s}.

5
Single-Step Electrode Reactions

5.1
The Overpotential, η

5.1.1
Definition and Physical Meaning of Overpotential

The standard potential for a given electrode reaction is related to the difference in Gibbs energy between reactants and products.

Its dependence on concentrations in solution is expressed by the Nernst equation, which has the general form

$$E_{rev} = E^0 + \frac{2.3\,RT}{nF}\log\frac{c_{Ox}}{c_{Red}} \tag{5.1}$$

where the activity coefficients were taken as unity, for simplicity.

When at open circuit, a nonpolarizable electrode assumes its reversible potential, whereas a polarizable electrode may deviate from it significantly. In either case, the overpotential, η, is defined as the difference between the actual potential measured (or applied) and the reversible potential.

$$\eta \equiv E - E_{rev} \tag{5.2}$$

This definition of overpotential is phenomenological and is always valid, irrespective of the reasons for the deviation of the potential from its reversible value. The overpotential is always defined with respect to a specific reaction, for which the reversible potential is known. When more than one reaction can occur simultaneously on the same electrode, there is a different overpotential with respect to each reaction, for any value of the measured potential. This situation is encountered most commonly during the corrosion of metals. When iron corrodes, for example, in a neutral solution, the overpotential may be $+0.4\,V$ with respect to metal dissolution and $-0.8\,V$ with respect to oxygen reduction. During metal deposition, hydrogen evolution often occurs as a side reaction. At any given potential the *overpotential* with

Physical Electrochemistry: Fundamentals, Techniques and Applications. Eliezer Gileadi
Copyright © 2011 WILEY-VCH Verlag GmbH & Co. KGaA, Weinheim
ISBN: 978-3-527-31970-1

respect to metal deposition and to hydrogen evolution will, of course, be different. However, in most fundamental studies of electrode kinetics, the experiment is set up in such a manner that only one reaction can take place in the range of potential studied, and the overpotential is unambiguously defined.

The overpotential is a measure of the distance of a reaction, on the Gibbs-energy scale, from its equilibrium state. Multiplied by the charge consumed per mol of reactants, nF, it is equal to the affinity, A, of a reaction, which is a measure of the thermodynamic driving force that exists to make the reaction occur,

$$A = nF|\eta| \tag{5.3}$$

Now, we have already seen that the electrochemical Gibbs energy of activation is linearly related to the applied potential, giving us a powerful tool to control the rate of electrode reactions over many orders of magnitude. At the other extreme, we can also use the potential to probe the reaction under conditions close to equilibrium, by applying small values of the overpotential in both directions around zero, and measuring the resulting current densities.

It will be remembered that equilibrium does not imply total freezing of the reaction. It is characterized by a dynamic situation, in which the reaction proceeds in both directions at a high rate. Equilibrium is reached when the forward and backward rates are equal and the *net rate*, which one would observe, say, as a change of concentration in a chemical reaction or a flow of current in the external circuit in an electrochemical reaction, is zero.

Consider the following simple chemical reaction at equilibrium, with reactants and products all in the gas phase.

$$I_2 + H_2 = 2HI \tag{5.4}$$

By definition, the affinity $A = 0$ and the net reaction rate $v = 0$. We can disturb the equilibrium slightly by adding a small amount of one of the molecules or by changing the temperature. The rate at which the reaction will proceed toward its new equilibrium is given by

$$v = v_{eq}(A/RT) \tag{5.5}$$

where v_{eq} is the exchange rate, namely, the rate at which the reaction proceeds back and forth at equilibrium.

For transfer of a unit charge (i.e. an electron or an ion) we can similarly write

$$j = j_0(\eta\,F/RT) \tag{5.6}$$

In this equation j_0 is called *the exchange-current density* and it represents, much as does v_{eq}, the current density in both the anodic and cathodic directions, when the system is at equilibrium.

Equations (5.5) and (5.6) are special cases of a general rule, according to which, whenever a system is perturbed to a small extent, the response is a linear function of the magnitude of the perturbation. But how small is "small" in the present context? This must be defined in dimensionless form, comparing affinity to the average thermal energy, or rate to the exchange rate. For Eq. (5.5) the perturbation is small if $A/RT \ll 1$. Likewise, for the case of charge transfer, the requirement is $|\eta|F/RT \ll 1$. Clearly, a small perturbation also leads to $v/v_{eq} \ll 1$ and $j/j_0 \ll 1$. However, this latter criterion is mainly relevant to electrode kinetics, since only there can one control the reaction rate (in a galvanostatic measurement) and observe the resulting affinity (i.e., overpotential).

To obtain a better feel for the quantities involved, we note that RT, the average thermal energy, is equal at room temperature to $0.0256\,\mathrm{eV} = 2.48\,\mathrm{kJ\,mol^{-1}}$. Thus, for $\eta = 5\,\mathrm{mV}$ we have

$$|\eta|F/RT \approx 0.2 \tag{5.7}$$

We conclude that a linear j/η relationship is to be expected for overpotentials in the range $-5 \leq \eta \leq +5\,\mathrm{mV}$. The value of the exchange-current density, j_0, can readily be obtained from the slope of such plots, using Eq. (5.6).

5.1.2
Types of Overpotential

In the previous section we viewed the overpotential as the stimulus (or perturbation), which causes the reaction to proceed in the desired direction, at a certain rate, represented by j. The ratio between the perturbation and the response is a measure of the effective resistance of the system, called the Faradaic resistance, R_F, (also referred to as the charge-transfer resistance, R_{ct}) In the example given earlier, which is applicable to small perturbations, the ratio (η/j) is constant, since the current is a linear function of the overpotential. However, in general, R_F is a function of potential, as implied by Figure 1.1. The units of R_F are $\Omega\,\mathrm{cm^2}$, since it is defined as the ratio between the overpotential and the current density. Multiplied by the current density, in units of $A\,\mathrm{cm^{-2}}$, it yields the overpotential

$$j\,R_F = \eta_F \tag{5.8}$$

In general, it is better to define the Faradaic resistance in differential form

$$R_F \equiv (\partial\eta/\partial j)_c \tag{5.9}$$

bearing in mind that R_F can be a function of η.

The overpotential observed experimentally can result from several unrelated physical phenomena, all hindering the reaction rate. Here we shall discuss the three main causes of overpotential.

Activation overpotential η_{ac}, (also called the charge-transfer overpotential, η_{ct},) is the kinetically significant quantity that acts on the electrochemical energy of activation, lowering it in one direction and increasing it in the opposite direction; η_{ac} is the change of potential across the interface, generated by an applied current density.

$$\eta_{ac} = {}^M\Delta^S\phi - {}^M\Delta^S\phi_{rev} = \delta\left({}^M\Delta^S\phi\right) \tag{5.10}$$

In proper kinetic studies one needs to isolate this type of overpotential. A mechanistic study is then based on determining the dependence of η_{ac} on current density, over a wide range of experimental conditions.

The second type of overpotential often mentioned in the literature is the *resistance overpotential*, η_{jR}, which results from the residual potential drop jR_S in the solution. In the research laboratory, this type of overpotential can be largely eliminated by proper positioning of the Luggin capillary, combined with electronic compensation for most of the remaining resistance, in real time. It does, however, have a major effect in industrial electrochemistry, with respect to its influence on the performance of batteries and on the energy consumption of electrolytic processes. About one third of the power consumption in chlorine production and two thirds in aluminum production are due to the voltage drop over the ohmic resistance in the cell[1]. This energy is transformed to heat, which must be removed by proper heat exchangers in order to maintain the desired temperature of operation. The immense magnitude of this effect may be appreciated when it is realized that the amount of electricity transformed into thermal energy in the chlorine and aluminum industries combined is about 2–3% of the total electric-power production in the United States, costing about $1.5 billion a year at an average price of ($0.05 per kW h). Being an electrochemist, one should, however, not be humbled by the staggering loss of energy and money due to the inefficiency of these electrochemical processes, when compared to the inherent inefficiency of electric-power generation. Consider the following: Thermal power stations (whether powered by fossil or nuclear fuel) operate at an average efficiency of about 45%, which means that 55% of the thermal energy of the fuel is wasted. This amounts to a direct loss of about $25 billion a year in the United States alone, not to mention the cost involved in the efficient removal of this heat, which is essential for the operation of the plants.

The third important type of overpotential is caused by mass-transport limitation. It is called *concentration overpotential*, η_{conc}, because, when the reaction is influenced by the rate of mass transport, the concentration at the surface is different from its bulk value (cf. Eq. (4.16)). Since the reaction rate depends only on the concentration *at the surface*, it is necessary to apply a higher overpotential to maintain the same rate when the concentration at the surface is lower than in the bulk.

In conclusion, one may be tempted to write the overpotential as the sum of three terms:

$$\eta = \eta_{ac} + \eta_{conc} + \eta_{jR} \tag{5.11}$$

This is conceptually correct, but it must be remembered that η_{ac} and η_{conc} (and to a lesser extent η_{jR}) are interrelated, and an attempt to obtain η by calculating the three contributions separately and adding them may lead to gross errors.

---

1) This is the total resistance between the two electrodes in an industrial cell, not the residual resistance measured in the laboratory between the working electrode and the tip of the Luggin capillary.

5.2
Fundamental Equations of Electrode Kinetics

5.2.1
The Empirical Tafel Equation

The Tafel equation was first written in 1905 in the form:

$$\eta = a - b \log j \tag{5.12}$$

as an empirical equation relating the observed overpotential to the current density during hydrogen evolution on mercury and lead cathodes. It shows a linear relationship between the overpotential and $\log j$, with a slope b, which came to be known as the *Tafel slope*. The numerical value of b observed by Tafel was about 0.10 ± 0.02 V per decade of current density. This is consistent with our previous discussion, in which we showed that the rate of an electrochemical reaction can be increased by many orders of magnitude by applying a relatively small overpotential to the electrode[2]. With present-day understanding of electrochemistry, one wonders whether Professor Tafel was lucky in choosing this particular system for his studies, or whether he had some deep insight, not reflected in his original papers: the conditions under which the Tafel equation can be observed, such as a high over-potential, no mass-transport limitation, uniform current distribution and no inter-ference by film formation, all happen to exist in the particular experimental systems chosen by him.

For many years the Tafel equation was viewed as an empirical equation. A theoretical interpretation was proposed only after Eyring, Polanyi and Horiuti developed the transition-state theory for chemical kinetics, in the early 1930s. Since the Tafel equation is one of the most important fundamental equations of electrode kinetics, we shall derive it first for a single-step process and then extend the treatment for multiple consecutive steps. Before we do that, however, we shall review very briefly the derivation of the equations of the transition-state theory of chemical kinetics.

5.2.2
Transition-State Theory

Consider the simple isotope-exchange reaction in which a hydrogen atom reacts with a deuterium molecule, D_2, to form a molecule of HD and a free deuterium atom:

$$H + D{-}D \rightarrow H{-}D + D \tag{5.13}$$

One could imagine this reaction occurring by D_2 first splitting into two atoms, followed by one of the atoms combining with the hydrogen atom. This is, however,

---

[2] The December 2005 issue of the journal "Corrosion" was dedicated to the 100th anniversary of the publication of the Tafel Equation. It contains mostly review articles, which can be useful for a deeper understanding of electrode kinetics.

a highly unlikely course of events, because a great deal of energy is required to break the D–D bond. A probable route is the formation of an intermediate, such as an unstable D---D---H species, in which both bonds are of about equal strength, and the dissociation of this intermediate to either the original reactants or to the products. The standard Gibbs energy needed to form the unstable intermediate, which is called *the activated complex*, is the standard Gibbs energy of activation for the reaction.

We shall now write the reaction in more general form as

$$A \rightarrow X^{\#} \rightarrow B \tag{5.14}$$

in which A and B are the reactant and product molecules, respectively, and $X^{\#}$ represents the activated complex. Two main assumptions are made in the framework of the transition-state theory: (i) the reaction rate is assumed to be proportional to the concentration of the activated complex, and (ii) the activated complex is assumed to be at equilibrium with the reactant. With these assumptions one can write, for the reaction rate

$$v = k_f c_A = \frac{k_B T}{h} c^{\#} \tag{5.15}$$

For equilibrium between the reactant and the activated complex we can write

$$\frac{c^{\#}}{c_A} = K^{\#} = \exp\left(-\frac{\Delta G^{0\#}}{RT}\right) \tag{5.16}$$

In these equations, k_f, k_B and h are the forward rate constant, Boltzmann's constant and Planck's constant, respectively, and $\Delta G^{0\#}$ is the standard Gibbs energy of activation. The latter equals, by definition, the standard Gibbs energy of the reaction in which the activated complex is formed from the reactants. The term $k_B T/h$ in Eq. (5.15) has units of frequency (s^{-1}) and follows from an added assumption, namely that the critical bond in the activated complex, which must be severed in order to form the product, is very weak and behaves "classically" (in the sense that the energy levels are so close to each other that there appears to be a continuum). We need not concern ourselves with the numerical value of this term, and it has been replaced by the Greek letter ω in all further equations. Substituting $c_{\#}$ from Eq. (5.16) into Eq. (5.15) we obtain

$$v = k_f \, c_A = \omega \, c_A \exp\left(-\frac{\Delta G^{0\#}}{RT}\right) \tag{5.17}$$

or

$$k_f = \omega \exp\left(-\frac{\Delta G^{0\#}}{RT}\right) \tag{5.18}$$

Equation (5.18) shows the chemical rate constant as a function of the Gibbs energy of activation. In order to use this equation in electrode kinetics, it is necessary to relate the standard Gibbs energy of activation to the potential difference, $\Delta\phi$, across the interface.

5.2.3
The Equation for a Single-Step Electrode Reaction

The relationship between the Gibbs energy and potential was discussed in Section 4.1.2. For the standard electrochemical Gibbs energy of a reaction, we wrote Eq. (4.6) and for the standard electrochemical Gibbs energy of activation, Eq. (4.7).

$$\Delta \overline{G}^0 = \Delta G^0 \mp z F \Delta \phi \tag{4.6}$$

$$\Delta \overline{G}^{0\#} = \Delta G^{0\#} \pm \beta F \Delta \phi \tag{4.7}$$

The symmetry factor β in Eq. (4.7) is discussed in Section 5.3 below. The positive and negative signs in Eq. (4.7) are applicable to cathodic and anodic reactions, respectively. When Eq. (4.7) is combined with the rate equation we have, for an anodic process

$$j = Fv = F \, \omega \, c_A \exp\left(-\frac{\Delta G^{0\#}}{RT}\right) \exp\left(\frac{\beta \, F\Delta\phi}{RT}\right) \tag{5.19}$$

There are two ways in which we wish to modify this equation. First, we simplify it by replacing the term $\omega \exp(\Delta G^{0\#}/RT)$ by a "chemical" rate constant, k_h^0, which is the value of the heterogeneous rate constant at $\Delta\phi = 0$, to yield

$$j = F \, k_h^0 c_A \exp\left(\frac{\beta F\Delta\phi}{RT}\right) \tag{5.20}$$

Second, we would like to replace the awkward term $\Delta\phi$ which, as we recall, cannot be measured. Now, we have gone to some length to show that, although $\Delta\phi$ cannot be measured, changes in it can be readily determined (cf. Section 2.1.2). Bearing this in mind we can write the overpotential as:

$$\eta = \Delta\phi - \Delta\phi_{rev} = \delta\Delta\phi \tag{5.21}$$

where $\Delta\phi_{rev}$ is the value of $\Delta\phi$ at the reversible potential $\eta = 0$. Substituting $\Delta\phi$ from this equation into Eq. (5.20), yields:

$$j = F \, k_h^0 \, c_A \exp\left(\frac{\beta \, \Delta\phi_{rev} \, F}{RT}\right) \exp\left(\frac{\beta\eta \, F}{RT}\right) \tag{5.22}$$

Equation (5.22) can also be written in the form

$$j = F \, k_h \, c_A \exp\left(\frac{\beta F}{RT}\eta\right) \tag{5.23}$$

in which k_h is the heterogeneous electrochemical rate constant at the reversible potential. Equation (5.23) is an expression for the current density (which, we recall, is the rate of the reaction in electrical units) in terms of the concentration and the overpotential, both of which can be measured and controlled experimentally. It should be noted that in deriving Eq. (5.23) we have tacitly assumed that the concentration at the surface is independent of the current density, in other words, we have neglected mass-transport limitations. Strictly speaking, we should therefore replace η by η_{ac} in these equations. For the sake of simplicity we shall leave it as it is, and refer specifically to mass-transport limitation separately.

Equation (5.23) represents the rate of an anodic oxidation reaction for which the overpotential is, by definition, positive. For a cathodic reduction we write a similar equation with a negative sign in the exponent, and take the overpotential, by definition, to be negative. In either case the current density increases exponentially with increasing absolute value of the overpotential, $|\eta|$.

The complete expression for the current density is obtained as the difference between the anodic and the cathodic current densities. This is the current measured in an external circuit. Now, if we have used Eq. (5.23) to represent the potential dependence of the electrochemical rate equation for the anodic direction, what would we use for the cathodic reaction? Clearly, the term including η must have the opposite sign, since the same potential that enhances the rate of the reaction in one direction must retard it in the opposite direction. What about the symmetry factor, β? We noted earlier (cf. Eq. (4.8)) that it must have a value between zero and unity. Also, we have seen in Eq. (4.9) that β represents the ratio of the effect of potential on the electrochemical Gibbs energy of activation to its effect on the electrochemical Gibbs energy of the overall reaction. If this fraction is β in the anodic direction, it must be $(1-\beta)$ in the cathodic direction. Thus we have, for the net current density measured in the external circuit:

$$ j = j_a - j_c = F\,k_{h,f}\,c_A \exp\left(\frac{\beta F}{RT}\eta\right) - F\,k_{h,b}\,c_B \exp\left[-\frac{(1-\beta)F}{RT}\eta\right] \qquad (5.24) $$

Consider now what happens at equilibrium. The overpotential is, by definition, zero and so is the net current density. This leads to

$$ F\,k_{h,f}\,c_A = F\,k_{h,b}\,c_B \equiv j_0 \qquad (5.25) $$

We can substitute this in Eq. (5.25) to obtain

$$ j = j_0\left\{\exp\left(\frac{\beta\,F}{RT}\right)\eta - \exp\left[-\frac{(1-\beta)\,F}{RT}\eta\right]\right\} \qquad (5.26) $$

The physical meaning of j_0, which is called the *exchange-current density*, should be clear from the definition given in Eq. (5.25). It represents the rate at which the electrochemical reaction proceeds back and forth at equilibrium, when the net reaction rate, observed as a current flowing through the external circuit, is zero. It is similar to the exchange rate discussed earlier in connection with Eq. (5.5). We also note that j_0 is the heterogeneous electrochemical rate constant at $\eta = 0$, multiplied by the appropriate concentration. One cannot overemphasize the fact that, while the measured current is zero, the reaction is not "frozen" and can occur at a high rate. The exchange-current density can be much larger than the current measured, which may be a small difference between large anodic and cathodic currents.

Some variations will be found in the sign convention used in different textbooks on electrode kinetics. In this book we shall consistently define anodic currents and anodic overpotentials as positive and the corresponding cathodic quantities as negative. The foregoing equations are consistent with this notation. Thus, if $\eta > 0$, the anodic partial current is larger than the cathodic partial current, the observed overall current will be positive, and vice versa.

Equation (5.26) is a general equation for a single-step charge-transfer process occurring under pure activation control. We shall now proceed to discuss some special cases of this equation, one of which leads to the Tafel equation.

5.2.4
Limiting Cases of the General Equation

Current–potential curves calculated from Eq. (5.26) are shown in Figure 5.1, for three values of β. The plots are linear near the equilibrium potential, and the current density increases exponentially at large overpotentials.

Consider the case of low overpotential, often referred to as the *micropolarization region* or the *linear current–potential* region. The exponential terms in Eq. (5.26) can be linearized, with the use of only the first two terms in the Taylor expansion of the exponential term

$$\exp x = 1 + x + \frac{x^2}{2} + \frac{x^3}{3!} + \ldots \ldots \frac{x^n}{n!} \tag{5.27}$$

This approximation is valid for $x \ll 1$. It yields the following linear relationship between the current density and the overpotential

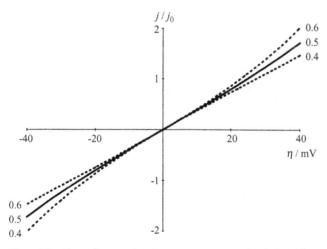

Figure 5.1 Plots of current density versus overpotential, calculated from Eq. (5.26), for a single-step charge-transfer process, for three values of β, at low overpotentials. Note that linearity is maintained longer for $\beta = 0.5$ than for other values of β.

$$j = j_0 \frac{\eta F}{RT} \tag{5.6}$$

Note that we have derived here Eq. (5.6), which was written intuitively in Section 5.1.1. It is also interesting that the symmetry factor, β, disappeared from Eq. (5.26) in the process of linearization. Thus, the rate of reaction a close to equilibrium does not depend on the detailed shape of the energy barrier for activation, (which determines the value of β). It does, however, depend on the magnitude of the energy of activation, which manifests itself in the value of j_0.

What is the range of the linear region represented by Eq. (5.6)? There is no single answer to this question, since it depends on the level of accuracy desired. On the other hand, we can readily calculate the range of overpotential for which the deviation from linearity of the j/η plot does not exceed, for example, 5% (or any other value); this turns out to be about ± 20 mV for $\beta = 0.5$. For other values of β the curve is no longer symmetrical and the linear region is shorter, as can be seen in Figure 5.1. Thus, while β does not appear in the equation for the low-overpotential linear approximation, its numerical value does influence the region over which this linear approximation is applicable.

It is interesting to note that in Section 5.1.1 we estimated the linear region to be approximately ± 5 mV, whereas here we find it to be ± 20 mV for $\beta = 0.5$. This discrepancy arises because in Section 5.1.1 we linearized the function $\exp(x)$, while here we linearize the difference between two exponential terms. For $\beta = 0.5$ this is given by $\sinh(x)$. Thus, Eq. (5.26) can be written as

$$j = 2 j_0 \sinh (\eta F / RT) \tag{5.28}$$

and the corresponding Taylor expansion is

$$\sinh (x) = 2 \left(x + \frac{x^3}{3!} + \frac{x^5}{5!} \right) \tag{5.29}$$

In Eq. (5.27) deviation from linearity is caused by the term $(x^2/2)$ while in Eq. (5.29) it is caused by the term $(x^3/6)$.

We turn our attention now to the case of high overpotential. One of the two exponential terms in Eq. (5.26) becomes negligible with respect to the other. For a large anodic overpotential one has

$$j = j_0 \exp \left(\frac{\beta F}{RT} \eta \right) \tag{5.30}$$

Written in logarithmic form, this equation becomes

$$\log j = \log j_0 + \frac{\beta F}{2.3 \, RT} \eta = \log j_0 + \frac{\eta}{b} \tag{5.31}$$

where the Tafel slope b is given as

$$b = \frac{2.3 \, RT}{\beta F} \tag{5.32}$$

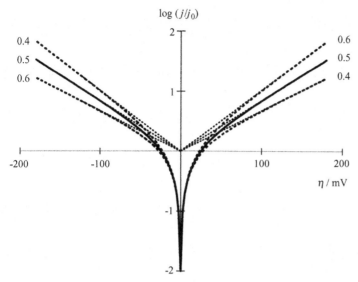

Figure 5.2 Semi-logarithmic j/η plots, calculated from Eq. (5.26) for three values of β. Note that the lines are symmetrical around $\eta = 0$ only for $\beta = 0.5$.

Equation (5.30) can be written in the form

$$\eta = b \log(j/j_0) \tag{5.33}$$

which is the same as the Tafel equation (Eq. (5.12)) introduced earlier in this chapter, except for a change of sign, because here we refer to an anodic reaction.

In Figure 5.1 we showed the j/η relationship in both the anodic and cathodic directions on a linear scale. In Figure 5.2 the same relationship is shown on a semi-logarithmic plot (Tafel plot) as η versus $b \log(j/j_0)$, for a larger range of over-potentials. A close look at this figure reveals several interesting points:

1) The Tafel plot is linear only at high values of the overpotential.
2) The extrapolated anodic and cathodic lines intersect at $\eta = 0$, irrespective of the value of β. This property could serve to determine the reversible potential, but it rarely does, because in most cases it is difficult to determine the current–potential relationship both at high anodic and high cathodic overpotentials. Usually, the reversible potential for the reaction studied is known, and the experimental line is extrapolated to this value to obtain j_0.
3) The Tafel plot is presented in terms of the current density, while the quantity determined experimentally is the total current. An uncertainty in the *real* surface area of the electrode (which often exists, as we shall see) causes a comparable uncertainty in the calculated value of j_0, but does not affect the Tafel slope. Similarly, if we do not know the reversible potential, we cannot determine j_0 but the Tafel slope can be calculated.
4) While j_0 represents the value of the current density extrapolated to $\eta = 0$, the parameter a in the Tafel equation is the overpotential extrapolated to a high

current density of $1.0\,A\,cm^{-2}$. It is determined by the heterogeneous rate constant k_h, the concentration of the reactant and the rate of change of current density with potential, given by the Tafel slope b.

5.3
The Symmetry Factor in Electrode Kinetics

5.3.1
The Definition of β

The symmetry factor has already been defined (Eq. (4.9)) in terms of the ratio between the effect of potential on the electrochemical Gibbs energy of activation and its effect on the electrochemical Gibbs energy of the reaction:

$$\beta \equiv \frac{(\delta \Delta \overline{G}^{0\#}/\delta \Delta \phi)}{(\delta \Delta \overline{G}^{0}/\delta \Delta \phi)} = \frac{\delta \Delta \overline{G}^{0\#}}{\delta \Delta \overline{G}^{0}} \tag{4.9}$$

It is appropriate to justify this definition here, and to compare it to some other ways in which this quantity has been defined in the literature. To do this, consider a simple reaction of the type

$$Ag^{+}_{soln} + e^{-}_{M} \rightarrow Ag^{0}_{M} \tag{5.34}$$

To be specific, we must note the positions of the various species before and after the reaction has taken place. The Ag^{+} ion is on the solution side of the interface, at a potential ϕ^{S}. The electron is in the metal, at a potential ϕ^{M}. The resulting silver atom will become part of the electrode and will be at the same potential ϕ^{M}, but since it is a neutral species, it is not affected by the potential, and we need not concern ourselves with it further. For the reaction to occur, a unit charge must cross the interface, a process that requires electrical energy amounting to $F(^{M}\Delta^{S}\phi)$. Any change in this potential difference will modify the electrochemical Gibbs energy of the reaction by the amount $F\,\delta(^{M}\Delta^{S}\phi)$. It does not matter, for the purpose of this argument, whether the reaction occurs by way of the positive ion crossing the interface (from ϕ^{S} to ϕ^{M}) to be neutralized by the electron in the metal, or if the electron jumps across, to neutralize the ion on the solution side of the interface, or the particles meet somewhere in between. The electrochemical Gibbs energy is *a function of state* defined thermodynamically as a function determined only by its position. Thus, changes in it depend only on the initial and final states, not on the route the system has taken to go from one to the other.

We do not know the nature of the activated complex in such a process, but fortunately we do not need to know, for the present purpose. All we have to assume is that it is an intermediate state, somewhere between the initial and the final states, which, by definition, it is. Thus we can say that the activated complex occurs at a point where the potential has an intermediate value, $\phi^{\#}$ between ϕ^{M} and ϕ^{S}, as shown schematically in Figure 5.3.

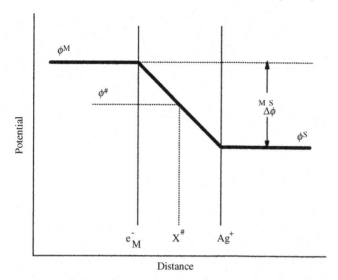

Figure 5.3 Charge transfer across the double layer. The reactant is at the potential ϕ^S, the product is at a potential ϕ^M and the activated complex is at an intermediate position, where the potential is $\phi^\#$, which is part way between ϕ^S and ϕ^M.

Since the formation of the activated complex is treated as an electrochemical reaction, we can write, for an anodic reaction:

$$\Delta\overline{G}^{0\#} = \Delta G^{0\#} - F\Delta\phi^\#$$ (5.35)

and for the overall reaction we have, as before

$$\Delta\overline{G}^0 = \Delta G^0 - F\Delta\phi$$ (4.6)

Hence the symmetry factor, β, can be written as the ratio

$$\beta = \frac{\delta\Delta\phi^\#}{\delta\Delta\phi}$$ (5.36)

Now, what is the physics behind these equations? We note that the potential difference $\Delta\phi$ across the interface affects the Gibbs energy of the reaction. A fraction of this potential affects the Gibbs energy of formation of the activated complex (i.e., the Gibbs energy of activation of the reaction). This fraction is defined as β. When we conduct an experiment, we can control $\delta\Delta\phi$, but we have no control over $\delta\Delta\phi^\#$, which amounts to a fraction β of the former.

It may be appropriate also to refer to the parameter β as an efficiency factor. Thus, we try to accelerate the rate of an electrochemical reaction by applying an amount of electrical energy $F\,\delta\Delta\phi$, but only a fraction β of this electrical energy is used to reduce the Gibbs energy of activation, the rest being "lost" as far as our endeavor to accelerate the reaction rate is concerned.

Why is it then that we call it the symmetry factor? It could be argued that a value of $\beta = 0.50$ corresponds to a symmetrical situation, in which the activated complex is formed exactly halfway between the reactant and the product. Also, if we look again at Figures 5.1 and 5.2 we note that the value of β determines the degree of symmetry of the j/η plots around the reversible potential. The important thing to remember is that β is determined by the shape of the Gibbs-energy barrier that the system must cross along the reaction coordinate, as it is transformed from reactants to products.

5.3.2
The Numerical Value of β

The determination of the numerical value of the symmetry factor β is a thorny problem in electrode kinetics. We might start with the conclusion, namely, that it is common practice to use the value of $\beta \approx 0.5$ in the study of electrode reactions. It is hard to come up with a satisfactory theory showing why this should be so, but there seems to be good evidence that it is, at least in some experimental systems.

The most reliable data are from studies of hydrogen evolution on mercury cathodes in acid solutions. This reaction has been studied most extensively over the years. The use of a renewable surface (a dropping-mercury electrode); our ability to purify the electrode material by distillation; the long potential range over which the Tafel equation is applicable and the relatively simple mechanism of the reaction in this system; all combine to give high credence to the conclusion that $\beta \approx 0.5$. This value has been used in almost all mechanistic studies in electrode kinetics, and has led to consistent interpretations of the experimental behavior. It is therefore reasonable to adopt this practice, in spite of the lack of solid theoretical evidence to support it.

5.4
The Marcus Theory of Charge Transfer

5.4.1
Outer-Sphere Electron Transfer

Electrode reactions can be divided into two major groups: those in which only charge is transferred across the interface, and those in which both charge and mass are transferred. Outer-sphere charge transfer is a good example of the former, while metal deposition and dissolution is an example of the latter.

A typical outer-sphere charge-transfer reaction is the ferricyanide–ferrocyanide redox couple

$$\left[\text{Fe(CN)}_6\right]^{3-}_{\text{soln}} + e^-_M \rightarrow \left[\text{Fe(CN)}_6\right]^{4-}_{\text{soln}} \tag{5.37}$$

This reaction occurs at the solution side of the interface, the electrode serves only as a source or sink for electrons. The important thing to note is that the close environment of the central ion is not changed significantly as a result of the transfer of charge. The configuration of the solvent molecules just outside the complex ion will change, but this is an effect on the "outer sphere". The inner sphere, that is, the six ligands around the central ion, remains intact and no chemical bonds are broken or formed as a result of charge transfer.

Although Eq. (5.37) does not show any change in the inner sphere of the ion, it should be noted that some changes must occur. Thus, as a first approximation, the ligands can be viewed as dipoles interacting electrostatically with the central ion. If the charge on this ion changes as a result of electron transfer, surely the positions of the ligands must change. A detailed approach must also take into account the covalent bonding of the ligands to the central ion, considering that the wavefunction of the central ion changes upon addition or removal of a unit charge. Thus, transfer of an electron must change the total solvation energy of the ion; that associated with the inner sphere, as well as that associated with the outer sphere. Recognition of the importance of this energy change is a central issue in the Marcus theory of charge transfer, to be discussed below.

5.4.2
The Born–Oppenheimer Approximation

The essence of the Born–Oppenheimer approximation is that the electronic and nuclear motions (following charge transfer) are separable in time. This can be understood tentatively, if one considers that the (CN) ligand is about 50 000 times heavier than an electron. It follows that the charge-transfer step presented by Eq. (5.37) should, in fact, be written in two steps, namely

$$\left[Fe(CN)_6 \right]^{3-}_{soln} + e^-_M \rightarrow \left[Fe(CN)_6 \right]^{\#4-}_{soln} \tag{5.38}$$

followed by

$$\left[Fe(CN)_6 \right]^{\#4-}_{soln} \rightarrow \left[Fe(CN)_6 \right]^{4-}_{soln} \tag{5.39}$$

In the first step, which is believed to occur very quickly, within a few femtoseconds, an electron has been transferred, but the nuclei of the ligand, as well as the water molecules in the outer solvation shell, did not move yet. Although the positive charge on the central atom was reduced from $+3$ to $+2$, the configuration of the ligand and the solvent around it is still in the equilibrium state corresponding to the Fe^{3+} ion. The superscripts on the complex $\left[Fe(CN)_6 \right]^{\#4-}_{soln}$ in the above two equations indicate that the product of electron transfer is not in its most stable state. In the second step, which may typically take a few picoseconds, that is, two or three orders of magnitude longer than the first step, the ligands around the newly formed Fe^{2+} ion are rearranged to their equilibrium position, corresponding to the lower charge. The variation of the Gibbs energy of the reactant and product along the reaction coordinate is shown schematically in Figure 5.4.

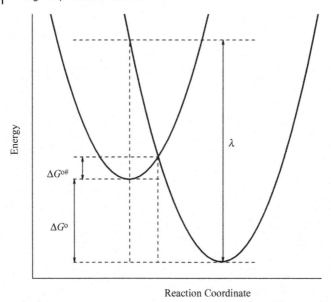

Reaction Coordinate

Figure 5.4 Schematic representation of "vertical" electron transfer, without movements of the nuclei, according to the Born–Oppenheimer approximation.

Transfer of the electron takes place without movement of the nuclei. The total energy of the system is increased (because the species formed in Eq. (5.38) is not in its most stable state). The energy released when the product of charge transfer decays to its equilibrium state is referred to as *the solvent reorganization energy, λ,* which is of central importance in the Marcus theory of charge transfer.

This course of events presents no problem from the thermodynamic point of view: when a negative overpotential is applied to the system, which was initially at equilibrium, the change in the electrochemical Gibbs energy is negative, and, hence, a driving force for the reaction to occur in the desired direction has been created. But the widely different timescales for the two steps associated with Eq. (5.38) and (5.39) does present a problem. The first step, which occurs in a few femtoseconds, creates an unstable intermediate, and hence increases the total Gibbs energy of the system. However, there is no mechanism that could supply the extra energy in this very short time; it could take as long as 10^3 fs for the unstable intermediate to relax to its stable final state. Thus, the sequence of steps represented by Eq. (5.38) and (5.39) violates the law of conservation of energy, and another route for charge transfer must be found. This problem is solved, in the framework of the Marcus theory, by assuming that charge transfer occurs only when the system is at the crossing point of the two curves shown in Figure 5.4. Since this point corresponds to the same energy level of the reactant and the product, transfer of an electron does not involve any change in the total Gibbs energy of the system. Further, it is assumed that the energy of the system is elevated to the crossing point of the two curves as a result of thermal fluctuations in the solvent.

5.4.3
The Calculated Energy of Activation

On the basis of the above physical model, we show the results of the Marcus theory, without showing the detailed calculations, which can be found in most textbooks of electrochemistry. The standard Gibbs energy of activation for an outer-sphere charge-transfer process is given approximately by

$$\Delta G^{0\#} = \frac{(\lambda + \Delta G^0)^2}{4\lambda} \tag{5.40}$$

where λ is the solvent reorganization energy and ΔG^0 is the standard Gibbs energy of the overall reaction. Equation (5.40) shows that solvent reorganization plays a major role in determining the rate of outer-sphere electron-transfer processes. For a symmetrical electron transfer step taking place in the bulk of the solution, in which an electron is transferred from the reduced to the oxidized form of the same species, $\Delta G^0 = 0$, because the reactants and the products are the same, and so Eq. (5.40) yields

$$\Delta G^{\#0} = \frac{\lambda}{4} \tag{5.41}$$

It is not easy to calculate or experimentally determine the value of λ. Estimated values given in the literature for aqueous solutions are in the range of approximately 50–200 kJ mol^{-1}. For reactions taking place in a solvent of lower polarity, this number could be much smaller, of course.

5.4.4
The Value of β and Its Potential Dependence

The results of the Marcus theory can be extended to electrode reactions by replacing the standard Gibbs energies by the standard *electrochemical* Gibbs energies, $\Delta \overline{G}^0$ and $\Delta \overline{G}^{0\#}$. Equation (5.40) is thus rewritten as

$$\Delta \overline{G}^{0\#} = \Delta G^{0\#} - \beta F|\eta| = \frac{(\lambda + \Delta G^0 - F|\eta|)^2}{4\lambda} = \frac{(\lambda + \Delta G^0)^2}{4\lambda}$$
$$- \frac{F|\eta|(2\lambda + 2\Delta G^0 - F|\eta|)}{4\lambda} \tag{5.42}$$

Combining Eq. 5.40 and 5.42 one has

$$\beta F|\eta| = \frac{F|\eta|(2\lambda + 2\Delta G^0 - F|\eta|)}{4\lambda} \tag{5.43}$$

and the symmetry factor, β, is given by

$$\beta = \frac{1}{2} + \frac{\Delta G^0}{2\lambda} - \frac{F|\eta|}{4\lambda} \tag{5.44}$$

For charge transfer occurring at the electrode/solution interface, ΔG^0 is not zero, since the reactant and the product are not identical. This could only lead to a value of $\beta \approx 0.5$ (commonly used in the analysis of the mechanism of electrode kinetics), for high values of the solvent reorganization energy, λ, or for a relatively short range of overpotentials, where the condition

$$\frac{\Delta G^0}{2\lambda} \approx \frac{F|\eta|}{4\lambda} \tag{5.45}$$

happens to apply. Moreover, according to Eq. (5.44), the Marcus theory predicts a potential-dependent symmetry factor, given by the expression

$$\frac{\partial \beta}{\partial |\eta|} = -\frac{F}{4\lambda} \tag{5.46}$$

which leads to a deviation from linearity of the Tafel plot

Certain experimental studies of outer-sphere charge-transfer processes seem to indicate that such a potential dependence of β does indeed exist, but this is based on measurements that were usually taken over a limited range of overpotential, so that the effect expected is very small, and the evidence is not compelling. On the other hand, highly accurate studies of the hydrogen evolution reaction on mercury showed values of $\beta = 0.50 \pm 0.01$, strictly independent of potential over a range of more than 0.7 V. But then, the hydrogen-evolution reaction is not an outer-sphere charge transfer process, so that this observation does not contradict the Marcus theory.

5.5
Time-Resolved Kinetics of Charge Transfer

5.5.1
Metal Deposition and Dissolution

In the previous section we discussed outer sphere charge transfer. That process is characterized by the following features:

1) Both reactant and product are on the solution side of the interface
2) The charge is carried by the electron, which is the only species crossing the interface
3) There are no chemical bonds broken and no new bonds formed as a result of charge transfer. Indeed the reactant and the product are very similar, except for their charge

Metal deposition is clearly *not an outer-sphere charge-transfer process*. The initial state is a solvated metal ion, while the final product is a neutral atom in the metal, for example[3].

---

3) The metal crystal consists of a network of positive ions immersed in a "sea" of delocalized electrons, but the total charge of these electrons equals the total charge of the ions, so that writing the product as a neutral atom is formally justified.

$$\left[Tl(H_2O)_m\right]^+_{soln} + e^-_M \rightarrow Tl^0_M + m(H_2O) \tag{5.47}$$

The Gibbs energy of solvation of ions in water is very high, of the order of

$$\Delta G_{solv} \approx 5 \times n^2 \text{ eV} \tag{5.48}$$

where n is the number of unit charges on the ion. The removal of all water molecules from the solvation shell of the metal ion, which is about 5 eV and 20 eV for monovalent and divalent metal ions, respectively, could lead to very high values of the electrochemical Gibbs energy of activation. In any case, one cannot expect the Marcus theory, which was developed for outer-sphere electron transfer processes, to apply to metal deposition and similar reactions, at least not without significant modification, which has yet to be developed.

The question then presents itself: is the charge carried across the interface by electrons, as in an outer-sphere charge-transport process, or by the positive metal ion? For the example shown in Eq. (5.47) we note that the Tl atom is about 3.75×10^5 times heavier than an electron, so one may be tempted to assume that it is the nimble electron that crosses the interface rather than the sluggish heavy ion. This has indeed been the practice in the discussion of the kinetics of metal deposition and dissolution, although reference to the fact that charge is carried across by the metal ions can be found in the literature. In this section it will be shown that it must be the positive metal ion that carries the charge across, not the electron. Consider first the three main features of outer-sphere charge transfer to metal deposition.

1) The metal ion is initially on the solution side of the interface but the product is on the metal side of the interface, so there is no doubt that mass transport across the interface has occurred.

2) The open question, at this point, is when is the ion neutralized: while it is still on the solution side of the interface, after it has landed on the metal surface, or somewhere in between?

3) There is no doubt that chemical bonds are broken during metal deposition. The energy needed to remove the solvation shell is of the order of 5 eV for a monovalent ion and about 20 eV for a divalent ion. Admittedly, this is the Gibbs energy of the overall reaction, not the energy of activation, but for a driven reaction (where the Gibbs energy of the product is higher than that of the reactant) the Gibbs energy of activation cannot be lower than the Gibbs energy of the overall reaction.

Consider now point 2 above, in order to show that charge must be carried across the interface by the metal cation, not the electron, during metal deposition. It is here that the different timescales for movement of mass and of electrons comes into play. It was noted above that electron transfer takes a few femtoseconds. Solvent reorganization may take a few picoseconds, but how long would it take for an ion or atom to diffuse across the interface, a distance of the order of 0.6 nm? This is hard to tell, since the water molecules at the interface do not have the same structure as those in the bulk of the solution, and hence the diffusion coefficient of ions or atoms may be different from the values known for bulk diffusion. Nevertheless, we can use the latter for an order-of-magnitude estimation, using the equation

$$t = \frac{\delta^2}{\pi D} \qquad (5.49)$$

and assuming $D \approx 8 \times 10^{-6}$ cm^2 s^{-1}, we find $t = 0.14$ ns or 1.4×10^5 fs, which is about five orders of magnitude longer than the time taken by the electron to cross the interface. In view of these widely different timescales for transfer of an electron and an ion, these processes cannot be considered to occur simultaneously, as a single step. Thus, metal deposition, represented, for example, by Eq. (5.47), would have to be considered as two separate steps

$$\left[Tl(H_2O)_m\right]^+_{soln} + e^-_M \xrightarrow{\approx 1\,fs} \left[Tl(H_2O)_m\right]^0_{soln} \xrightarrow{\approx 1\,ps} Tl^0_{soln} + m(H_2O) \qquad (5.50)$$

followed by

$$Tl^0_{soln} \xrightarrow{\approx 0.1\,ns} Tl^0_M \qquad (5.51)$$

But a neutral species in solution, formed in the fast electron transfer (Eq. (5.50)) is highly unstable, and thermal fluctuations of the solvent could not bring the reactant and the product to the same energy level, as assumed in the Marcus theory for outer-sphere charge transfer. It must be concluded, therefore, that the sequence shown in Eq. (5.50) and (5.51) cannot be the path followed, and charge must be carried across the interface by the metal ions. Another way to look at it is to consider the mechanism of anodic dissolution of a metal. Consider Eq. (5.47) written in the anodic direction, we have

$$Tl^0_M + m(H_2O) \rightarrow \left[Tl(H_2O)_m\right]^+_{soln} + e^-_M \qquad (5.52)$$

If electron transfer is assumed, then the electron would have to move from the solution side of the interface to the metal, but the species that is oxidized here would have to be a neutral thallium atom, which is very unstable in solution. So, if charge is transferred by the electrons, Eq. (5.52) would need to be rewritten as

$$Tl^0_M \rightarrow Tl^0_{soln} \qquad (5.53)$$

followed by

$$Tl^0_{soln} + m(H_2O) \rightarrow \left[Tl(H_2O)_m\right]^+_{soln} + e^-_M \qquad (5.54)$$

But, besides the neutral atom being highly unstable, there is no driving force for Eq. (5.53) to occur, and no reason for it to be influenced by the applied potential. So, metal dissolution cannot occur by a mechanism in which charge is transferred by the electron. Moreover, the principle of microscopic reversibility states that a reaction going forward and backward (as in metal deposition and dissolution) must follow the same pathway. Therefore, if we have proven that in metal dissolution the charge must be carried across the interface by the metal ions, not by the electrons, the same must apply for metal deposition, and vice versa.

What is the mechanism of charge transfer by ions? Unfortunately, little has been done so far to evaluate this mechanism. It would seem that, as the solvated ion moves from one side of the interface to the other, it must lose its solvation shell and its effective charge is reduced gradually until it is neutralized. Interestingly, the heterogeneous rate constant for deposition of different metals is quite high, comparable to and sometimes exceeding the rate constants observed for outer-sphere charge transfer. This is consistent with a mechanism in which the ion moves in many small steps, increasing its chemical interactions with the surface, while its charge interacts with the delocalized electrons in the metal. The solvation shell is gradually distorted and then lost continuously along its path. If the path across the interface consists of many small paths, each representing a minor change in configuration and in effective charge, the activation energy could be very small all along the way, from the initial to the final state. The electrostatic field applied by applying an overpotential may play a major role in enhancing the rate of such ion transfer.

As noted above, this mechanism has not yet been evaluated in a satisfactory manner, and any further discussion is outside the scope of this book.

6
Multi-Step Electrode Reactions

6.1
Mechanistic Criteria

6.1.1
The Transfer Coefficient, α, and Its Relation to the Symmetry Factor, β

So far, we have discussed the current–potential relationship for a simple electrode reaction, which occurs in one step, in which a single electron is transferred. Usually, it is assumed that there can be only one electron transferred in each elementary step, so that all single-step reactions involve the transfer of only one electron, although simultaneous two-electron transfer can occur in certain cases. Real systems are, however, more complex. Most electrode reactions occur in several steps, and the transition from reactants to products requires the transfer of several electrons. When one considers a complex reaction, such as the oxidation of methanol in a fuel cell, the overall reaction is:

$$CH_3OH + H_2O \rightarrow CO_2 + 6H^+ + 6e_M^- \tag{6.1}$$

which must occur in several electron- and proton-transfer steps, as it proceeds from reactants to products.

We shall not discuss here the mechanism of this reaction. It is mentioned only to show the complexity of an electrode reaction, which happens to be of great interest in the context of the development of direct-methanol fuel cells. It is evident, then, that the equations of electrode kinetics derived in Chapter 5 must be generalized to describe multi-step electrode processes, which involve the transfer of several electrons.

We recall that the Tafel relation was originally observed for the hydrogen-evolution reaction on mercury. We should therefore have an equation similar to Eq. (5.26) to describe the current–potential relationship. For a single cathodic step we wrote:

$$j = j_0 \exp\left(-\frac{\beta_c F}{RT}\eta\right) \tag{5.30}$$

Physical Electrochemistry: Fundamentals, Techniques and Applications. Eliezer Gileadi
Copyright © 2011 WILEY-VCH Verlag GmbH & Co. KGaA, Weinheim
ISBN: 978-3-527-31970-1

But the symmetry factor β has been defined strictly for a single step and is related to the shape of the energy barrier and to the position of the activated complex along the reaction coordinate. To describe a multi-step process, β_c must be replaced by an experimental parameter, which we call the *cathodic transfer coefficient* α_c. Instead of Eq. (5.26) we then write:

$$j = j_0 \exp\left(-\frac{\alpha_c F}{RT}\eta\right) \tag{6.2}$$

It is now evident that the Tafel slope will take the form:

$$b_c = \left(\frac{\partial\eta}{\partial\log j}\right)_c = -\frac{2.3RT}{\alpha_c F} \tag{6.3}$$

This can be rearranged and written as follows:

$$\alpha_c = -\frac{2.3RT}{F}\frac{1}{b_c} \tag{6.4}$$

Either Eq. (6.3) or (6.4) can be considered to be a definition of the transfer coefficient α_c. Equation (6.3) relates it directly to the measured quantity $(\partial\eta/\partial\log j)_c$, while Eq. (6.4) can be regarded as a formal definition, indicating that the transfer coefficient is simply the reciprocal Tafel slope in dimensionless form.

Although the definition of α and the difference between it and β should be quite clear from the preceding explanation, there has been a fair amount of confusion in the literature concerning these two parameters. It is, therefore, appropriate to state again the definitions of these quantities and to highlight their different physical meanings.

The symmetry factor, β, is a fundamental parameter in electrode kinetics. It must always be discussed with respect to a specific *single* step in a reaction sequence, and its value (which must be between zero and unity, by definition) is related to the shape of the energy barrier and to the position of the activated complex along the reaction coordinate. When the potential across the interface is changed (by application of an external current or potential), both the electrochemical Gibbs energy of activation and the electrochemical Gibbs energy of the reaction are altered. The symmetry factor, β, represents the ratio between the changes in these quantities, as given by Eq. (4.9) namely:

$$\beta \equiv \frac{(\delta\Delta\bar{G}^{0\#}/\delta\Delta\phi)}{(\delta\Delta\bar{G}^{0}/\delta\Delta\phi)} = \frac{\delta\Delta\bar{G}^{0\#}}{\delta\Delta\bar{G}^{0}} \tag{4.9}$$

In contrast, the transfer coefficient, α, is an experimental parameter obtained from the current–potential relationship. Just that and nothing more! It will be shown later that the relationship between α and β depends on the mechanism of the reaction. The transfer coefficient is therefore *one of the parameters* that allows us to evaluate the mechanism of electrode reactions or to distinguish between different plausible mechanisms.

It cannot be overemphasized that one can measure only the transfer coefficient, α, not the symmetry factor, β. The latter can be inferred from the former by making a suitable set of assumptions. For example, for the hydrogen-evolution reaction on mercury, it is commonly *assumed* that $\alpha_c = \beta_c$. One often tends to refer in this case to the measurement of the symmetry factor, but even in this simple case only the transfer coefficient can be measured, and the symmetry factor must be calculated from it (even if in this case "calculation" simply means assuming that they are equal), on the basis of certain assumptions.

There is an additional difference of great importance between β and α, which is often overlooked. The way in which β is defined requires that the sum of the symmetry factors in the anodic and cathodic directions must be unity; if it is β for the cathodic reaction, it must be $(1 - \beta)$ for the anodic reaction, and vice versa. The same is not true with respect to the transfer coefficient. To begin with, α is a parameter obtained from experiment. One must therefore *find* its value, not *assume* what it should be. When we wish to write a rate equation, such as Eq. (5.26), for a multi-step reaction, the correct form will be

$$j = j_0 \left[\exp\left(\frac{\alpha_{an} F}{RT} \eta \right) - \exp\left(-\frac{\alpha_c F}{RT} \eta \right) \right] \tag{6.5}$$

in which one can usually write $\alpha_{an} + \alpha_c = n$, (while $\beta_{an} + \beta_c = 1$). Indeed, one can readily write mechanisms for which the transfer coefficient is greater than unity, as we shall see below.

6.1.2
Steady State and Quasi-Equilibrium

Consider a simple anodic reaction, such as the oxidation of Cl^- to Cl_2, which can occur in the following two steps:

$$Cl_{soln}^- \rightarrow Cl_{ads}^0 + e_M^- \tag{6.6}$$

$$Cl_{soln}^- + Cl_{ads}^0 \rightarrow Cl_2 + e_M^- \tag{6.7}$$

The concentration of the adsorbed intermediate can best be expressed in terms of the partial surface coverage, θ, which is defined as the surface concentration, Γ, (mol cm^{-2}) divided by the maximum surface concentration, Γ_{max}, (expressed, of course, in the same units):

$$\theta \equiv \frac{\Gamma}{\Gamma_{max}} \tag{6.8}$$

The net rate of formation of the adsorbed intermediate can be written as

$$\frac{\partial \theta}{\partial t} = k_1 (1-\theta) c_{b,Cl^-} - k_{-1}\theta - k_2\theta\, c_{b,Cl^-} + k_{-2}(1-\theta) c_{b,Cl^-} \tag{6.9}$$

in which k_i are the potential-dependent electrochemical rate constants. This can be solved readily under steady-state conditions by setting $\partial\theta/\partial t = 0$. It is, however, easier to treat multi-step reaction sequences in the framework of the *quasi-equilibrium* assumption, as shown below.

When a reaction occurs in several consecutive steps, the rate of all steps must be equal at steady state (otherwise the system would not be at steady state). This rate is determined by the slowest step in the sequence, which we refer to as *the rate-determining step* (rds). In the preceding example, if k_1, the specific rate constant for the step shown by Eq. (6.6), is much smaller than k_2, the specific rate constant for the step shown by Eq. (6.7), the rate of the second step will effectively be limited by the supply of adsorbed intermediates Cl_{ads}, namely, by the rate of the first step. To visualize this situation, consider a potential applied to several resistors in series. The current is determined by the overall resistance, which is simply the sum of the resistances in series $R = R_1 + R_2 + R_3 + \ldots.R_n$. If one of these resistances is much larger than all others, it will be dominant, and one has $R \approx R_i$.

To use this simple equivalent circuit, it must be realized that the rate constants of the various steps can be represented by the *inverse* of the corresponding resistances; the higher the resistance, the lower the rate constant, and vice versa. Thus, the overall effective rate constant is given by

$$\frac{1}{k} = \frac{1}{k_1} + \frac{1}{k_2} + \ldots.\frac{1}{k_n} \tag{6.10}$$

It is clear, then, that the overall rate constant is determined by the *lowest* individual rate constant. We note, in passing, that Eq. (6.10) is similar to Eq. (6.15), which correlates the overall current to the activation- and mass-transport-controlled currents.

$$\frac{1}{j} = \frac{1}{j_{ac}} + \frac{1}{j_L} \tag{1.5}$$

This should not be surprising, since the activation and mass-transport processes always occur in series and should combine to determine the overall rate in the same way as do several activation-controlled steps in series.

How do we know that there is a distinct rate-determining step, one for which k_1 is much smaller than all other rate constants? The fact is that we do not know *a priori*, and, indeed, one could envisage situations in which several rate constants in a sequence would be comparable in magnitude. However, it will be recalled that in electrode kinetics, the rate constant is a function of potential, and in a reaction sequence the rate constants could depend differently on potential. Thus, in electrode kinetics, more than in other fields of chemical kinetics, one is likely to observe a single rate-determining step, at least over a certain potential range. It is therefore common (and sensible) to treat the mechanisms of electrode reactions on the assumption that there is a well-defined rate-determining step in each potential region studied, with possible transitions from one rate-determining step to another, as the potential is changed.

We now proceed to the concept of quasi-equilibrium. If there is a distinct rate-determining step in a reaction sequence, then *all other steps before and after it must be effectively at equilibrium.* This comes about because the overall rate is, by definition, very slow compared to the rate at which each of the other steps could proceed by itself, and equilibrium in these steps is therefore barely disturbed. In order to better understand this, consider the specific example given above for chlorine evolution. Assume, for the sake of argument, that the values of the exchange-current density j_0 for steps (6.6) and (6.7) are 250 and $1\,\text{mA cm}^{-2}$, respectively. Now consider the application of a current density of $0.5\,\text{mA cm}^{-2}$. We can calculate the overpotential corresponding to each step in the sequence, from Eq. (5.6) namely

$$j = j_0 \left(\frac{\eta F}{RT} \right) \tag{5.6}$$

we find for step (6.6) $\eta = (RT/F)(0.5/250) = 0.05\,\text{mV}$ and for step (6.7) $\eta = (RT/F)(0.5/1) = 12.8\,\text{mV}$. The total overpotential is the sum of these values, namely $\eta = 12.85\,\text{mV}$, of which only $0.05\,\text{mV}$ (0.4%) is associated with the first step. Hence, we can consider this step to be effectively at equilibrium. In this way we proceed to calculate the kinetic parameters for the reaction sequence, assuming that all steps other than the rate-determining step are at equilibrium.

Why call it the *quasi*-equilibrium assumption? This has nothing to do with the mathematical treatment that follows; it is used only to soothe our conscience. Strictly speaking, the preceding assumption is self-contradictory. Equilibrium is defined as the state in which no net reaction takes place. There is an exchange reaction (the rate of which is represented in electrode kinetics by j_0) but the rates of the forward and backward reactions are equal. How can we say that a step in a reaction sequence is at equilibrium while it is proceeding at a finite rate in one direction? We use the term *quasi-equilibrium* in recognition of the fact that this is only an approximation that serves our purpose well, even as we remain fully aware of its logical limitations.

6.1.3
Calculation of the Tafel Slope

Equipped with the assumption of quasi-equilibrium, we can now proceed to calculate the Tafel slopes and some other kinetic parameters for a few very simple cases, to show how such calculations are made. In Chapter 7 we shall discuss the kinetics of several reactions that either have been important in the development of the theory of electrode kinetics or are of current practical importance.

Consider again the chlorine evolution reaction and let us assume first that step (6.6) is rate-determining

$$Cl^-_{soln} \rightarrow Cl^0_{ads} + e^-_M \tag{6.6}$$

The rate of this step is given by

$$j = F k_1 c_{b,Cl^-} (1-\theta) \exp \left(\frac{\beta F}{RT} E \right) \tag{6.11}$$

This equation is applicable at high overpotentials, where the reverse reaction can be ignored. Also, it is assumed in writing this equation (and all following kinetic equations) that mass-transport limitation is either negligible or has been corrected for in a quantitative manner. It should be noted that we have used here the symmetry factor, β, not the transfer coefficient α. We do this because we are referring to a *specific step* in the reaction sequence and not the overall reaction.

Considering Eq. (6.11), we have good reason to assume that the partial coverage θ is very small, since the intermediate is formed in the rate-determining step and is removed by the following step, which is much faster. Taking $\theta \ll 1$ and hence $(1-\theta) \approx 1$, we can rewrite Eq. (6.11), to a very good approximation, as:

$$j_1 = F k_1 \, c_{b,Cl^-} \exp\left(\frac{\beta_{an} F}{RT} E\right) \tag{6.12}$$

For the overall reaction we can then write:

$$j = nF \, k_1 c_{b,Cl^-} \exp\left(\frac{\alpha_{an} F}{RT} E\right) = j_0 \exp\left(\frac{\alpha_{an} F}{RT} E\right) \tag{6.13}$$

It is important to understand the logic of the transition from Eq. (6.12) to (6.13). In the former we used the partial current density, j_1, for a step in the reaction sequence; in the latter we used the total current density, j. In the present case $j = 2j_1$, since for every electron transferred in the first step, another is transferred in the second step $(n = 2)$. In addition, we write α_{an} instead of β_{an}, since Eq. (6.12) refers to an elementary step in the reaction sequence, while Eq. (6.13) refers to the overall reaction. In this particular case, we find that $\alpha_{an} = \beta_{an}$, but that is beside the point. It should also be noted here that the exchange-current density is related to the rate constant, the concentration of the reactant and to the metal–solution potential difference at the reversible potential (since $E = E_{rev} + \eta$).

$$j_0 = nF \, k \, c_{b,Cl^-} \exp\left(\frac{\alpha_{an} F}{RT} E_{rev}\right) \tag{6.14}$$

We can readily calculate the Tafel slope for this case, if we assign a numerical value to the symmetry factor an β_{an}. This, as we said before, is commonly taken to be 0.5. The Tafel slope can then be obtained either from Eq. (6.5),

$$b_{an} = \frac{2.3RT}{\alpha_{an} F} = \frac{2.3RT}{\beta_{an} F} = 118 \, \text{mV decade}^{-1} \, (\text{at } 25\,^\circ\text{C})$$

or directly from Eq. (6.4), by calculating the value of $(\partial \eta / \partial \log j)_{c,T,P}$.

We can now proceed to obtain the kinetic equation for the same reaction in a somewhat more complicated case, when the second step (Eq. (6.7)) is assumed to be the rate-determining step, and the first step (Eq. (6.6)) is at quasi-equilibrium.

For quasi-equilibrium in step (6.6) we can write

$$k_1 c_{b,Cl^-} (1-\theta) \exp\left(\frac{\beta F}{RT} \eta\right) = k_{-1} \theta \exp\left(-\frac{(1-\beta) F}{RT} \eta\right) \tag{6.15}$$

from which it follows that

$$\frac{\theta}{1-\theta} = K_1 c_{b,Cl^-} \exp\left(\frac{F}{RT} E\right) \tag{6.16}$$

in which $K_1 = k_1/k_{-1}$ is the equilibrium constant. The absence of the symmetry factor β from this equation is not an error, and it does not depend on the numerical value of β. The symmetry factor is strictly related to the shape of the Gibbs-energy barrier and the position of the activated complex along the reaction coordinate. Equation (6.16) describes an equilibrium that is independent of the preceding considerations and is related only to the difference in Gibbs energy between the initial and the final states.

Let us work out the kinetic equations for this case. Assuming that the step shown in Eq. (6.7) is the rds, we write

$$j_2 = Fk_2 c_{b,Cl^-} \theta \exp\left(\frac{\beta F}{RT} E\right) \tag{6.17}$$

in which we have to substitute the potential-dependent value of θ from Eq. (6.16). To simplify things, we consider two extreme cases:

1. For $\theta \ll 1$, we can write $(1-\theta) \approx 1$. Then, by combining Eqs. (6.16) and (6.17) we have

$$j = nFk_2 K_1 c_{b,Cl^-}^2 \exp\left(\frac{(1+\beta_{an})F}{RT} E\right) \tag{6.18}$$

from which it follows that, under these assumptions,

$$\alpha_{an} = (1+\beta_{an}) \text{ and } b_{an} = 2.3RT/(1+\beta_{an})F$$

2. When the surface coverage approaches unity and can no longer change significantly with potential, we can substitute $\theta = 1$ in Eq. (6.17), to obtain

$$j = nFk_2 c_{b,Cl^-} \exp\left(\frac{\beta_{an} F}{RT} E\right) \tag{6.19}$$

which leads to $\alpha_{an} = \beta_{an}$ and $b_a = 2(2.3RT)/F = 118$ mV. This result seems to be identical to that derived by assuming that the first step in the reaction sequence is rate-determining. There is, however, an important difference. Equation (6.19) was derived for the case of high surface coverage, while Eq. (6.12) applies for low coverage, where $(1-\theta)$ can be replaced by unity. To distinguish between these possibilities, we must make independent measurements of the surface coverage.

The two limiting cases just discussed are approximately applicable for $\theta \leq 0.1$ and $\theta \geq 0.9$, respectively. In the intermediate region, one could readily solve the equation by substituting θ from Eq. (6.16) into Eq. (6.17), but the result (besides being cumbersome) leads to a transfer coefficient that decreases gradually with increasing overpotential, from $\alpha_{an} = 1+\beta_{an}$ to $\alpha_{an} = \beta_{an}$. This also means that the Tafel slope changes with potential. In other words, the Tafel *plot* (which is the plot of log j versus E is not linear in this intermediate range of coverage. It should be noted that the potential dependence of α_{an} shown here is due to the variation of the partial coverage with potential, *not* because β_{an} itself is potential dependent.

We have stated that Eq. (6.16) is equivalent to the Nernst equation. This can be proved in the following simple way. We write this equation in logarithmic form as:

$$\log\left[\frac{\theta}{1-\theta}\right] = \log K_1 + \log c_{b,Cl^-} + \frac{F}{2.3RT}E \tag{6.20}$$

which can be rearranged to

$$E = -\frac{2.3RT}{F}\log K_1 + \frac{2.3RT}{F}\log\left[\left(\frac{1}{c_{b,Cl^-}}\right)\left(\frac{\theta}{1-\theta}\right)\right] \tag{6.21}$$

The fractional coverage, θ, is proportional to the concentration of the product in this reaction, Cl_{ads}, and $(1-\theta)$ is proportional to the concentration of one of the reactants (free sites on the surface), making the second term in Eq. (6.21) a typical ratio of concentrations of products to reactants. Also, we know from thermodynamics that the equilibrium constant is related to the standard Gibbs energy of the reaction, and the latter is related to the standard potential:

$$\frac{2.3RT}{F}\log K = -\frac{\Delta G^0}{F} = E_\theta^0 \tag{6.22}$$

Substituting in Eq. (6.21) we thus have

$$E = E_\theta^0 + (2.3RT/F)\log\left[\left(\frac{1}{c_{b,Cl^-}}\right)\left(\frac{\theta}{1-\theta}\right)\right] \tag{6.23}$$

where E_θ^0 is the standard potential for a reaction involving the formation of an adsorbed species by charge transfer, such as described by Eq. (6.6). We note that the standard state for the adsorbed species is chosen as $\theta = 0.5$. Hence the term $\theta/(1-\theta)$ will be unity. Equation (6.23) is the Nernst equation, written for a single electron transfer. There is nothing unusual in this result, of course. We treated Eq. (6.6) as an electrochemical equilibrium and obtained the correct equation for an electrochemical reaction at equilibrium. The point is that the Nernst equation is commonly derived from thermodynamic considerations. It is reassuring to be able to start from kinetic equations and reach the same result, in the limit where equilibrium can be assumed.

6.1.4
Reaction Orders in Electrode Kinetics

The reaction order is defined in chemical kinetics by the partial derivative

$$\rho = (\partial \log v/\partial \log c_i)_{c_{j\neq i}} \tag{6.24}$$

which measures the dependence of the reaction rate, v, on the concentration of one species in solution, with the concentrations of all other species (as well as the temperature and pressure) maintained constant.

In electrode kinetics, the reaction order is defined in a similar manner, but, in addition to keeping the temperature and the pressure constant, a constant potential is maintained. As a result, there are two reaction orders in electrode kinetics, one taken at constant *potential*.

$$\rho_1 = (\partial \log j / \partial \log c_i)_{E, c_j \neq i} \qquad (6.25)$$

and

$$\rho_2 = (\partial \log j / \partial \log c_i)_{\eta, c_{j \neq i}} \qquad (6.26)$$

It is important to distinguish between these two parameters, since the reversible potential changes as we change the concentration of the reactant, and the over-potential can thus change while the applied potential remains constant.

In Eq. (6.25), the potential E, which we keep constant, is that measured with respect to a fixed reference electrode. It does not matter which reference electrode is used, as long as it is the same throughout the experiment. Actually, we would like to obtain the reaction order while keeping the metal–solution potential difference, $^S\Delta^M\phi$, constant. This would seem to be impossible, since $^S\Delta^M\phi$ cannot be measured, as explained in detail at the beginning of the book. But the *absolute* metal–solution potential difference differs from the *measured* potential only by a constant! Thus, by keeping the potential E constant, we can be assured that the metal–solution potential difference $^S\Delta^M\phi$ is also maintained constant, even though its value is not known.

We recall that the current is a very sensitive measure of the rate of an electrochemical reaction. It is therefore quite easy to determine the current–potential relationship without causing a significant change in the bulk concentration of either reactants or products. Thus, measurements in electrode kinetics are conducted effectively under *quasi-zero-order* kinetic conditions. It would be wrong to infer from this that electrode reactions are independent of concentration. To determine the concentration dependence (i.e., the reaction order), one must obtain a series of j/E or j/η plots as shown in Figure 6.1a and derive from them the dependence of $\log j$ on $\log c_i$ at different potentials, as shown in Figure 6.1b. The slopes in Figure 6.1b yield the parameter ρ_1, since they are measured at constant potential E. Here, and in all further equations, we shall assume that T, p, and the concentration of all species in solution $c_{j \neq i}$ other than the one being studied, are kept constant, in order to permit us to write the equations in a more concise form.

Which of the two reaction-order parameters should one prefer? When measurement is made with a constant reference electrode (e.g., calomel), the reaction order at constant potential, ρ_1, is obtained. If, however, an indicator-type reference electrode is employed (e.g., a reversible hydrogen electrode, often used in the study of the hydrogen-evolution reaction), the overpotential can be kept constant and the parameter ρ_2 is the one directly obtained. In either case, both parameters can be calculated

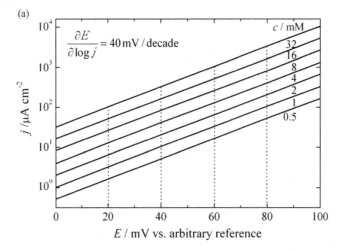

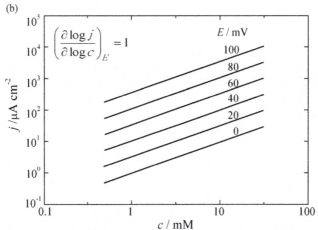

Figure 6.1 Calculation of the reaction order from Tafel plots. (a) Tafel lines for a series of concentrations of the electroactive species. (b) Reaction-order plots derived from (a) for different values of the potential.

from the same experimental data, since the dependence of the reversible potential on concentration is well known.

To show the difference between these two reaction-order parameters, we return to the equations derived in Section 6.1.3 for the chlorine-evolution reaction. Consider the case in which the first charge-transfer step (cf. Eq. (6.6)) is rate-determining. Equation (6.13) in logarithmic form is:

$$\log j = \log \left(n F k_1 \right) + \log c_{b,Cl^-} + \frac{\alpha_{an} F}{RT} E \tag{6.27}$$

hence

$$\rho_1 = \left(\frac{\partial \log j}{\partial \log c_{b,Cl^-}}\right)_E = 1 \tag{6.28}$$

To calculate ρ_2, the reaction order at constant overpotential, we substitute $E = E_{rev} + \eta$ in Eq. (6.27) and express the reversible potential in terms of the Nernst equation, namely:

$$E_{rev} = E^0 + \frac{2.3RT}{2F}\log\frac{c_{Cl_2}}{(c_{b,Cl^-})^2} \tag{6.29}$$

Assuming that the solution is saturated with respect to Cl_2, we can rewrite Eq. (6.27) as

$$\log j = \log(nF k_1) + \log c_{b,Cl^-} + \frac{\alpha_{an} F}{RT}E^0 - \alpha_{an}\log c_{b,Cl^-} + \frac{\alpha_{an} F}{RT}\eta \tag{6.30}$$

Differentiating with respect to $\log c_{b,Cl^-}$ while keeping η constant yields:

$$\rho_2 \equiv \left(\frac{\partial \log j}{\partial \log c_{b,Cl^-}}\right) = 1 - \alpha_{an} \tag{6.31}$$

We note that the reaction order at constant overpotential is a little more complex and depends also on the transfer coefficient α_{an}. One may be tempted to obtain both ρ_1 and ρ_2 from the experimental data, in order to calculate the transfer coefficient from the difference. This can be done, in principle, but the measurement of reaction orders is inherently much less accurate than the measurement of the Tafel slope, hence this approach is not expedient, unless the transfer coefficient cannot be measured directly.

In the literature one often encounters two additional types of reaction orders, which are actually derived from the two we have just discussed. The first relates to the variation of the *exchange-current density* with concentration, namely $(\partial \log j_0/\partial \log c)$. It is easy to see that this parameter is equal to ρ_2, since by using j_0, one has actually specified a constant overpotential of $\eta = 0$. This parameter is useful for the study of fast electrode reactions, where measurements can be taken only at low overpotentials, (as a result of mass-transport limitation) and the Tafel region is not experimentally accessible. The second reaction-order parameter occasionally used is $(\partial E/\partial \log c_i)_{c_{j \neq i}}$. However, applying the rule of partial derivatives, we can write:

$$\left(\frac{\partial E}{\partial \log c}\right)_j \left(\frac{\partial \log c}{\partial \log j}\right)_E \left(\frac{\partial \log j}{\partial E}\right)_c = -1 \tag{6.32}$$

Hence

$$\left(\frac{\partial E}{\partial \log c}\right)_j = -\rho_1 b \tag{6.33}$$

There is nothing wrong with using any of these reaction-order parameters, as long as we remember that only two of them (any two, of course) are independent, and the third can always be derived from these.

6.1.5
The Effect of pH on Reaction Rates

The influence of pH on reaction rates may be looked upon as just another concentration effect, which can be dealt with in terms of the reaction orders just discussed. It merits special attention, however, for two reasons: first, because it allows us to change the concentration of the reactant (H_3O^+ or OH^-) over many orders of magnitude, without encountering solubility limitations at the high concentration end and mass transport limitations at the low concentration end (as long as the solution is buffered). The second reason is that, in aqueous solutions, the solvent itself can be the reactant or the product in the reaction being studied.

A few words of caution might be appropriate in regard to the use of this concept, particularly for readers who did not major in chemistry. The pH of an aqueous solution is formally defined as

$$pH \equiv -\log a_{H_3O^+} \tag{6.34}$$

in which $a_{H_3O^+}$ is the activity of the hydronium ion. Single-ion activity cannot be measured, but in dilute solutions, the activity can be approximated by the concentration. This is so often assumed that one may forget that it applies only to dilute solutions, where the physical properties of the solvent are not perceptibly affected by the addition of solute. In contrast, the pH of a 7.0 M solution of KOH is about 16, far from the value of 14.85 calculated from Eq. (6.34), employing concentration instead of activity. Also, if one uses a mixed solvent (e.g., methanol and water), the concept of pH becomes somewhat more complicated, and must be employed with caution. In such cases, even if the pH can be measured by a conventional method (e.g., with a glass electrode), the result may represent quite different concentrations of H_3O^+ ions in different solvent mixtures.

When the temperature is changed, the equilibrium constant for the dissociation of water is also changed. As a result, pH 7.0 represents the point of neutrality only at 25 °C. At higher temperatures the point of neutrality moves to lower pH values. Similarly, in mixed solvents the point of neutrality is not necessarily at pH 7.0.

Let us now return to the effect of pH on electrode kinetics, using concentrations instead of activities. Consider the hydrogen-evolution reaction, and assume that it proceeds in the following two steps, with the second step being rate determining.

$$H_3O^+ + e_M^- H_{ads} + H_2O \tag{6.35}$$

$$2H_{ads} \xrightarrow{rds} H_2 \tag{6.36}$$

For (quasi)-equilibrium in the first step we can write:

$$\frac{\theta}{1-\theta} = K_1 c_{b,H_3O^+} \exp\left(-\frac{F}{RT}E\right) \tag{6.37}$$

This is identical to Eq. (6.16), derived for the first step in chlorine evolution, except for the change in the sign of the exponent, which is necessary because we are now dealing here with a cathodic reaction.

For the rate-determining step we write:

$$j_2 = Fk_2\theta^2 \tag{6.38}$$

This is an interesting case, which may need some clarification. Equation (6.36), as written, is not an electrochemical step in the sense that it does not involve charge transfer. Nevertheless we are justified in expressing the rate of this step in terms of a current density, since every time step (6.36) occurs, two electrons must have been transferred in the preceding step.

If we assume low coverage ($\theta \ll 1$) we can substitute θ from Eq. (6.37) into Eq. (6.36), to obtain

$$j_2 = nFK_1^2 k_2 (c_{b,H_3O^+})^2 \exp\left(-\frac{2F}{RT}E\right) \tag{6.39}$$

from which we obtain directly

$$\rho_1 = (\partial \log j / \partial \log c_{b,H_3O^+})_E = -(\partial \log j / \partial pH)_E = 2 \tag{6.40}$$

Also, since we can write the Nernst equation in this case as:

$$E_{rev} = E^0 + \left[\frac{2.3RT}{2F}\right]\log\left[\frac{(c_{b,H_3O^+})^2}{p_{H_2}}\right] \tag{6.41}$$

we have, from Eq. (6.39), (taking $p_{H_2} = 1$)

$$\log j_2 = \log(n\,FK_1^2 k_2) + 2\log(c_{b,H_3O^+}) - 2\left(\frac{2F}{2.3RT}E^0\right)$$

$$-2\log(c_{b,H_3O^+}) - 2\left(\frac{2F}{2.3RT}\right)\eta \tag{6.42}$$

Hence

$$\log j_2 = \log(nF\,K_1^2 k_2) - \frac{2F}{2.3RT}(E^0 + \eta) \tag{6.43}$$

We note that, in this particular case, the reaction order at constant overpotential, ρ_2, becomes independent of pH. When the pH is lowered, the effect of increasing concentration of the hydronium ion is exactly compensated by the influence of the variation of the reversible potential on the reaction rate.

It is interesting to note that the symmetry factor β does not appear in any of these equations. This is because the rate-determining step assumed here does not involve charge transfer. The current depends indirectly on potential, through the potential dependence of the fractional coverage, θ. The transfer coefficient $\alpha_c = 2.0$ *exactly*, as can be seen in Eq. (6.39), and this corresponds to a Tafel slope of $b_c = 29.5$ mV at room temperature.

Note that the transfer coefficient obtained here is not in any way related to the symmetry factor. It arises from the quasi-equilibrium assumption and should therefore be a true constant, independent of potential, as long as the assumptions leading to Eq. (6.39) are valid.

6.1.6
The Enthalpy of Activation

To discuss the enthalpy of activation in electrode kinetics, we make use of the fundamental rate equation

$$j = nFc_i\omega\exp\left(-\frac{\Delta G^{0\#}}{RT}\right)\exp\left(\frac{\alpha_{an}F}{RT}\right)E \tag{6.44}$$

which is equivalent to Eq. (5.19) discussed earlier, except that it is written for a multi-step anodic reaction. We split the Gibbs energy of activation into its enthalpic and entropic parts, and rewrite Eq. (6.44) in logarithmic form as:

$$\log j = \log(nFc_i\omega) - \frac{\Delta H^{0\#}}{2.3RT} + \frac{\Delta S^{0\#}}{2.3R} + \frac{\alpha_{an}F}{2.3RT}E \tag{6.45}$$

from which it follows that a plot of $\log j$ versus $1/T$ should be a straight line with a slope of

$$[\partial \log j/\partial(1/T)]_E = -\frac{\Delta H^{0\#} - \alpha_{an}E\,F}{2.3R} = -\frac{\Delta \bar{H}^{0\#}}{2.3R} \tag{6.46}$$

This is similar to the usual treatment in chemical kinetics, except that the enthalpy of activation is found to be a function of potential, as shown in Figure 6.2.

Often, reference is made in the literature to the *energy* of activation, instead of the *enthalpy* of activation. It follows from elementary thermodynamics that the former applies if the reaction is conducted at constant volume, whereas the latter is applicable to conditions of constant pressure. In aqueous solutions, the difference between the two is negligible, and when measurements are made at ambient pressure, the terms can be used interchangeably.

To be exact, it should be remembered that the frequency term, ω, in Eqs. (5.19) and (6.44), is generally considered to be a function of temperature

$$\omega = k_B T/h \tag{6.47}$$

where k_B and h are the Boltzmann and the Planck constants, respectively. This should lead to a nonlinear dependence of $\log j$ versus $1/T$. The effect is small, however, and

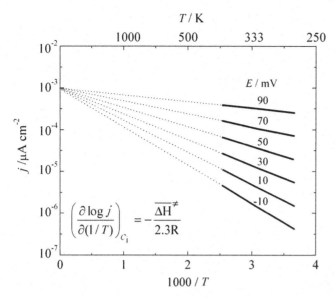

Figure 6.2 Schematic Arrhenius plots, showing the variation of the apparent electrochemical enthalpy of activation as a function of potential. All lines extrapolate to the same point at $1000/T = 0$, that is, at infinite temperature.

can be neglected in the narrow temperature range accessible to experiment in most solutions.

An inherent source of uncertainty in the calculation of the enthalpy of activation, unique to electrochemistry, is related to the temperature-dependence of the potential of the reference electrode. Thus, in order to obtain $\Delta \bar{H}^{0\#}$, we determine $\log j$ versus $1/T$ at a constant metal–solution potential difference, $\Delta \phi$. Now, at any given temperature, $\Delta \phi$ is constant, as long as the potential with respect to a given reference electrode is constant. When the temperature is changed, this is no longer true, since the metal–solution potential difference at the reference electrode has changed by an unknown amount.

There are two ways to approach this problem: measurement can be conducted *isothermally* (i.e., keeping the working and the reference electrodes always at the same temperature) or *nonisothermally* (i.e., keeping the temperature of the reference electrode constant while, that of the working electrode is scanned). In the latter case, the reference and working electrode compartments are connected through a salt bridge, and there must be a temperature gradient somewhere along this bridge, causing a small thermal-junction potential.

There is some merit to each of these methods, and both have been used. In isothermal mode, we estimate the change in the value of $\Delta \phi$ with T for the reference electrode. In nonisothermal mode, we estimate the additional potential drop generated in the salt bridge, as a result of the thermal gradients. Neither can be measured directly. Consequently, there is always some uncertainty in the value of the enthalpy of activation of electrode reactions.

We note (cf. Eq. (6.47)) that the enthalpy of activation decreases with potential (or rather with increasing overpotential) and its rate of change with potential is proportional to the transfer coefficient, α_{an}

$$-\Delta \bar{H}^{0\#} = -(\Delta H^{0\#} - \alpha_{an} E\, F) \tag{6.48}$$

Thus, α_{an} could, in principle, be obtained from Eq. (6.48), but since direct measurement of α_{an} (from the current–potential relationship) is much more accurate, it is better to introduce its value, determined from the Tafel slope, into Eq. (6.48) in order to test the accuracy of the measured values of the electrochemical enthalpy of activation, $\Delta \bar{H}^{0\#}$.

7
Specific Examples of Multi-Step Electrode Reactions

7.1
Experimental Considerations

7.1.1
Multiple Processes in Parallel

In all the discussions so far, it has been tacitly assumed that we know the reaction taking place and that *only one reaction* is occurring in the potential range of interest. Unfortunately, this is not always the case. In the electroplating industry, for example, one must specify the so-called *Faradaic efficiency*, which is the fraction of the current utilized for metal deposition (the rest is usually taken up by hydrogen evolution). The measured current in itself does not yield any information regarding the fraction of the current supporting each reaction, of course. If two or more reactions occur simultaneously, the current–potential relationship can become rather complicated, since the different partial currents may depend on potential in different ways. In the case of metal deposition, one can obtain the partial current density by measuring the weight of metal deposited or the volume of hydrogen evolved (or both, for double checking). However, in that case, the high sensitivity and ease of determination of the rate of electrode reactions, offered by the simple measurement of the current density, is lost.

In other cases, it may not be obvious how far the reaction can proceed under a given set of conditions. Oxygen, for instance, can be reduced either to hydrogen peroxide ($n = 2$) or to water ($n = 4$), depending on the nature of the electrode, the composition and purity of the solution and the range of potential studied. There could also be a region in which both reactions occur simultaneously, and their relative rates depend on potential. In such cases, analysis of the j/E relationship could become rather complicated and extracting the mechanistic parameters may not be possible.

Another tacit assumption is that the surface of the electrode remains unchanged during the experiment. Changes in the catalytic activity of the surface (which would change the specific rate constant of the reaction) could occur during metal deposition or dissolution, particularly when alloy deposition is concerned, where

Physical Electrochemistry: Fundamentals, Techniques and Applications. Eliezer Gileadi
Copyright © 2011 WILEY-VCH Verlag GmbH & Co. KGaA, Weinheim
ISBN: 978-3-527-31970-1

the atomic composition of the surface could change during the course of measurement, or the surface roughness could increase or decrease. Adsorption of impurities could cause major changes in the catalytic activity of the electrode, particularly if it is nonuniform, and contains active sites where most of the reaction takes place. This problem can be partially alleviated when a renewable surface, such as a dropping-mercury electrode is used, as shown below. On solid electrodes it is practically impossible to maintain the desired high level of surface purity, unless the surface is held in a state where adsorption of impurities is inherently prevented. For example, oxygen evolution on platinum can be conducted without interference from most impurities, since the oxide-covered surface resists contamination by most impurities. On the other hand, low-frequency measurements of electrochemical impedance spectroscopy (EIS), which will be discussed below, take a long time, and the results obtained may be corrupted by variations of the surface during the course of the measurement.

7.1.2
The Level of Impurity That Can Be Tolerated

The above arguments give rise to the need for extreme conditions of purity in electrochemical measurements. Why is high purity necessary and how clean should the system be in order that reliable results will be obtained? While it may be a good idea to conduct *any* experiment in a clean system, it must be remembered that purity has a price, and for extreme purity the price may be quite high, in terms of both the cost of electrode materials and other chemicals, and the time and effort required for each experiment. Impurities can be divided into two groups: those that are electroactive in the potential range of interest, and those that may interfere with measurement by adsorbing on the surface and poisoning it (a more general, and perhaps less ominous, phrase would be "changing its catalytic activity").

For the former group, the allowed level of impurity is relatively easy to assess. It should be several orders of magnitude less than the concentration of the material being studied, so that the mass-transport-limited current density at which the impurity can react will be smaller than the smallest current we wish to measure.

Let us clarify this argument by an example – the need to remove oxygen from solution during the study of the hydrogen-evolution reaction on mercury. The concentration of oxygen in a dilute aqueous solution at equilibrium with air is about 0.25 mM (8 ppm). For a dropping-mercury electrode, which is renewed every second, this should yield a limiting current of about 3 µA. This is the current calculated for the four-electron reduction of oxygen to water, which is the reaction taking place at high overpotentials with respect to oxygen reduction, over the range in which hydrogen evolution is studied. (cf. the Ilkovic Equation, (4.19)). For the hydrogen-evolution reaction on mercury, one typically tries to obtain the j/E relationship for current densities from $1 \times 10^{-2} \rightarrow 1 \times 10^{4}\ \mu A\ cm^{-2}$ It is necessary, therefore, to reduce the concentration of oxygen by two to three orders of magnitude, so that the limiting current for this reaction will be lower than the lowest current we wish to measure.

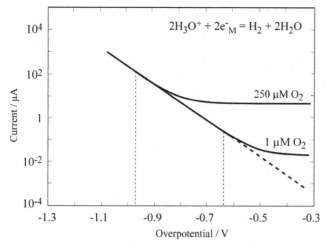

Figure 7.1 The current calculated for hydrogen evolution at a DME, assuming $I_0 = 3.3 \times 10^{-12}$ A ($j_0 = 1 \times 10^{-10}$ A cm^{-2}) in the presence of two concentrations of dissolved oxygen. The dotted line shows the current that would be observed in the total absence of oxygen

This can be done by bubbling purified nitrogen or argon though the solution for some minutes before the experiment begins. The partial current for oxygen reduction and for hydrogen evolution are shown in Figure 7.1, for two concentrations of oxygen.

This figure shows that reducing the concentration of oxygen by a factor of 250, should extend the range over which the hydrogen-evolution reaction (HER) can be studied by about 0.28 V.

It should be added that the experimental situation could be better than indicated in this figure, because it is the *fluctuation* in the background current, not its absolute value, which determines the lowest current for hydrogen evolution that can be reliably evaluated. Thus, if the experiment is conducted on a rotating-disc electrode, where the limiting current for oxygen reduction is stable and can be measured accurately, the range over which the HER can be studied could be extended, by subtracting the oxygen-reduction current from the total current measured.

Now, consider the effect of an impurity that is adsorbed on the surface and alters its catalytic properties. We recall that it takes a very small amount of material (about 2×10^{-9} mol cm^{-2}) to form a monolayer. Moreover, very often a fraction of a monolayer is enough to change the properties of the surface significantly. In a typical experiment there will be about 25 cm^3 of solution per cm^2 of electrode area. If the concentration of impurity is 1.0 μM, the total amount of impurity in solution will be 25 nmol, for each cm^2 of surface area, enough to form 10–20 molecular layers, depending on the size of the species being adsorbed. This is too high, even if we accept the fact that only a fraction of these impurity molecules reaches the surface and adheres to it during measurement. Thus, a higher level of purity is required, if the solution is to be in contact with the electrode for a long time. A relatively simple way to ease the requirement for very high purity is to decrease the volume of the solution per unit surface area. Reducing this ratio to $1\ cm^3$ of solution per cm^2 of surface area can readily be achieved. Moreover, thin-layer cells have been built with a gap of $10\ \mu m$, corresponding to a volume of $1 \times 10^{-3}\ cm^3$ per cm^2. With an impurity level of $1.0\ \mu M$, this corresponds to a mere $1 \times 10^{-12}\ mol\ cm^{-2}$ of surface area, namely 0.05% of a monolayer. Thus, by employing a thin-layer cell, one could relax the purity requirement by one or two orders of magnitude and still maintain a clean surface during measurement.

The other aspect of allowed impurity level relates to the maximum rate of adsorption, as compared to the duration of the experiment, or rather to the time interval between successive renewals of the surface. We shall explain this in relation to studies on the dropping-mercury electrode, which is typically renewed every second, although the argument can be applied also to solid electrodes, under certain favorable conditions, as we shall see.

The rate of adsorption of an impurity may be kinetically controlled, in which case it depends on the specific system being studied. This cannot be discussed in general terms, but we can calculate *the maximum rate of adsorption*, which is controlled by mass transport, as a function of the concentration of impurity. This rate is given by the flux of the impurity molecules reaching the surface. The logic behind this argument is very simple. If every molecule of impurity reaching the surface is instantaneously adsorbed, the concentration of these molecules in the solution nearest to the surface (i.e., at $x = 0$) will be zero. The rate of adsorption will then depend on the rate of supply of molecules to the surface, that is, it will be totally mass-transport controlled. This situation is similar to the case in which an electroactive material is oxidized or reduced at the limiting current. In the case of adsorption, charge transfer may not take place, but the adsorbed molecules are removed from the solution, just as if they had reacted at the surface. Clearly, this is a worst-case scenario, since the rate of diffusion is the maximum rate for a given concentration, and the assumption that each species reaching the surface is adsorbed may not be valid. Thus, the degree of coverage could be less than the value calculated in this manner, but it could not exceed it.

In the case of the dropping-mercury electrode, the limiting current for a one-electron reduction is on the order of $3\ \mu A\ mM^{-1}$, or a current density of about $100\ \mu A\ cm^{-2}\ mM^{-1}$ for a typical surface area of $0.03\ cm^2$. The flux of the impurity reaching the surface, in units of $mol\ cm^{-2}\ s^{-1}$, is equal to the diffusion-limited current density, divided by the charge per mole, nF. Assuming, as before, an impurity concentration of $1.0\ \mu M$, we obtain a flux of

$$\frac{j_d}{F} = \frac{(1 \times 10^{-4}\ A\ cm^{-2}\ mM^{-1})(1 \times 10^{-3}\ mM)}{9.65 \times 10^4\ C\ mol^{-1}} \approx 1 \times 10^{-12}\ mol\ cm^{-2}\ s^{-1}$$

$$(7.1)$$

Considering that a monolayer amounts to about $2\ n\ mol\ cm^{-2}$, it would take about $2000\ s$ to deposit a monolayer of impurity. Since the drop is renewed every second, the maximum coverage by an impurity that exists in solution at a concentration of $1.0\ \mu M$ cannot exceed $\theta = 5 \times 10^{-4}$.

The foregoing calculation shows that one can relax the requirements for purification substantially. Thus, allowing a tenfold increase in the impurity level would still limit the maximum impurity coverage during the lifetime of a drop to $\theta \leq 5 \times 10^{-3}$.

These conditions can be realized in the simplest way on a dropping-mercury electrode, but they are not totally limited to liquid metals. The surface of a solid electrode could also be cleaned or effectively renewed periodically, for example, by a suitable series of pulses of potential, by mechanical abrasion *in situ*, or by sudden heating with a laser pulse.

On the other hand, when a mercury pool or a solid electrode is employed and the surface is not renewed periodically, the electrode may be in contact with the solution for 10^3 to 10^4 s, allowing ample time for the impurity to diffuse to the surface and be adsorbed on it. The requirements for solution purity are much more stringent in this case, of course.

So far we have discussed only the *maximum* extent and rate of adsorption. This represents the worst case, and one cannot go wrong if the preceding requirements are satisfied. In general, however, adsorption depends on potential. This can best be treated in the context of electrosorption of organic materials, discussed below. We shall anticipate that discussion here by saying that adsorption depends mostly on the charge density on the metal. Impurities that carry a negative charge will be adsorbed when the excess charge density on the electrode surface is positive, and vice versa. It is also true, although not obvious from first principles, that neutral species tend to be adsorbed mainly in the region in which the excess charge density is the lowest. The so-called *potential of zero charge*, E_z, where the excess charge density on the metal is zero, can be measured for most metals. It can be said qualitatively that a region of about 0.5 V on either side of this potential is the most susceptible to interference by adsorption of neutral impurities.

When very high solution purity is required, the last stage of purification is often *pre-electrolysis*, employing a large-surface-area electrode, composed of the same metal as the working electrode. The idea behind this procedure is simple: if there is an electroactive impurity in solution, let it be consumed during pre-electrolysis (which is typically conducted for a long time, compared to the duration of the experiment) so that none will be left to interfere with the reaction to be studied in the purified electrolyte. If the impurity is not electroactive, there will be sufficient time for it to be adsorbed on the surface during pre-electrolysis. The electrode used for pre-electrolysis can then be removed from solution, carrying the impurities with it. One could estimate the time required for pre-electrolysis, but it is probably best to develop the procedure by trial and error, choosing a time beyond which further pre-electrolysis does not help.

We conclude this discussion by stating that maintaining high-purity conditions is essential in measurements of electrode kinetics. Experimental results obtained without due control of the impurity level cannot be trusted. Yet, the level of impurity required in each case depends on the system being studied and on the method of measurement. One must prudently combine the use of high-purity solvents and chemicals with suitable measuring techniques, to reach the desired level of purity *during measurement* at the lowest cost in materials and effort.

7.2
The Hydrogen-Evolution Reaction

7.2.1
Hydrogen Evolution on Mercury

Mercury electrodes have been studied more than any other type of electrode, because of their ease of purification and the high degree of reproducibility attainable with them. All aspects of hydrogen evolution on mercury have probably been studied at one time or another. On the basis of all experimental evidence, it is commonly accepted that in this case, the first charge-transfer step is rate-determining, and is followed by fast ion–atom recombination

$$H_3O^+_{soln} + e^-_M \xrightarrow{rds} H_{ads} + H_2O \qquad (7.2)$$

$$H_3O^+_{soln} + H_{ads} + e^-_M \xrightarrow{fast} H_2 + H_2O \qquad (7.3)$$

The coverage by adsorbed hydrogen atoms must be very low, since none has ever been detected, even at the highest overpotentials measured. This also rules out atom–atom recombination as the fast second step, since the rate of the reaction

$$H_{ads} + H_{ads} \rightarrow H_2 \qquad (7.4)$$

is proportional to θ^2, while the rate of step (7.3) is proportional to the first power of θ.

The Tafel slope for this mechanism is $2.3RT/\beta_c F$, and this is one of the few cases offering good evidence that $\alpha_c = \beta_c$, namely, that the experimentally measured transfer coefficient is equal to the symmetry factor. A plot of $\log \chi$ versus E, where χ is the dimensionless rate constant, given by

$$\chi = [(12/7)(t/D)]^{1/2} k_{s,h} \qquad (7.5)$$

obtained on a dropping-mercury electrode in a dilute acid solution, is shown in Figure 7.2. The activation-controlled current density j_{ac}, corrected for mass-transport limitation, is proportional to the parameter χ. The accuracy shown here is very high and can be achieved only on a HDME. On solid electrodes, one must accept a significantly lower level of accuracy and reproducibility. The best values of the symmetry factor obtained in this kind of experiment are close to, but not exactly equal to, 0.50. It should be noted, however, that the Tafel lines are very straight; that is, β_c is strictly independent of potential over a range 0.5–0.6 V, corresponding to four to five orders of magnitude of current density.

The hydrogen-evolution reaction on the HDME is the best-known case in which β_c is experimentally shown to be independent of potential and to have a value close to 0.50. This is probably the best experimental evidence favoring the use of this value of β_c in the analysis of more complex electrode reactions.

The exchange-current density for this system depends on the composition of the solution, but generally it is in the range 10^{-12}–10^{-10} A cm^{-2}. Mercury is often

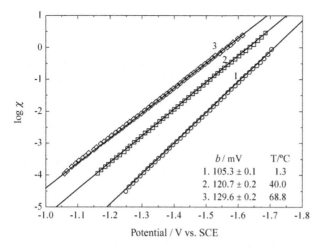

Figure 7.2 Tafel plots for the HER at a hanging dropping-mercury electrode (HDME) in 3 mM HCl and 0.8 M KCl. χ is the dimensionless rate constant given by $\chi = [(12/7)(t/D)]^{1/2}k_{s,h}$. Based on data from E. Kirowa-Eisner, M. Schwarz, M. Rosenblum and E. Gileadi, *J. Electroanal. Chem.* 381 (1995) 29.

referred to in the literature, rather loosely, as a "high-overpotential metal". This is a poorly chosen term, since the overpotential clearly depends on the current density and cannot be said to have a specific value for a particular metal. It would be better to describe the situation by saying that mercury is "a low-exchange-current-density metal". Of course, if overpotentials are compared at a given current density, then it is higher for mercury and similar "high-overpotential metals" than for platinum and other "low-overpotential metals".

Which metals are similar to mercury in this respect? It turns out that most of the soft metals in Groups 13–15 of the periodic table (including Pb, Bi, Cd, In and Sn) behave rather similarly to mercury as cathodes for the hydrogen-evolution reaction. It would be presumptuous to claim that we could have predicted this similarity from theory, but on being confronted with the facts, we can reasonably well explain this result on the basis of the poor catalytic activity of these metals, as shown in Section 7.7.

7.2.2
Hydrogen Evolution on Platinum

It is much more difficult to study the kinetics of the hydrogen-evolution reaction on platinum than on mercury. To begin with, the exchange-current density is found to be many orders of magnitude higher, in the range $j_0 = 10^{-4}-10^{-2}$ A cm^{-2}, depending on the composition of the solution and its purity. It may be recalled that the linear Tafel region (in which the rate of the reverse reaction can be neglected), starts only when $j/j_0 \geq 10$ (corresponding to $\eta/b \geq 1$). Since it is necessary to make measurements over at least two decades of current density in order to obtain a reliable Tafel slope, the measurement on platinum must be extended to rather high current densities, where interference by the uncompensated solution resistance and by mass-transport limitations can be significant. Also, it is found that platinum, being a catalytic metal, adsorbs most impurities very well. Moreover, the potential region in which the HER is studied is close to the potential of zero charge on platinum, where adsorption of neutral species is favored. Thus, a very high level of purification is required in order to obtain reliable data for the HER on platinum electrodes.

The results obtained in acid solutions indicate that there are two distinct mechanisms. At low overpotentials, the atom–atom recombination step (7.4) is believed to be rate-determining. This should yield a Tafel slope of $b = 2.3RT/2F = 29.5$ mV and a reaction order (at constant potential) of $\rho_1 = 2$, in agreement with experiment. As the overpotential is increased, the fractional coverage θ must also increase. This increases not only the rate of the atom–atom recombination step, but also that of ion–atom recombination, which occurs in parallel. As θ approaches unity, the rate of step (7.4) can no longer increase, but the rate of the ion–atom recombination step can grow, because it depends on potential (cf. Eq. (7.3)). This step then becomes rate-determining. The resulting rate equation (for $\theta \to 1$) is:

$$j_2 = nFk_2 c_{b,H_3O^+} \exp\left(-\frac{\beta_c F}{RT} E\right) \tag{7.6}$$

This leads to a Tafel slope of $b = 2.3RT/\beta_c F = 118$ mV for $\beta_c = 0.5$, and a reaction order (at constant potential) of $\rho_1 = 1$. The transfer coefficient α_c is equal to the symmetry factor β_c, as in the case of mercury, but we recall that the Tafel slope is calculated here on the assumption of essentially full coverage, whereas that on mercury was obtained for very low coverage.

Is the assumption of high coverage borne out by experiment? Platinum is a catalytic metal, on which molecular hydrogen is adsorbed spontaneously from the gas phase, dissociating and forming adsorbed atoms on the surface. The surface coverage can be readily measured electrochemically, and indeed it is found to be high at high overpotentials. The difficulty with the foregoing interpretation is that θ is found to be close to unity already at low overpotentials, making it hard to justify the low value of the Tafel slope observed in this region. One way out of this dilemma is to assume that there are two types of adsorbed hydrogen on platinum: a first monolayer, which is already observed at the reversible potential, is partially *ab*sorbed and becomes, in a sense, a part of the metal surface. This layer can be removed by anodic oxidation, yielding a measure of the coverage, but it does not participate in the HER at low negative overpotentials. The precursors of the HER are hydrogen atoms adsorbed on top of the first layer at low partial coverage. At high overpotentials, the first layer of adsorbed hydrogen becomes electroactive, and the observed kinetics are as expected on a complete monolayer of adsorbed hydrogen. This interpretation is tentative, in the sense that it has not been confirmed directly. There is, however, indirect evidence that this kind of behavior is possible and even plausible, from the study of the oxidation of iodide in thin-layer cells. In that case, it was shown experimentally that the first layer of iodide ions adsorbed on the surface of platinum is not electroactive and iodine is formed from ions in solution, which presumably can be adsorbed on top of the first layer. It was also found that the first adsorbed layer can be oxidized at higher overpotentials, as proposed here for the HER.

The region of micro-polarization, where the j/η plot is linear, can extend to about $\eta/b \leq 0.2$ whereas the linear Tafel region starts at about $\eta/b \geq 1$. As a result, the intermediate region of $0.2 \leq \eta/b \leq 1$ would be left unused, as far as the evaluation of kinetic parameters is concerned. For fast reactions, such as the HER on platinum, this represents a loss of crucial data, since it may be difficult to extend the measurements to overpotentials much above $\eta/b = 1$, because of mass-transport limitations. Fortunately, modern microcomputers allow us to make use of this intermediate region. To do this, we write the full equation for an activation-controlled electrode reaction as follows:

$$j/j_0 = \exp\left(\frac{\alpha_{an} F}{RT} \eta\right) - \exp\left(-\frac{\alpha_c F}{RT} \eta\right) \qquad (7.7)$$

assuming that mass transport has been corrected for. The values of j_0, α_{an} and, α_c are then calculated by parameter-fitting of the data, according to Eq. (7.7), over the whole range of overpotentials accessible experimentally. Moreover, for fast reactions, it is often possible to make measurements both at positive and at negative overpotentials, extending even further the range of data from which the kinetic parameters could be obtained.

Metals that are known to be highly catalytic for hydrogenation and dehydrogenation reactions are similar to platinum with respect to the HER. These include palladium, iridium, rhenium, nickel, and cobalt. In the case of palladium, the situation is further complicated by the tendency of atomic hydrogen to diffuse *into* the metal and dissolve in it. Although this effect is most pronounced for palladium, which can adsorb hydrogen to the extent represented by the formula $PdH_{0.6}$, hydrogen is also absorbed in iron, nickel, titanium, aluminum, platinum and many other metals. The hydrogen-evolution reaction in these cases should be written in the form:

$$H_3O^+ + e_M^- \rightarrow H_2O + H_{ads} \qquad (7.8)$$

The *adsorbed* hydrogen atom can then follow two parallel paths: form molecular hydrogen or be *absorbed* in the metal. Penetration of hydrogen into the metal is a side reaction for the HER. If it is substantial, as in the case of palladium, the j/E plot cannot be analyzed properly, unless the relative rates of the two parallel reactions are determined as a function of potential and time. This could be performed, for example, by employ a ring–disc electrode. Molecular hydrogen would be formed on the disc, and its rate of formation could be determined by oxidizing it back to H_3O^+ on the ring, but such measurements have rarely, if ever, been reported.

It should be noted here that Pt and some of its alloys are of great importance in the area of fuel cells and water electrolyzers and in the general context of the so-called "*hydrogen economy*". In these fields the high value of the exchange current density is a great advantage, and the fact that it hinders our efforts to determine the mechanism is quite irrelevant. Apart from being an excellent catalyst for oxidation of molecular hydrogen (as well as some prospective organic fuels, such as methanol), its stability in acid solutions under usual operation conditions is an important asset. Great efforts have been made to reduce the cost of Pt employed as anodes in fuel cells, by using small amounts of the metal in highly dispersed form, usually on large-surface-area carbon. This is discussed in Chapter 20.

7.3
Hydrogen Storage and Hydrogen Embrittlement

Two interesting fields of technology are related to the hydrogen-evolution reaction. These are the storage of hydrogen in the form of a solid, either as a hydride or as hydrogen dissolved in the metal, and hydrogen embrittlement.

7.3.1
Hydrogen Storage

Hydrogen is in some respects the ideal fuel. It is clean (leaving only water as the product of its oxidation), it is renewable in unlimited amounts (provided there is a method of producing energy that does not depend on burning fossil fuel (such as solar, wind, hydroelectric or nuclear). In addition, because of its low equivalent weight, it has the highest energy density per unit weight, (about 3.8 times the energy density of gasoline). Much of this advantage is lost, however, as a result of the difficulty of storage. In reality, one should consider the *effective equivalent weight* of hydrogen, which is the weight of 1 g stored, including the weight of the medium in which it is stored. In the chemical industry hydrogen is usually stored under high pressure or cryogenically. Both methods are expensive and consume a lot of energy. Thus, the amount of energy needed to compress hydrogen to 700 atm can be as high as 15–25% of the energy in the hydrogen stored[1]. Cryogenic storage consumes at least 30% of the energy stored. In addition, the weight of the container, particularly for compressed hydrogen, could reduce its effective equivalent weight significantly. There is therefore a clear incentive to develop better methods to store hydrogen, to permit the gas to be used at room temperature and at moderate pressures.

Hydrogen could be stored in palladium as $PdH_{0.6}$ at low pressure, and be easily retrieved from it, but this is not practical because its effective equivalent weight in this medium is 177, and the price of palladium is too high.

An alloy of iron with titanium can absorb and release hydrogen reversibly. With a composition of $Ti Fe H_{1.8}$ at maximum loading, this has an effective

---

1) This pressure is often used in the literature because it is equal to 10 000 psi, which is a nice round number. Moreover, it is probably the highest pressure that could be used in transportation with presently available high-strength materials.

equivalent weight of 58.6, not including the weight of the container in which it is stored.

Another alloy developed for the same purpose is made of lanthanum and nickel. The composition at maximum loading is $La\,Ni\,H_{6.7}$, which corresponds to an effective equivalent weight of 65.5. Indeed, a very similar alloy, in which Ni is replaced partially by one or more transition metals and La may be replaced by a mixture of lanthanides, is used commercially in so-called Ni/metal hydride batteries. Other hydrides, such as $Mg\,H_2$ or AlH_3 (effective equivalent weights of 13 and 10, respectively) have been suggested, but have not been widely used so far, mainly because charging and discharging with hydrogen can be done only at an elevated temperature, and the cost of preparing these hydrides, both in terms of energy and money is prohibitive.

Attempts have been made in recent years to use nanotechnology for better hydrogen storage. Carbon nanotubes and other forms of highly dispersed carbon have been tried, with only partial success. It should be realized that storage of hydrogen by bonding it chemically (as in AlH_3) or physically (as in carbon nanotubes) must satisfy two criteria that are inherently opposing each other. The thermodynamic criterion is that bonding should be strong, so that hydrogen could be stored at or close to ambient temperature and pressure. On the other hand, the kinetics of releasing it on demand should be fast and consume as little energy as possible. Now, the standard Gibbs energy of a reaction is given by the difference between the standard Gibbs energies of activation of the forward and reverse direction.

$$\Delta G_f^{0\#} - \Delta G_b^{0\#} = \Delta G^0 \tag{7.9}$$

where adsorption (storage) and desorption (release) are considered to be the "forward" and "backward" reactions, respectively. A good medium for hydrogen storage should have a high negative value of ΔG^0. But it follows from Eq. (7.9) that the standard Gibbs energy of activation for releasing the adsorbed or absorbed hydrogen is given by

$$\Delta G_b^{0\#} = \Delta G_f^{0\#} - \Delta G^0 \tag{7.10}$$

Considering that the energies of activation are, by definition, positive while ΔG^0 is negative, it follows that $\Delta G_b^{0\#} \geq -\Delta G^0$. Thus, increasing the capacity for storage, by finding a medium that strongly binds hydrogen, decreases the rate of its delivery. Catalysis cannot increase the rate of release of hydrogen because $-\Delta G^0$ in the above inequality is a thermodynamic quantity, which is not influenced by catalysis. On the other hand, increasing the surface area per unit weight could improve hydrogen storage capacity, and this is where nanotechnology may eventually lead to a suitable solution.

It should be noted that the numbers given above for total weight per unit weight of hydrogen (namely the effective equivalent weight) do not tell the whole story. Even if hydrogen is stored chemically in the manner discussed above, it still must be kept in a suitable container and heat exchangers must be provided to remove the heat evolved during storage and to provide heat needed when hydrogen is discharged. Thus, the *effective equivalent weight* in an actual storage device could be significantly higher.

The most interesting application of hydrogen as a fuel at present is for electrical cars operated with H_2/O_2 fuel cells, in the context of the hydrogen economy, although hydrogen could also be used directly as the fuel in internal combustion engines. In either case, the ability to store hydrogen cheaply and safely in a vehicle is critical. However, at this time (2010) it can be stated that this will not be practical until a major breakthrough in developing more suitable storage systems is made.

7.3.2
Hydrogen Embrittlement

The best-known case of hydrogen dissolved in a metal in small quantities, causing major changes in its bulk properties, is that of steel. Hydrogen absorbed in steel can cause embrittlement. As a rule, the higher the tensile strength of the steel, the more susceptible it is to hydrogen embrittlement. It is interesting to note that the amount of hydrogen that can cause severe embrittlement is minute, of the order of 0.01 atom% or less.

How can hydrogen get into iron and its alloys? Iron is not a particularly good catalyst with respect to hydrogenation and dehydrogenation reactions. When placed in contact with *molecular* hydrogen at room temperature and moderate pressure, nothing much happens. The surface of iron is not a good enough catalyst to split the H_2 molecules and form H_{ads} on the surface – a necessary precursor for diffusion of hydrogen into the metal. The rate of hydrogen evolution on iron and its alloys is intermediate between that on mercury and on platinum, with j_0 values in the range $10^{-5}-10^{-7}$ A cm^{-2}, depending on the type of alloy and the solution used. The kinetics of the HER on iron has been studied extensively, and it has been established that atomic hydrogen is adsorbed on the surface at a substantial level, enough to support the entry of hydrogen atoms into the metal. Naturally, one does not cause hydrogen embrittlement on purpose, by evolving hydrogen on a steel part, but this can occur as a side reaction in several industrial processes. For example, steel parts are often plated with a thin layer of cadmium to prevent corrosion, with hydrogen evolution occurring as a side reaction. Hydrogen evolution can also occur during corrosion, when the sample is exposed to the environment. Furthermore, in an attempt to protect the metal against corrosion by applying a negative potential to it (a technology referred to as *cathodic protection*, cf. Section 18.3.2), hydrogen is evolved and atomic hydrogen is formed on the surface as an intermediate. Cleaning the surface in preparation for electroplating or painting often involves the dissolution of a thin layer of the surface in acid (a process called *pickling*). During such processes hydrogen is evolved, and some of it may penetrate the bulk of the metal, causing severe embrittlement.

Hydrogen embrittlement affects not only ferrous alloys. It is also important in many other alloys of great engineering importance, in particular titanium- and aluminum-containing alloys. In all cases it reduces the mechanical strength of the metal and may cause catastrophic failure if left undetected and untreated.

7.4
Possible Paths for the Oxygen-Evolution Reaction

Oxygen evolution is a more complex reaction than hydrogen evolution, involving the transfer of four electrons. The overall reaction in alkaline solution is

$$4OH^- \rightarrow O_2 + 4e_M^- + 2H_2O \tag{7.11}$$

whereas in neutral or acid solutions it is

$$6H_2O \rightarrow O_2 + 4\,H_3O^+ + 4e_M^- \tag{7.12}$$

Let us write a mechanism (one of many possible) for this reaction. The first step would be charge transfer

$$OH^- \xrightarrow{k_1} OH_{ads} + e_M^- \tag{7.13}$$

This could be followed by a further charge-transfer step, such as

$$OH^- + OH_{ads} \xrightarrow{k_2} O_{ads} + e_M^- + H_2O \tag{7.14}$$

Followed, for example, by atom–atom recombination, to yield

$$2O_{ads} \xrightarrow{k_3} O_2 \tag{7.15}$$

Let us evaluate the kinetic parameters corresponding to this reaction sequence. For the first charge transfer as the rate-determining step, we already know the result, since it is equivalent to step (7.1) for hydrogen evolution on mercury. If the second step is assumed to be rate-determining, we also know the result, (cf. Eqs. (6.18) and (6.19)). The Tafel slope changes from $b_{an} = [2.3RT/(1+\beta_{an})F]$ to $b_{an} = 2.3RT/\beta_{an}F$ as the partial coverage increases. The reaction order at constant potential, ρ_1, changes from 2 to 1, as θ approaches unity.

Now let us consider the third case, assuming the atom–atom recombination (Eq. (7.15)) to be rate-determining. In this case the two preceding steps are at quasi-equilibrium. The corresponding equations are:

$$\left[\frac{\theta_{OH} + \theta_O}{1 - (\theta_{OH} + \theta_O)} \right] = K_1 c_{OH^-} \exp\left(\frac{F}{RT} E \right) \tag{7.16}$$

$$\left[\frac{\theta_O}{\theta_{OH}} \right] = K_2 c_{OH^-} \exp\left(\frac{F}{RT} E \right) \tag{7.17}$$

Hence, for the rate-determining step one has

$$j_3 = nFk_3\theta_O^2 = nFk_3 K_2^2 \theta_{OH^-}^2 (c_{b,OH^-})^2 \exp\left(\frac{2F}{RT} E\right) \tag{7.18}$$

At low values of the total coverage, $[1-(\theta_O + \theta_{OH})] \approx 1$, which leads to

$$j_3 = nFk_3 K_1^2 K_2^2 (c_{b,OH^-})^4 \exp\left(\frac{4F}{RT} E\right) \tag{7.19}$$

For this mechanism, the transfer coefficient α_{an} is 4, the Tafel slope b_{an} is 15 mV and the reaction order at constant potential, ρ_1, is 4.

One could write many other pathways and rate-determining steps, and calculate the kinetic parameters for each, following the same line of reasoning. It is important to note that, in a complex reaction sequence, there can be more than one type of adsorbed intermediate on the surface, and some steps may involve the transformation of one kind of adsorbed species to another, by either an electrochemical or a chemical route.

Two aspects of the oxygen-evolution reaction are common to all electrodes studied so far: (i) the exchange-current density is low, of the order of 10^{-10} A cm^{-2} or less, and (ii) a reversible oxygen electrode operating at or near room temperature has not yet been found. At sufficiently high temperatures (say, in molten salts, at about 600 °C or with high-temperature solid electrolytes operating at around 1000 °C) the kinetics of the reaction can be sufficiently accelerated to make reversible oxygen electrodes operate as well as do reversible hydrogen electrodes at room temperature.

An important point that is often ignored, or at least not included explicitly in the interpretation of the experimental findings, is that oxygen evolution never occurs on the bare metal surface. By the time the reversible potential for oxygen evolution is reached, an oxide layer has been formed on all metals. At more anodic potentials, where measurements can actually be conducted, (remember that j_0 is small, and it takes a high overpotential to drive the reaction at a measurable rate), the oxide film may be several molecular layers thick. On the other hand, oxygen reduction may well occur on the bare metal surface, or one that is covered only by a fraction of a monolayer. Thus, even when Pt is used, it is well to remember that the oxidation and reduction of oxygen on the same metal occur at different surfaces and may therefore follow entirely different pathways.

The above statement might appear to violate the well-established *principle of microscopic reversibility*, which states in effect that the forward and reverse reaction follow the same pathway along the reaction coordinate, and the same step has the highest Gibbs energy of activation in both directions. This principle is implicit in the commonly used equation relating the forward and reverse rate constants to the equilibrium constant of the reaction

$$\vec{k}/\overleftarrow{k} = K \tag{7.20}$$

from which it follows that the difference between the standard Gibbs energies of activation should be equal to the standard Gibbs energy of the overall reaction

$$(\Delta G_f^{0\#} - (\Delta G_f^{0\#}) = \Delta G^0 \tag{7.21}$$

Applied to electrochemistry it implies that the exchange current density must be the same in both directions and the Tafel lines extrapolated from high anodic and high cathodic overpotentials should intersect at the reversible potential, yielding the numerical value of the exchange current density. It is important to emphasize that this does not apply to oxygen evolution and reduction, because they occur on different surfaces, and, hence, could follow different pathways.

Oxides on metals can be divided into three groups: those having a high electronic conductivity, such as RuO_2 and the oxides formed on platinum and iridium; those that are semiconducting, such as NiO, and the oxides formed on tungsten and molybdenum; and those that are insulators, such as Al_2O_3 and the oxides formed on the valve metals (Ti, Ta, and Nb). Oxygen evolution can occur readily on the electronically conducting oxides, and these are the best catalysts for it. On semiconducting oxides, the reaction can still occur, but it may be associated with pitting on the one hand and further build-up of the oxide layer on the other, causing poor reproducibility and making the interpretation rather dubious. On valve metals, oxide formation is the main reaction occurring during anodic polarization and the current either decays to zero with time or reaches a constant value, at which the rate of dissolution of the oxide is equal to the rate of its electrochemical formation.

On noble metals (e.g., Pt, Ir, Au), the region of potential in which the oxygen-evolution reaction is studied is far removed from the potential of zero charge in the positive direction. Thus, there is little danger of adsorption of neutral or positively charged impurities, but negatively charged impurities will be heavily adsorbed. One may expect, therefore, that the nature of the anion in the electrolyte will influence the measurements substantially, while the type of cation will have little effect. Also, most organic molecules are rapidly oxidized at these potentials, so that the requirements for solution purification are much less severe than in the case of hydrogen evolution.

The preceding statement does not hold true for the oxygen-*reduction* reaction. Using platinum as an example, we note that oxygen evolution is typically studied in the range 1.5–2.0 V, vs. a reversible hydrogen electrode (RHE) in the same solution, while oxygen reduction is studied in the range 1.0–0.4 V on the same scale. In most of the latter range, the surface is free of oxide (and very sensitive to impurities) if approached from low potentials, whereas it may be covered partially with oxide (and less sensitive to impurities) when approached from higher potentials. This is due to the high degree of irreversibility of formation and removal of the oxide layer on most noble metals.

A discussion of oxygen evolution and, even more, of oxygen reduction, cannot be complete without mentioning the efforts that have gone into improving the catalytic activity of metals by the use of metal-organic catalysts of the phthalocyanine group.

These are large planar molecules, with a metal atom in the center, resembling the structure of the porphyrin molecule, which is involved in the breathing processes of living organisms. The reasoning behind these experiments is to try to imitate nature. If molecular oxygen can be reduced efficiently in living organisms at or near ambient temperatures, perhaps the molecule involved in this reaction, or other metal–organic compounds having a similar structure, could also act as good catalysts for oxygen reduction occurring *in vitro*, in an electrochemical device.

Unfortunately, the analogy is rather poor. Nature operates in an intricate manner, and the porphyrin molecule is only one part of a complex system, which includes various enzymes and other protein molecules. Nevertheless, some progress has been made along these lines, and phthalocyanines and similar molecules may eventually be developed to become potent catalysts for oxygen evolution and reduction.

7.5
The Role and Stability of Adsorbed Intermediates

Intermediates are commonly formed in chemical reactions, as well as in electrode reactions. The preferred mechanism is that which involves the most stable intermediates, since this is the path of lowest Gibbs energy of activation. For reactions taking place in the gas phase or in the bulk of the solution, the stability of different species can be calculated, or at least estimated, from existing thermodynamic data. This is not the case for electrode reactions. For the HER discussed earlier, a hydrogen atom was assumed to be an intermediate. The standard potential for the formation of this species *in solution*, that is, for the reaction

$$H_3O^+_{soln} + e^-_M \rightarrow H^0_{soln} + H_2O \tag{7.22}$$

is about -2.1 V vs. SHE. Thus, this species could not be formed in the potential range over which the hydrogen-evolution reaction is studied, even on mercury electrodes. The real reaction we are looking at, however, is not represented by Eq. (7.22), but rather by Eq. (7.1):

$$H_3O^+_{soln} + e^-_M \rightarrow H^0_{ads} + H_2O \tag{7.2}$$

The hydrogen atom formed as an intermediate is stabilized by adsorption. To clarify this point, we should perhaps write Eq. (7.2) in explicit form, as follows:

$$H_3O^+_{soln} + e^-_M + Pt \rightarrow Pt\text{-}H + H_2O \tag{7.23}$$

implying that a chemical bond is formed between the hydrogen atom and a metal atom on the surface. The reversible potential for this reaction is clearly different from that for reaction (7.22). The difference between them depends on the Pt–H bond energy on the surface. This should not be confused with the bond energy of bulk platinum hydride, or with the energy of formation of a Pt–H species in the gas phase, since platinum atoms on the surface are energetically different from atoms in the bulk, which, in turn, are different from isolated atoms in the gas phase. It is found that the standard reversible potential for reaction 7.23 is $E_0^0 \approx +0.2$ V vs. RHE. For the hydrogen-evolution reaction

$$2H_3O^+ + 2e_M^- \rightarrow H_2 + 2H_2O \tag{7.24}$$

the standard reversible potential on the same scale is, of course, zero, by definition. We note that the adsorption energy of hydrogen on platinum is so high that it is easier (i.e., it requires a less negative potential) to form the adsorbed species Pt–H than to form H_2 molecules in solution. In comparison, the Hg–H bond is very weak, and consequently a hydrogen atom is not stabilized on the surface of mercury. The reversible potential for reaction 7.23 – with platinum replaced by mercury – would be close to that for reaction 7.22. This is why θ_H on mercury cathodes is below the detection limit, even at the highest overpotentials studied.

The situation is similar in the case of oxygen evolution, where OH_{ads} and O_{ads} are postulated as adsorbed intermediates, even though these radicals are very unstable in the bulk of the solution. The same type of argument has been employed to justify the existence of adsorbed intermediates in some complex organic reactions, such as $RCOO_{ads}$ in the Kolbe reaction of decarboxylation of some organic acids and formation of hydrocarbons

$$2CH_3COO_{soln}^- \rightarrow 2CH_3COO_{ads}^0 + 2e_M^- \rightarrow C_2H_6 + 2CO_2 \tag{7.25}$$

7.6
Catalytic Activity: The Relative Importance of j_0 and b

When the catalytic activity of different metals for a given reaction is considered, there may be some question regarding the way the comparison should be made. In the study of heterogeneous catalysis in chemistry, the obvious choice would be to compare the specific rate constant observed for different catalysts. The reaction rate is, of course, proportional to the concentration of the reacting species, but this is a trivial relation, not related to the catalytic properties of the surface. It would seem that the same procedure might be followed in electrode kinetics, by comparing either the rate constants or the exchange-current densities at a given concentration of the reacting species. However, the rate constant is potential-dependent, and its variation with potential is different for different mechanisms of the reaction being studied. Two cases exemplifying this situation are shown schematically in Figure 7.3. The Tafel slopes were taken as 30 mV and 120 mV, and the exchange-current densities are 10^{-10} and 10^{-5} A cm^{-2}, respectively. Following the example of chemical catalysis, the reaction with the higher exchange-current density would seem to be more catalytic. This applies at low overpotentials up to $\eta = 0.2$ V, where the two lines cross and the current density has a value of $j = 0.46$ mA cm^{-2}. However, at higher overpotentials, the reaction having the lower exchange-current density yields the higher reaction rates for any given overpotential. So how should one make the choice?

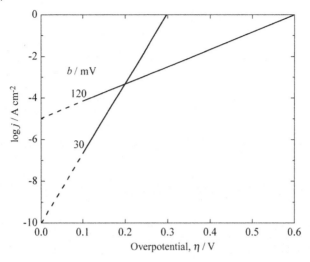

Figure 7.3 Schematic Tafel plots having different slopes, showing the catalytic activity from the engineering point of view.

From the practical point of view, it is easy to choose, since most industrial electrolytic processes are conducted at high current densities, in the range $0.1-1.0 \, A \, cm^{-2}$, and a process occurring at a rate of less than $0.5 \, mA \, cm^{-2}$ is of no practical interest. Thus, in the example shown in Figure 7.3, the reaction having the lower value of the Tafel slope turns out to be more catalytic in the context of industrial application, in spite of the fact that its exchange current density is five orders of magnitude lower.

7.7
Adsorption Energy and Catalytic Activity

When one tries to correlate the electrocatalytic activity of metals with some fundamental property of the system, the result is often a "volcano-type" plot, as shown in Figure 7.4. Such behavior is easy to understand qualitatively, although a quantitative relationship may be hard to derive. The role of a heterogeneous catalyst is to adsorb the reactant or intermediate and transform it to a species that can undergo the desired chemical reaction more readily. If the bonding energy for adsorption is very low, the extent of adsorption will be very small. Moreover, the adsorbed molecules are weakly bound to the surface and are not affected by it. As the bonding energy increases, the fractional surface coverage also increases, and the adsorbed species can be modified and activated by its bond to the surface. But this can be overdone. Beyond a certain point, the coverage approaches saturation, and the rate of reaction can no longer increase with increasing energy of adsorption. Also, if the bonding energy to the surface is too high, the adsorbed intermediate or its product may stick to the surface,

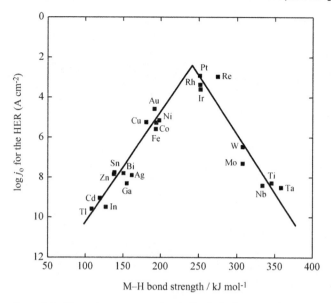

Figure 7.4 The exchange-current density for the hydrogen-evolution reaction as a function of the M–H bond energy. Reprinted with permission from S. Trasatti, *J. Electroanal. Chem.* **39**, 163. Copyright 1972, Elsevier Sequoia.

effectively poisoning it. In short, the energy of adsorption must be high enough to attach the reactant to the surface in sufficient amount, yet low enough to allow it to react there and release the product to the solution. This leads to a "volcano-type" relationship, where the best catalyst is one giving rise to an intermediate value of the bonding energy for adsorption.

A word of caution is appropriate here in relation to making such correlations. The proper comparison would be one in which the energy of adsorption was changed, while all other properties of the metal were kept constant. This cannot be done, of course. When we compare, say, the rate of hydrogen evolution on gold and on platinum, the heat of adsorption of hydrogen for the two metals is different, *but so is every other property*: the electronic work function, the heat of sublimation of the metal, the potential of zero charge, the free-electron density, the crystal structure and dimension, and any other property one may think of. Often some of these properties are correlated (e.g., the work function, the potential of zero charge, and the heat of adsorption are nearly linearly related). Finding a numerical correlation between any two quantities does not necessarily prove that one is caused by, or is directly related to, the other.

Support for the volcano-type relationship is also found in the gas-phase catalysis of the Haber process, namely the synthesis of NH_3 from H_2 and N_2. On the other hand, the data given on the right-hand side of Figure 7.4, where the exchange-current density declines with increasing M–H bond strength, may be questioned, because the metals involved (from W to Ta in this figure) all have an oxide layer that is very hard to remove, so that the rate of hydrogen evolution may have been measured on the oxide rather than on the bare metal, in which case, the M–H bond strength may not be relevant.

7.8
Electrocatalytic Oxidation of Methanol

A very interesting approach to electrocatalysis has been developed in recent years, for the oxidation of methanol in the so-called direct-methanol fuel cell (DMFC). Employing a Pt–Ru alloy was found to increase the rate of oxidation of methanol very substantially, compared to the performance of each of the two metals in their pure form. It turns out that an OH_{ads} species is formed on Ru at a less positive potential, where the platinum surface is still bare and highly active for adsorbing methanol. The rate-determining step is assumed to be the interaction between adjacent adsorbed species

$$Ru\text{-}OH + Pt\text{-}CH_3OH \rightarrow Ru + Pt\text{-}CH_2OH + H_2O \qquad (7.26)$$

Further interaction between Ru–OH and the partially oxidized organic molecule on the surface leads eventually to the total oxidation of methanol to CO_2 and H_2O.

It is interesting to note that in this and other complex reactions, which involve the overall transfer of many electrons per molecule, the Tafel slopes often turn out to be close to 0.12 V, which is characteristic of a sequence in which the first charge-transfer step is rate-determining. A commonly accepted explanation for this observation is not yet available, but it may be surmised that this comes about because the first charge-transfer step transforms a stable molecule (in the present case CH_3OH) to a much less stable intermediate (in this case CH_2OH_{ads}), which is then transformed to a series of further unstable intermediates. It is as if, following the first step, the floodgates were opened, and the rest just flows down smoothly on the Gibbs energy slope.

8
The Ionic Double-Layer Capacitance C_{dl}

8.1
Theories of Double-Layer Structure

8.1.1
Phenomenology

It was pointed out at the very beginning of this book that the impedance of the metal/ solution interface is partially capacitive. In simple cases, the equivalent circuit is that shown in Figure 8.1a. The double-layer capacitance C_{dl} and the Faradaic resistance R_F are inherent properties of the interface, which we measure experimentally and interpret theoretically. The solution resistance R_S is not a property of the interface. It can be viewed as an "error term" arising from the fact that the potential in solution is always measured far from the interface on the molecular scale, typically at a distance of 10^6–10^7 nm.

If the interface is ideally polarizable, the Faradaic resistance approaches infinity, and the equivalent circuit shown in Figure 8.1a can be simplified to that shown in Figure 8.1b. If it is ideally nonpolarizable, the Faradaic resistance tends to zero, and the equivalent circuit shown in Figure 8.1c results. Real systems never behave ideally, of course; they may approach one extreme behavior or the other, or be anywhere in between. It is also important to remember that both C_{dl} and R_S depend on potential and should be defined in differential form as follows:

$$C_{dl} \equiv \left(\frac{\partial q_M}{\partial E}\right)_\mu \text{ and } R_F \equiv \left(\frac{\partial E}{\partial j}\right)_\mu \tag{8.1}$$

where q_M is the excess surface-charge density on the metal. The double-layer capacitance can be measured in a number of ways, some of which are discussed in this chapter.

The system that has been most widely studied is mercury. In addition to being highly reproducible and easy to purify, the interface it forms in solution is almost ideally polarizable over a relatively wide range of potential, making both experimental

Physical Electrochemistry: Fundamentals, Techniques and Applications. Eliezer Gileadi
Copyright © 2011 WILEY-VCH Verlag GmbH & Co. KGaA, Weinheim
ISBN: 978-3-527-31970-1

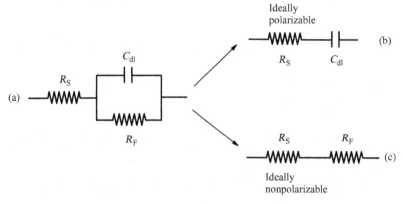

Figure 8.1 Equivalent circuit for an activation-controlled process, showing the three basic circuit elements: the double-layer capacitance, C_{dl}, the Faradaic resistance, R_F, and the residual solution resistance, R_S. (a) general (b) ideally polarizable (c) ideally non-polarizable.

measurement and theoretical interpretation easier. The best results on solid electrodes have been obtained on single-crystal surfaces of gold, which are the most reproducible of solid surfaces. Typical results for mercury are shown in Figure 8.2. The effect of a strongly adsorbed organic molecule on the capacitance–potential curve is also shown. The reader will note that the potential scale is reversed, negative to the

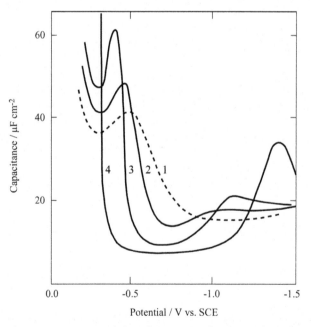

Figure 8.2 Experimental plots of C_{dl} versus E. on Hg in 1 M KCl, showing the effect of addition of N-propylamine concentration: (1) zero, (2) 5×10^{-5} M, (3) 1×10^{-4} M, (4) 5×10^{-4} M. Data from M. Lorenz, F. Möckel and W. Müller, Z. *Phys. Chem., N.F. Frankfurt* **25**, (1960) 145.

right and positive to the left. This unfortunate choice was made in the early days of electrochemical research, when the field was dominated by polarography on mercury, and it is still in use in most papers and textbooks, when dealing with the mercury electrode. In order to avoid confusion, we have chosen, somewhat reluctantly, to follow the same convention when studies of the double-layer capacitance and of adsorption on mercury are discussed.

One of the most important features seen in Figure 8.2 is the nearly constant value of the capacitance at the far negative end, in the absence of adsorption (curve 1). This value, of about $16\,\mu F\,cm^{-2}$, is essentially independent of the electrolyte used. This observation played an important role in the development of our understanding of the structure of the double layer at the metal/solution interface, as discussed below.

The other point of great interest is the effect of the organic material. First, we note that the capacitance is much smaller when organic matter is adsorbed. Secondly, we observe that this occurs only in a certain range of potential around the potential of zero charge, E_{pzc} (compare curve 1 with curve 2–4), disappearing both at more negative and at more positive potentials. This result is far from being self-evident, considering that the organic species used here is not charged. Moreover, similar behavior is observed for many neutral organic molecules of widely different structure. These observations form a further cornerstone of our understanding of the structure of the double layer, and, in particular, of the factors controlling the adsorption of neutral molecules.

8.1.2
The Parallel-Plate Model of Helmholtz

The first attempt to explain the capacitive nature of the interface is credited to Helmholtz (1853). In his model, the interface is viewed as a parallel-plate capacitor – a layer of ions on its solution side and a corresponding excess of charge on the surface of the metal. It should be noted here that electroneutrality must be maintained *in the bulk* of all phases, but not at the interface. Here, there can be an excess charge density on the metal, which we denote q_M, and an excess charge density, q_S, on the solution side of the interface. The interface as a whole must be electroneutral. It follows, then, that at any metal/solution interface we can write

$$q_M + q_S = 0 \qquad (8.2)$$

While the Helmholtz model can explain the existence of a capacitance at the interface, it can explain neither its dependence on potential nor its numerical value at any potential. The capacitance of a parallel-plate capacitor, per unit surface area, is given by:

$$C_H = \frac{\varepsilon_o \varepsilon}{d} \qquad (8.3)$$

where $\varepsilon_o = 8.8542 \times 10^{-12}\,J^{-1}\,C^2\,m^{-1}$ is the permittivity of free space, ε is the ratio between the permittivity of the solvent and that of free space (usually referred to as the dielectric constant) and d is the distance between the plates. The latter is equal to the radius of the solvated cations. The capacitance calculated for this model, according to

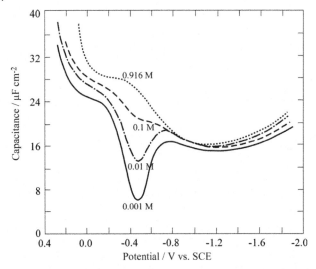

Figure 8.3 Double-layer capacitance on Hg in solutions of different concentrations of NaF at 0 °C. D.C. Grahame and B.A. Soderberg, ONR Tech. Rep. N° 14 (1954). The minima observed in the curves for 0.01 and 0.001 M NaF show the position of the potential of zero charge $E_{pzc} = -0.472$ vs. SCE.

Eq. (8.3), should be independent of potential, contrary to the experimental observation shown in Figure 8.2. Moreover, it should be independent of charge; hence of concentration in solution, while in reality it depends strongly on concentration, as shown in Figures 8.2 and 8.3.

The numerical value obtained from Eq. (8.3), with $\varepsilon = 78$ for the bulk dielectric constant of water at room temperature and a typical value of 2×10^{-8} cm for d, is 340 μF cm^{-2}, more than 20 times the experimentally observed value of 16 μF cm^{-2} shown in Figure 8.3 at negative potentials, where the solution side of the interface is dominated by cations.

If a theory has failed so utterly in describing the experimental results, how could it withstand the test of time, and still merit mentioning a century and a half later?

A hint may be found by observing the extreme negative end of the plots in Figures 8.2 and 8.3. Over about 1 V, from − 0.8 to − 1.8 V vs. SCE, the double-layer capacitance in this region is almost independent of concentration and only slightly dependent on potential, as expected from the simple parallel-plate model. The numerical value is off by more than an order of magnitude, but this can be remedied by a better choice of the values of ε and d in Eq. (8.3), as seen below.

8.1.3
The Diffuse-Double-Layer Theory of Gouy and Chapman

A different approach to interpret the behavior of the double-layer capacitance was taken by Gouy (1910) and later by Chapman (1913). The fundamental premise of the *diffuse-double-layer* model they proposed is that the ions constituting the charge q_S, on the solution side of the interface, are acted upon by two forces: They interact electrostatically with the excess surface charge on the surface of the metal, q_M, but are also subject to random thermal motion, which acts to equalize the concentration throughout the solution. The equilibrium between these two opposing tendencies is expressed by the well-known *Boltzmann equation*

$$c_i(x) = c_{b,i} \exp\left(-\frac{z_i F}{RT} \phi_x\right) \qquad (8.4)$$

where ϕ_x represents the potential at a distance x from the surface of the metal (with respect to the potential in the bulk of the solution, ϕ_s, which is taken as zero), and z_i is the charge on the ion. The concentration of any ionic species at a distance x from the surface, $c_i(x)$, is determined by the exponent of the ratio between the electrostatic energy, $z_i F \phi_x$, and the thermal energy, RT. From here on, the derivation is equivalent to the better-known derivation of the Debye–Hückel limiting law, giving the mean activity coefficients of ions in solution as a function of the ionic strength, which was, however, published about 10 years later.

The potential ϕ_s is related to the charge density per unit volume, $\rho(x)$, by the *Poisson equation*. Since the changes in ϕ_x and $\rho(x)$ are considered only in the direction perpendicular to the surface, this equation takes the simple, one-dimensional, form:

$$\left(\frac{\partial^2 \phi_x}{\partial x^2}\right) = -\frac{\rho(x)}{\varepsilon_0 \varepsilon} \qquad (8.5)$$

The volume-charge density $\rho(x)$ is related to the concentration of all the ions in solution by

$$\rho(x) = F \sum z_i c_i(x) \qquad (8.6)$$

One additional equation that is needed to solve this problem is the *Gauss theorem*, which relates the excess surface-charge density to the gradient of the potential

$$q_M = -\varepsilon_0 \varepsilon \left(\frac{\partial \phi_x}{\partial x}\right)_{x=0} \qquad (8.7)$$

We shall skip the details of the derivation, which is described in many textbooks, and present the final results, which are valid for symmetrical electrolytes (e.g. NaF or MgSO$_4$). Note that in Eq. (8.8) the potential ϕ_x is substituted by ϕ_0, since the equations were solved for the potential at the surface, not at some distance x from it.

$$q_M = (8RT \, \varepsilon_0 \, \varepsilon \, c_b)^{1/2} \sinh\left(\frac{|z| F}{2RT} \phi_0\right) \qquad (8.8)$$

The diffuse-double-layer capacitance is readily derived from this equation by differentiating q_M with respect to potential, leading to

$$C_{G.C.} = \left(\frac{\partial q_M}{\partial \phi_0}\right)_\mu = |z| F \left(\frac{2\varepsilon_0 \, \varepsilon}{RT} c_b\right)^{1/2} \cosh\left(\frac{|z| F}{2RT} \phi_0\right) \qquad (8.9)$$

From Eqs. (8.8) and (8.9) it can be seen that, at the potential of zero charge, E_{pzc} (where $q_M = 0$), the potential ϕ_0 at the surface also equals zero, and the differential capacitance has its minimum value, which is proportional to the square root of the bulk concentration:

$$C_{G.C.}(\text{min}) = |z| F \left[\frac{2\,\varepsilon_0 \varepsilon}{RT} c_b \right]^{1/2} \tag{8.10}$$

This is a beautiful theory, in that it allows us to calculate the excess surface-charge density and the double-layer capacitance from well-known principles of electrostatics (the Poisson equation and the Gauss theorem) and thermodynamics (the Boltzmann equation). It has, however, one major drawback: it does not predict the correct experimental results! Perhaps it would be more accurate to state that agreement between theory and experiment is found only in dilute solutions and over a limited range of potentials, near the potential of zero charge, as seen in Figure 8.4.

Now, we could come up with several reasons to explain why this theory would deviate *to some extent* from experiment:

1) In Eq. (8.4) we made the simplifying assumption that the only energy involved in bringing an ion from infinity to a distance x from the surface is the electrostatic energy, $z_i F \phi_x$, neglecting ion–ion interactions, which are bound to be important at higher concentrations.
2) In the derivation of Eq. (8.9) it was tacitly assumed that the dielectric constant, ε, is independent of the distance from the electrode. This is only an approximation. Theoretical calculations indicate that ε changes from 8 to10 in the first layer of water molecules at the metal surface to its bulk value of 78, over a distance of 1–2 nm.
3) Perhaps the largest error is introduced by using the potential ϕ_0 at $x = 0$, which is equivalent to treating the ions as point charges, for which the distance of closest approach to the surface is taken as zero.

All these are valid objections, and the theory could be improved, at least in principle, by deriving appropriate correction terms. However, none of these could explain the total disagreement between theory and experiment at a potential of, say, 0.5 V on either side of the potential of zero charge, as well as in concentrated solutions. As in the discussion of the Helmholtz model, we might ask ourselves why a theory that is in almost total disagreement with experiment is discussed at all! It turns out that a clever combination of the Helmholtz and the Gouy–Chapman models, resulting from a clear physical grasp of the situation at the interface, yields quite good agreement between theory and experiment, as we shall see next.

8.1.4
The Stern Model

The puzzle was assembled by Stern (1924), who showed that good agreement between theory and experiment could be achieved, once it was realized that both

(a)

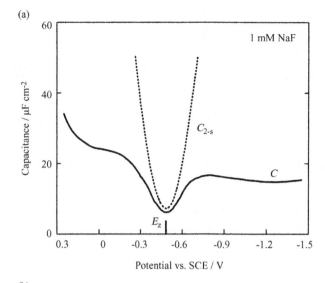

(b)

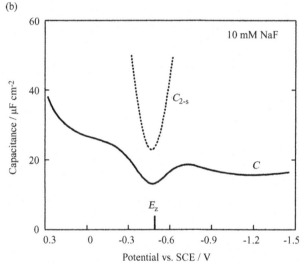

Figure 8.4 A comparison between $C_{G.C.}$, calculated from the Gouy–Chapman Theory (dashed lines) and experiment (solid lines) for (a) 1 mM NaF and (b) 10 mM NaF at 0 °C, based on data from D.C. Grahame, *J. Am. Chem. Soc.* **76** (1954) 4819.

the Helmholtz and the Gouy–Chapman models are valid and exist simultaneously. Stern postulated a layer of ions at the surface, which give rise to the Helmholtz (compact) part of the double layer, written as C_H or C_{M-2}. Outside this layer there is an ionic space charge, which constitutes the Gouy–Chapman (diffuse) double layer, written as $C_{G.C.}$ or C_{2-S}. For electroneutrality across the interface, one must still have $q_M + q_S = 0$, but the charge on the solution side is partly in the compact layer and partly in the diffuse layer.

With this model in mind, we can think of the potential drop between the metal and the solution as being divided into two segments:

$$\phi_M - \phi_S = (\phi_M - \phi_2) + (\phi_2 - \phi_S) \tag{8.11}$$

The potential ϕ_2 is called the potential of the *outer Helmholtz plane*. The diffuse double layer starts at the outer Helmholtz plane, where the potential is ϕ_2. It is this value of the potential, rather than ϕ_o that must be used in Eqs. (8.8) and (8.9), to relate the surface-charge density and the diffuse-double-layer capacitance to the potential. It should be noted here that the distance of the outer Helmholtz plane from the metal surface is determined by the radius of the hydrated cations in solution, and the potential in that plane is ϕ_2.

There is a similar quantity called the inner Helmholtz plane, where the potential is ϕ_1. This is determined by the radius of the anions in solution, which tend to be specifically adsorbed on the metal surface. A discussion of this phenomenon and its effect on the double-layer capacitance is outside the scope of this book.

Differentiating Eq. (8.11) with respect to q_M (while setting $\phi_S = 0$), one has:

$$\left[\frac{\phi_M}{q_M}\right]_\mu = \left[\frac{(\phi_M - \phi_2)}{q_M}\right]_\mu + \left[\frac{\phi_2}{q_M}\right]_\mu \tag{8.12}$$

This can also be written in the form

$$\frac{1}{C_{dl}} = \frac{1}{C_{M-2}} + \frac{1}{C_{2-S}} \tag{8.13}$$

Here C_{dl} is the experimentally measured double-layer capacitance. Equation (8.13) has the usual form for two capacitors connected in series, in agreement with the model postulated by Stern, in which the two parts of the double layer are consecutive in space.

The important thing to note in Eq. (8.13) is that in a series combination of capacitors, it is always the *smallest* capacitor that will predominantly determine the overall capacitance observed. This can explain qualitatively the observation shown in Figures 8.3 and 8.4. Agreement between the Gouy–Chapman theory and experiment is found only in dilute solutions and only in the vicinity of the potential of zero charge. Since C_{2-S} increases linearly with $(c_b)^{1/2}$ and is proportional to $\cosh(|z|F\phi_2/2RT)$, it follows from Eq. (8.13) that it can no longer be observed at high concentration and at potentials far removed from the potential of zero charge.

It is evident now why the Helmholtz and Gouy–Chapman models were retained. While each one alone fails completely when compared with experiment, a series combination of the two yields reasonably good agreement. There is room for improvement and refinement of the theory, but we shall not deal with that here. The model of Stern brings theory and experiment close enough for us to believe that it does describe the real situation at the interface. Moreover, later work of Grahame shows that the diffuse-double-layer theory, used in the proper context, yields consistent results and can be considered to be correct, within the limits of the approximations used to derive it.

We have failed to discuss so far the numerical value of the capacitance of the compact layer C_{M-2} and its dependence on potential (or charge), both of which are in disagreement with the simple parallel-plate-capacitor model proposed originally by Helmholtz. These issues, and the important effect of the solvent at the interface, are discussed in the next section.

8.1.5
The Role of the Solvent at the Interface

The solvent can influence the structure of the interface in many ways:

1) Different ions in solution are solvated to different degrees, depending on the nature of the solvent. The interaction of ions with the surface depends on their degree of solvation. In aqueous solution most cations are strongly hydrated. As a result, their interaction with the surface is mainly electrostatic. Most anions are not hydrated (or may have, on average, only one water molecule attached to them). As a result, anions can be in direct contact with the surface, allowing specific chemical interactions to play a role. Therefore, their adsorption is called *specific* or *contact* adsorption.

2) The surface itself is solvated. It can be viewed as a giant ion, having a large number of charges. The field near the surface (inside the compact part of the double layer) is comparable to the field around an ion in the bulk of the solution. Thus, solvation of the electrode surface accompanied by breakdown of the bulk structure of the solvent is expected to occur, just as it does in the inner solvation shells of ions. Since the solvent molecules are in direct contact with the surface, short-range chemical interactions with the metal electrode cannot be ignored. The chemical part of the adsorption energy of water on mercury is not very large, but it is enough to explain the experimentally observed lack of symmetry of the plots of q_M versus E around the potential of zero charge. It may be much larger on solid electrodes, particularly on catalytic metals such as platinum and nickel.

3) The solvent also acts as a dielectric medium, which determines the electrostatic field $d\phi/dx$ and the energy of interaction between charges. Now, the dielectric constant ε depends on the inherent properties of the molecules (mainly their permanent dipole moment and polarizability) and on the structure of the solvent as a whole. Water is unique in this sense. It is highly associated in the liquid phase and so has a dielectric constant of 78 (at 25 °C), which is much higher than that expected from the properties of the individual molecules. When it is adsorbed on the surface of an electrode, inside the Helmholtz layer, the structure of bulk water is destroyed, and the molecules are essentially immobilized by the high electric field. Consequently, the appropriate dielectric constant to be used in this region has been estimated to be about 8. Farther out in the solution, the bulk structure of water is rapidly regained. The dielectric constant is estimated to reach its bulk value within about 1–2 nm of the surface. In the discussion of the parallel-plate model of Helmholtz, a capacitance of $0.34\,\mathrm{mF\,cm^{-2}}$ was calculated

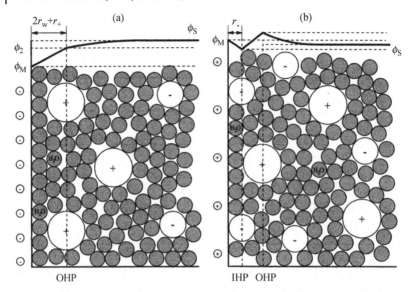

Figure 8.5 Schematic representation of the structure of the double layer. (a) At very negative potentials (with respect to E_z). (b) At positive potentials, where specific adsorption of anions occurs. The outer Helmholtz plane is shown at a distance of $(2r_W + r_+)$ from the surface, where r_W and r_+ represent the radii of a water molecule and the solvated cation, respectively.

using $\varepsilon = 78$ and $d = 0.2$ nm. This value is at variance with the capacitance of about 16 μF cm^{-2} observed experimentally at extreme cathodic potentials (with respect to E_{pzc}). If, however, we use $\varepsilon = 8$ instead, we find agreement with experiment for $d = 0.44$ nm, which is of the right order of magnitude.

What is the "correct" value of the thickness of the parallel-plate capacitor in Eq. (8.3)? This may be seen by reference to Figure 8.5a, which shows the surface of a negatively charged electrode covered with a layer of water molecules. The distance of closest approach of a cation is the sum of the diameter of a water molecule (0.27 nm) and the radius of the hydrated ion, the so-called Stokes radius, (calculated from electrolytic conductivity data), which are in the range 0.1–0.3 nm for most cations. The thickness of $d = 0.44$ nm calculated above agrees with the model, assuming a solvation radius of $d = 0.17$ nm, confirming the validity of the model of the parallel-plate capacitor given here.

Figure 8.5b shows the structure of the interface at positive potentials, in the presence of specifically adsorbed anions. Here the physical meaning of the inner Helmholtz plane (IHP) is illustrated. The distance of this layer is always smaller than that of the outer Helmhotz plane OHP, since there is no layer of water molecules between the anions and the surface. The absence of such a layer can qualitatively explain why the value of C_{M-2} is higher on the positive side of E_{pzc} than on the negative side, taking the parallel-plate model one step closer to the experimental observations.

From an examination of the C_{dl}/E curves in Figures 8.2–8.4, it is clear that not all the experimental details have been accounted for. There are further effects resulting from lateral ion–ion, ion–dipole and dipole–dipole interactions. For example, the "hump" observed on the positive side of E_{pzc} can be explained in terms of these interactions, which we shall not discuss, since they are not essential for a basic understanding of the structure of the double layer.

Which ions are specifically adsorbed? It depends, of course, on the metal, but detailed and accurate data available for mercury show that anions, which are generally not hydrated, tend to be specifically adsorbed. This includes most of the anions, but not F^-. Also, some highly symmetrical anions such as ClO_4^-, BF_4^-, and PF_6^- are not specifically adsorbed on mercury. In contrast, most cations are not specifically adsorbed on mercury. Cesium, which was found to be specifically adsorbed to some extent, is an exception. Also, large organic cations of the tetra-alkyl ammonium type (for example $[(C_2H_5)_4N]^+$) are found to be specifically adsorbed. While this information has been most valuable for the understanding of the double-layer structure, it should be borne in mind that mercury is a rather inert, non-catalytic surface. Thus, the structure of the double layer at the surface of catalytic metals could be quite different, and should be considered for each solid metal separately.

8.1.6
Simple Instrumentation for the Measurement of C_{dl}

In a later chapter we shall discuss the technique of electrochemical impedance spectroscopy, (EIS), which can provide the most accurate measurements of the double-layer capacitance over a wide range of experimental conditions. Here we shall discuss two methods, requiring relatively simple instrumentation.

1) The charge on the double layer is related to the potential as:

$$q_M = C_{dl}\bar{E} = C_{dl}(E - E_z) \tag{8.14}$$

where $\bar{E}$ is the so-called rational potential, which is measured with respect to the potential of zero charge E_{pzc}. Hence, the charging current can be written as

$$j = \left(\frac{\partial q_M}{\partial t}\right) = C_{dl}\left(\frac{\partial E}{\partial t}\right) \tag{8.15}$$

Although the double-layer capacitance is a function of potential, this dependence can be ignored, if the perturbation is small. If a current step is applied to an interface, the change of potential with time is given by

$$E = E_\infty\left[1 - e^{-t/\tau}\right] + jR_S \tag{8.16}$$

where E_∞ is the steady-state potential and τ is a characteristic time constant in the equivalent circuit (cf. Figure 4.3), given by

$$\tau = R_s \times C_{dl} \tag{8.17}$$

When $t/\tau \ll 1$ the exponent in Eq. (8.16) can be linearized to yield:

$$E = E_\infty \frac{t}{\tau} \tag{8.18}$$

Thus, the potential varies linearly with time in this region, and the capacitance can be calculated from the slope according to Eq. (8.18). Usually a repetitive square-wave pulse is applied, as seen in Figure 8.6, from which both the capacitance and the solution resistance can be determined.

2) The technique of cyclic voltammetry, to be discussed later (Section 15.1) can also be used to determine the double-layer capacitance. In this case, a relatively slow triangular potential waveform is applied, and the current is determined as a function of potential. If the interface is highly polarizable, the result will be that shown in Figure 8.7(a). If there is a significant Faradaic current, a plot such as shown in Figure 8.7(b) will be observed.

The best way to calculate the capacitance from such experiments is to conduct measurements over a range of sweep rates, obtaining C_{dl} from the plot of $\Delta j = j_{an} - j_c$ versus $(\partial E/\partial t)$, as shown in Figure 8.8.

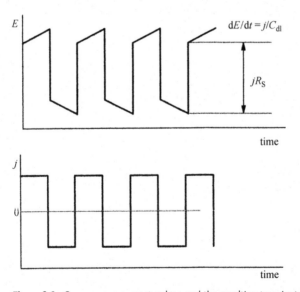

Figure 8.6 Square-wave current pulses and the resulting transients of potential.

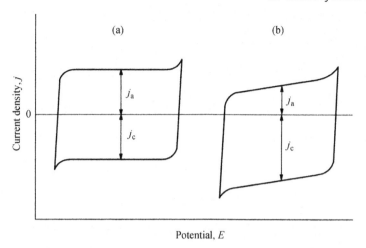

Figure 8.7 Current response to a triangular potential sweep. (a) ideally polarizable interface, (b) real interface, with a finite Faradaic current.

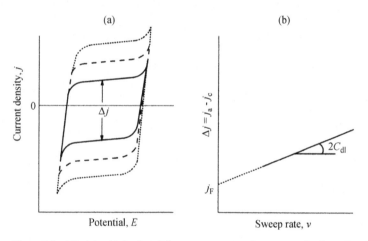

Figure 8.8 j/E plots obtained at different sweep rates, for a system having a Faradaic current in addition to the double-layer charging current. Using the difference, Δj, rather than the individual values of j_{an} or j_c, eliminates most of the error due to the Faradaic reaction.

For any method to be used prudently, it must be remembered that, as a rule, the interface consists of at least a resistor in series with a parallel combination of a capacitor and a resistor, but more complicated equivalent circuits are often needed for a true description of the interface. If both resistances are high, one should conduct measurements at low frequencies, to increase the impedance of the capacitor compared to the solution resistance in series with it. If the Faradaic resistance is low, the experiment should be conducted at high frequency, so that the impedance of the capacitor is small compared to the Faradaic resistance in parallel with it. If the series resistance is high and the parallel resistance low, one faces a difficult experimental situation. In such a case, measurements should be conducted over a wide range of frequencies, with the highest possible accuracy, and the optimum conditions for the experiment should be carefully chosen. Under such conditions, electrochemical impedance spectroscopy apparatus may be indispensable.

9
Electrocapillarity

9.1
Thermodynamics

9.1.1
Adsorption and Surface Excess

The creation of an interface is a symmetry-breaking process. Even if the two phases in contact are entirely homogeneous, the molecules at the interface experience a different average force in each of the two directions perpendicular to the interface. As a result, the Gibbs energy of a molecule at the surface is different from its value in the bulk phase. This effect is not restricted to a single layer of molecules, it can extend to a distance of several molecular diameters into the solution. Indeed, it is possible to envision a region on either side of the interface in which the energy of the molecules is different from that in the two bulk phases. This region is referred to as the *interphase*. When only neutral species are involved, the interphase cannot be more than a few molecular layers thick, since the forces involved are "chemical" and decay quickly with distance (cf. Figure 2.1). When charged particles (i.e., ions in solution at the metal/electrolyte interface) are considered, the interphase can extend much further, since the electrostatic interactions between charges decrease linearly with the distance. The best example of such behavior is the diffuse double layer discussed in Section 8.1.3. In the discussion of the relevant theory, we want to determine the variation of the potential ϕ_2 at the OHP with the excess surface-charge density q_M, or the rational potential $\bar{E}$. The resulting equation is

$$\phi_x = \phi_2 \exp(-\kappa x) \tag{9.1}$$

Where κ is the so-called *reciprocal Debye length*, given by

$$\kappa = \left[\frac{2z^2 F^2}{\varepsilon_o \, \varepsilon \, RT} \right]^{1/2} (c_b)^{1/2} \tag{9.2}$$

Substituting the numerical values of the constants, we find that for a 1 mM solution of a symmetrical 1-1 electrolyte, $\kappa^{-1} = 10$ nm. Looking again at Eq. (9.1), it is noted

Physical Electrochemistry: Fundamentals, Techniques and Applications. Eliezer Gileadi
Copyright © 2011 WILEY-VCH Verlag GmbH & Co. KGaA, Weinheim
ISBN: 978-3-527-31970-1

that at a distance $x = \kappa^{-1}$ from the metal, the potential is $\phi_\kappa = \phi_2/e = 0.37\phi_2$, and this is often taken to be "the thickness" of the diffuse double layer, but should more correctly be referred to as "the characteristic length" of the diffuse double layer. A better choice for defining the thickness might be to use $x = 3\kappa^{-1}$, since at this distance ϕ_2 has decayed to 5% of its original value, but this does not matter, from the conceptual point of view.

The thickness of the diffuse double layer is seen to be a function of concentration. In a 1.0 M solution, $\kappa^{-1} = 0.3$ nm, which is of the order of magnitude of the thickness of the compact part of the double layer. This shows that the diffuse double layer no longer exist in 1.0 M solutions.

The asymmetry at the surface can affect the energy of different species in solution to different degrees. If, for example, equal amounts of methanol and n-butanol are dissolved in water, the concentration of n-butanol in the interphase will be higher. This comes about because the transfer of n-butanol from the bulk to the interphase decreases the total Gibbs energy of the system more than does the transfer of methanol.

It is necessary, at this point, to define a new quantity, called the *surface excess*, Γ, such that

$$\Gamma \equiv \int_0^\infty (c - c_b)\,dx \tag{9.3}$$

The surface excess, Γ, represents the total amount of the relevant species in a cylinder of unit cross-section, extending from the interface into the bulk of the solution, less the amount that *would have been* in the same volume, had there been no interface. The definition of the surface excess is shown graphically in Figure 9.1.

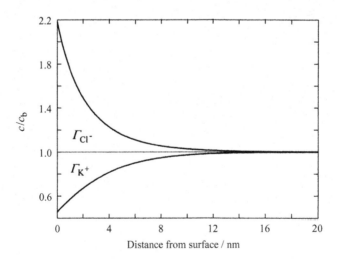

Figure 9.1 Illustration of the concept of the surface excess, Γ. The lines represent the dimensionless concentrations of positive and negative ions as a function of distance from the mercury surface, calculated for 0.01 M KCl, $q_M = +10\ \mu C\ cm^{-2}$, in the absence of specific adsorption.

Fortunately, we do not have to determine how far exactly the interphase extends. The function $(c-c_b)$ is integrated to "infinity" – far enough into the bulk so that its value will have become negligible. It should be noted that on the scale of interest for interphases, "infinity" is not very far. In fact, it is less than $1\,\mu m$ in the direction perpendicular to the interface! For example, for $\kappa^{-1} = 10\,nm$ and a distance of $1.0\,\mu m$ from the surface, one has, according to Eq. (9.1)

$$\phi_x = \phi_2 \exp(-100) = 3.7 \times 10^{-44} \tag{9.4}$$

We recall, however, that every potential must be referred to some (arbitrarily chosen) zero. When considering the metal/solution interface in electrochemistry one usually chooses $\phi_s = 0$. Thus, Eq. (9.4) could be written as

$$(\phi_x - \phi_s) = (\phi_2 - \phi_s)\exp(-100) = 3.7 \times 10^{-44} \tag{9.5}$$

It follows that, at a distance of $1.0\,\mu m$, the potential caused by the diffuse-double layer is essentially equal to that in the bulk of the solution. Infinity is less that $1.0\,\mu m$ from the interface!

The surface excess is an integral quantity. This has the advantage of relieving us of the need to define the boundary of the interphase. On the other hand, measurement of Γ cannot yield any information concerning the variation of the concentration *inside* the interphase, although this can be calculated on the basis of a suitable model. Another point to remember is that the surface excess, as defined here, can have either positive or negative values, as seen in Figure 9.1.

In the case of adsorption of an intermediate in electrode kinetics, we have discussed the extent of adsorption in terms of the fractional surface coverage, θ, which was defined as:

$$\theta \equiv \frac{\Gamma}{\Gamma_{max}} \tag{6.8}$$

In this equation Γ is the surface concentration of the adsorbed species, and Γ_{max} is the value of Γ needed to form a complete monolayer. We recall that the intermediates discussed (such as an H atom or an OH radical) are unstable in solution, and can be formed only on the surface. In this case, the surface concentration is equal to the surface excess, and it is appropriate to use the same symbol for both. It should be noted, though that Eq. (9.3) does not show a clear value of Γ_{max} and, in fact, Γ could exceed, (at least in principle, although this is rarely the case) the amount corresponding to a monolayer, because the material is distributed over a thickness that can correspond to several molecular layers.

9.1.2
The Gibbs Adsorption Isotherm

The most fundamental equation governing the properties of interphases is *the Gibbs adsorption isotherm*:

$$-d\gamma = \sum_i \Gamma_i \, d\mu_i \tag{9.6}$$

The *surface tension*, γ, is given in units of force per unit length $(10^3 \, \text{dyne cm}^{-1} = 1 \, \text{N m}^{-1})$. It is related to the *two-dimensional surface pressure*, Π, by the simple equation

$$-d\gamma = d\Pi \tag{9.7}$$

It is also the *excess surface Gibbs energy* per unit area, namely the extra Gibbs energy added to a system as a result of formation of the interface. $(10^3 \, \text{erg cm}^{-2} = 1 \, \text{J m}^{-2})$. When we speak of surface tension, we have a mechanical model in mind, as though there was an imaginary membrane on the surface, pulling it together with a force (which can readily be measured using a Langmuir–Blodgett trough). The notion of an excess surface Gibbs energy is, of course, purely thermodynamic.

We noted earlier that the driving force in chemistry is the decrease in Gibbs energy (cf. Eq. (2.1)). Thus, a system will change spontaneously in the direction of decreasing surface tension. This leads to two observations:

1) A pure phase always tends to assume a shape that creates the minimum surface area per unit volume. This is why droplets of a liquid are almost spherical (they are completely spherical in the absence of gravity, in a spacecraft orbiting the Earth, for example).
2) When a solution is in contact with another phase, the composition of the interphase differs from that of the bulk in such a manner as to minimize the total excess surface Gibbs energy of the system.

The second observation represents the essence of the physical meaning of the Gibbs adsorption isotherm. The adsorption of any species in the interphase ($\Gamma_i) > 0$) must always cause a decrease in the Gibbs energy of the surface ($d\gamma < 0$), since it is the reduction in this Gibbs energy that acts as the driving force for adsorption to occur.

The Gibbs adsorption isotherm is derived in many standard textbooks of physical chemistry and surface chemistry. We shall not repeat this derivation here. Rather, we show how this isotherm is modified when it is applied in electrochemistry.

9.1.3
The Electrocapillary Equation

The surface tension of an electrode in contact with an electrolyte depends on the metal–solution potential difference, $^M\Delta^S\phi$. The equation describing this dependence is called the *electrocapillary equation*. It follows by simple logic from the Gibbs adsorption isotherm. Thus, the sum $\sum_i \Gamma_i \, d\mu_i$ in Eq. (9.6) should represent the surface excess (or deficiency, i.e. negative surface excess) of all the species in the interphase. On the solution side there are terms of the type $(\Gamma_{Cl^-} \, d\mu_{Cl^-})$ and $(\Gamma_{RH} \, d\mu_{RH})$ for charged and neutral species, respectively, where the subscript "RH" stands for an unspecified organic molecule). On the metal side, the surface excess is expressed by q_M and the chemical potential is replaced by the electrical potential, E, measured with respect to a chosen reference electrode. This potential differs from the metal–solution potential difference only by a constant.

A general form of the electrocapillary equation can hence be written as:

$$-d\gamma = q_M dE + \sum \Gamma_i d\mu_i \tag{9.8}$$

Several equations follow directly from this equation. The partial derivative of γ with respect to potential, at constant composition of the solution, yields the excess surface charge density:

$$-(\partial\gamma/\partial E)_{\mu_i} = q_M \tag{9.9}$$

and the second derivative with respect to potential yields the double-layer capacitance:

$$-(\partial^2\gamma/\partial E^2)_{\mu_i} = (\partial q_M/\partial E)_{\mu_i} = C_{dl} \tag{9.10}$$

These equations relate the surface tension, or excess surface Gibbs energy, to two very important electrical characteristics of the interface: the charge density q_M and the double-layer capacitance, C_{dl}. It should be noted that these are purely thermodynamic relationships, not based on a model. The only assumption made in the derivation of Eqs. (9.8)–(9.10) (from Eq. (9.6)) is that the interface is ideally polarizable, that is, charge cannot cross the interface.

It follows from Eq. (9.9) that at the maximum of the electrocapillary curve, $q_M = 0$. In other words, the potential of zero charge, E_{pzc}, coincides with the potential of the electrocapillary maximum. The double-layer capacitance can be obtained by double differentiation of the surface tension with respect to potential, and the surface tension can be obtained by double integration of the dependence of C_{dl} on E. The situation is not entirely symmetrical, however. For double differentiation, all one needs is very accurate data of γ versus E. For double integration, one also needs two constants of integration. These are the coordinates of the electrocapillary maximum, namely E_{pzc} and γ_{max}. For liquid electrodes (e.g., mercury and some amalgams) both can be readily measured with high accuracy. For solid electrodes, it is possible to measure E_{pzc}, but reliable values of the surface tension as a function of potential are hard to come by. Thus, for solid electrodes, one can integrate the double-layer capacitance data to obtain q_M as a function of potential, but the second integration, needed to obtain the electrocapillary curve, cannot be readily performed.

The third important relationship that follows from the electrocapillary equation is:

$$\Gamma_i = -(\partial\gamma/\partial\mu_i)_{E,\mu_{j\neq i}} = -\frac{1}{2.3RT}(\partial\gamma/\partial\log a_i)_{E,\mu_{j\neq i}} \tag{9.11}$$

This equation can be used to determine the surface excess (in effect, the extent of adsorption in the interphase) of any one species, from the variation of the surface tension with the activity of this species in solution, maintaining the activities of all other species, as well as the potential, constant.

Consider now the application of Eq. (9.8) to a specific system. We choose for this presentation the simple case of a mercury electrode in contact with an aqueous solution of KCl, and an Ag/AgCl reference electrode. This cell can be represented by:

$$Cu'||Ag||AgCl||KCl, H_2O||Hg||Cu''$$

where the double vertical lines represent separation between different phases (which are in physical contact with each other) and Cu' and Cu'' are copper wires connected to the terminals of a suitable source of variable potential. Sparing the reader the tedium of going through a long derivation, we arrive at the final equation, which is

$$-d\gamma = q_M dE_- + \Gamma'_{K^+} d\mu_{KCl} \tag{9.12}$$

where the new symbol, E_-, has been introduced to show that this specific equation applies to the case in which the reference electrode is reversible with respect to the anion in solution. Had we used a reference electrode that is reversible with respect to the positive ion in solution, Eq. (9.12) would have been replaced by

$$-d\gamma = q_M dE_+ + \Gamma'_{Cl^-} d\mu_{KCl} \tag{9.13}$$

The symbols Γ'_{K^+} and Γ'_{Cl^-} stand for the *relative surface excess* of the marked species, defined as

$$\Gamma'_{K^+} \equiv \Gamma_{K^+} - \Gamma_{H_2O}\left(\frac{X_{KCl}}{X_{H_2O}}\right) \tag{9.14}$$

The need to introduce this function arises from the fact that the derivative given by Eq. (9.11) represents an experiment that is, strictly speaking, impossible to perform. Thus, in a system containing *n* components, it is impossible to change the concentration of only one of these components, leaving *all other concentrations* constant. However, from the definition of Γ'_{K^+} it follows that it is very close to Γ_{K^+} in fairly dilute solution, and this refinement can usually be ignored. (For example, in a 0.1 M solution of KCl the ratio of mole fractions in Eq. (9.14) is about 0.002). It must, however, be taken into account in a rigorous treatment of the problem, particularly in more concentrated solutions or when mixed solvents of comparable concentrations are employed.

9.2
Methods of Measurement and Some Results

9.2.1
The Electrocapillary Electrometer

The measurement of surface tension is an old trade in science. There are consequently many methods of determining this quantity. In chemistry, the surface

(a)   (b)

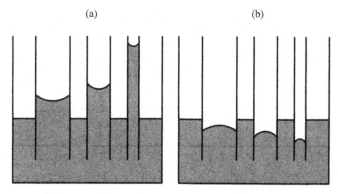

Figure 9.2 (a) Capillary rise observed for a wetting liquid, (water) and (b) capillary depression observed for a non-wetting liquid, (Hg). Note that this figure is not drawn to scale.

tension is measured as a function of the solvent, the composition of the solution, and the nature of the two phases. In electrochemistry the potential is an added variable. The *electrocapillary curve* is a plot of the surface tension of a liquid-metal (usually mercury) electrode versus potential, at a given composition of the solution. This type of measurement is then repeated in solutions of different composition, to obtain the surface excess of the appropriate species, employing Eq. (9.11).

To be more exact, we should be talking about the *interfacial tension*, which is the surface tension between two specified phases. In electrochemistry it is customary to use the term *surface tension* to refer to the interfacial tension at the metal/solution interface, or more generally, at the interface between an electronic and an ionic conductor.

We recall from our elementary science classes the phenomenon of *capillary rise*, shown in Figure 9.2a. When a series of glass capillaries is inserted into water, the water will rise in the capillaries to a height that is inversely proportional to the radius of the capillary. The equation describing this behavior is the *Young–Laplace* equation, which can be written as:

$$\Delta p = \frac{2\gamma}{r} \qquad (9.15)$$

This equation shows the pressure difference Δp needed to keep two phases in mechanical equilibrium, if the interface has a radius of curvature r. One should distinguish here between *thermodynamic* equilibrium and *mechanical* equilibrium. The former is defined by the condition of minimum Gibbs energy of the system. The latter represents the condition that the vector sum of all forces acting on the interface is zero. When the interface is flat, r goes to infinity and $\Delta p \to 0$. In the present example, the pressure in each capillary is related to the height of the liquid, since

$$\Delta p = \frac{\text{force}}{\text{area}} = \frac{\pi r^2 h \rho g}{\pi r^2} = h \rho g \qquad (9.16)$$

where ρ is the density of the liquid and g is the acceleration due to gravity. The height of the liquid in the capillary is hence given by:

$$h = \frac{\Delta p}{\rho g} = \frac{2\gamma}{r} \frac{1}{\rho g} \tag{9.17}$$

Thus, the capillary rise is determined by the surface tension and the density of the liquid on the one hand, and on the other hand by the radius of curvature, which in this case equals the radius of the capillary.

If we replace water by mercury, a *capillary depression* will be observed, instead of a capillary rise, as shown in Figure 9.2b. The simple physics behind this is that water wets glass whereas mercury does not. In everyday language we could say that there is an affinity between water and glass, causing the liquid to "crawl up" the capillary. To force the water in the capillary back to its level in the vessel, it would be necessary to apply a pressure (given by the Young–Laplace equation) that is inversely proportional to the radius of the capillary.

There is no affinity between mercury and glass. Thus mercury must be forced to enter the capillary. One would have to apply a pressure in the opposite direction to bring the mercury in the capillary to its level in the vessel. Clearly, the physics controlling capillary rise and capillary depression must be the same. To accommodate the difference between the two in the framework of the same equations, we note the difference in the form of the meniscus in Figure 9.2a and b. For water the meniscus is concave, while for mercury it is convex. If we agree to define the radius as being *negative* for the convex interface and choose the level of the liquid in the outer vessel as zero (leading to negative values of h for capillary depression), we find that the Young–Laplace equation is applicable to wetting as well as non-wetting liquids, and Eq. (9.17) correctly describes both phenomena.

The reader may object to the use of an artificial concept such as a negative radius, but this is not an uncommon practice in science. For instance, the capacitive impedance of the metal/solution interface is described as the "imaginary" part of the impedance, although it is a very real impedance indeed!

We have obviously taken some shortcuts to keep the foregoing discussion short and, hopefully, clear. For example, we ignored the angle of contact between the liquid and the solid at the edge of the meniscus. This is tantamount to considering water to be an *ideally wetting* liquid (with a contact angle of zero) and mercury to be an *ideally non-wetting* liquid (having a contact angle of 180°. One may also question the exact definition of the height h in Figure 9.2a. Is it measured to the bottom of the meniscus, to the point at which the meniscus contacts the glass, or somewhere in between? Is the capillary ideally circular, leading to a hemispherical meniscus, or should we define two radii of curvature? These are important points when one conducts research in this field, but they are not relevant to the basic physical understanding of the phenomena observed. Although Figure 9.2 is not drawn to scale, the capillary rise is intentionally drawn larger than the capillary depression, to show the effect of the high density of Hg (cf. Eq. (9.16)).

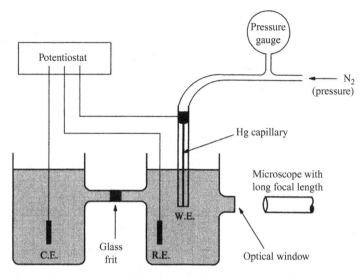

Figure 9.3 The electrocapillary electrometer. For each potential the pressure of nitrogen is adjusted to bring the mercury inside the capillary back to the same fixed position.

In the classical electrocapillary electrometer, the configuration is inverted. Mercury is placed in a glass tube that ends with a fine capillary, as shown in Figure 9.3. Since we need pressure to force mercury into a fine capillary, there will be a certain height of mercury column supported by the capillary in this configuration. This is the exact equivalent of the capillary depression shown in Figure 9.2b, and the height of the column is also given by Eq. (9.17). In this equation we note that the height depends on the surface tension, which depends on potential. Hence, the height of the mercury column above the capillary is a function of potential.

Measurement of the electrocapillary curve consists of changing the potential stepwise and determining the pressure required to return the mercury meniscus to the same location in the fine capillary. A plot of this pressure as a function of potential is nothing but the electrocapillary curve, multiplied by a constant, according to Eq. (9.17). Although this equation allows an absolute determination of the surface tension, the best way to determine it is by calibration with a known system. This requires one accurate determination of γ by an independent method. Very careful experiments were performed by Gouy around the beginning of the 20th century. The highest value he measured at the electrocapillary maximum in a solution of NaF is 0.426 N m^{-1}. This value is used even now as the *primary standard* for electrocapillary measurements. There are several ways to measure the electrocapillary curve. One of them is shown schematically in Figure 9.3.

The potential of the mercury electrode is controlled against a suitable reference electrode. It is not essential to use a three-electrode system in this case, since the mercury/solution interface is studied in the range of potentials where it behaves essentially as an ideally polarizable interface. A reference electrode is commonly used, nevertheless, to ensure stability and reproducibility of the measured potential.

The position of the mercury in the capillary is observed either though a long-focal-length microscope, or by a video camera connected to a high-resolution monitor. We shall not dwell on technical details, except to note that building and operating an electrocapillary electrometer at the desired level of accuracy is a delicate matter requiring both skill and experience.

Let us make a small detour here to discuss a minor point, which appears to be purely technical, but may help us to better understand the physics behind electrocapillary measurements. The question we want to address is, should one make an effort to use a perfectly cylindrical capillary, or is a slightly tapered capillary satisfactory? The answer cannot be found in Eq. (9.17), which relates the height of the mercury column, or the pressure difference, to the radius. Consider, however, the situation in a perfectly cylindrical tube. If we move the meniscus to a different position, it will stay there, since the radius has not been changed. Thus, the system is at equilibrium with the mercury meniscus *anywhere* in the cylindrical tube. What we have is a *neutral* equilibrium. Its mechanical equivalent is a perfect sphere on a perfectly flat and horizontal surface, as shown in Figure 9.4a.

Now consider a tapered tube, wider at the top. This represents a stable equilibrium. For a given pressure applied, the mercury is stable in only one position in the capillary. If the pressure is increased momentarily (or the surface tension is decreased, say, by a fluctuation in the applied potential), the mercury meniscus will move to a lower point, where the radius is a little smaller, to establish a new equilibrium, in accordance with Eqs. (9.15)–(9.17). The mechanical equivalent of this configuration is a sphere at the bottom of a concave surface, shown in Figure 9.4b. Stability is attained by negative feedback. If the system is perturbed, a force is created, acting in the direction opposite to the direction of the perturbing force, bringing it to a new equilibrium position.

Consider now the same tapered capillary, but placed in the inverted position, namely, wider at the bottom. This represents an *unstable equilibrium*. The mechanical equivalent is a sphere on the top of a convex surface, as shown in Figure 9.4c. It is

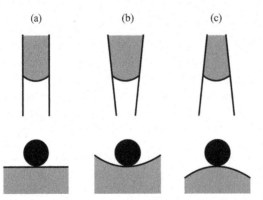

Figure 9.4 Neutral, stable and unstable equilibria (a, b and c, respectively) in mechanics and in an electrocapillary electrometer.

theoretically possible to find an equilibrium position for the sphere, but it will be impossible to maintain it in practice. The slightest disturbance will cause it to roll down. Unstable equilibrium is the result of positive feedback. Removing the system even very slightly from its state of equilibrium creates a force in the direction of removing it even further away from equilibrium.

What is the effect of a small increase in pressure on the mercury in such an inverted capillary? The meniscus will move downwards slightly, but in its new position the radius is larger, so that the same pressure will tend to move it farther down, where the radius is even larger, and so on. Thus, while the general equations allow the existence of an equilibrium position in this situation, it will not be possible, in practice, to maintain a stable position of the mercury meniscus in the capillary.

Fortunately, the normal manufacturing process of glass tubing produces slightly tapered capillaries, and the problem does not arise, unless one connects the capillary in the electrocapillary electrometer in the wrong direction.

The information one can derive from measurements of the surface tension as a function of potential is specified in Eqs. (9.9)–(9.14). It includes the dependence of charge and the double-layer capacitance on potential and the relative surface excess of all the species in solution, (except the solvent), as a function of potential and of the composition of the solution. It should be noted, however, that all these quantities must be obtained by numerical differentiation of the experimental results. This requires very high accuracy in measurement, since differentiation inherently amplifies experimental errors (while integration tends to smooth them out). The range of values of γ in most cases is $(0.250-0.426)\,\mathrm{N\,m^{-1}}$. The best measurements recorded claim an accuracy of $\pm 0.1\,\mathrm{mN\,m^{-1}}$, which amounts to about $\pm(0.04-0.02)\%$.

9.2.2
Some Experimental Results

9.2.2.1 The Adsorption of Ions
Figure 9.5 shows electrocapillary curves obtained in NaF solutions at different concentrations. The most remarkable feature of these curves is that both E_{pzc} and γ_{max} are practically independent of the concentration of the electrolyte. Also, there is hardly any adsorption of the anion at negative potentials or of the cation at positive potentials. This observation led Grahame and others to conclude that neither Na^+ nor F^- are specifically adsorbed in the mercury/electrolyte interphase.

Distinctly different behavior is shown in Figure 9.6 for KBr. The potential of zero charge is shifted in the negative direction as the concentration of the electrolyte increases. The anion is strongly adsorbed, as can be seen from the depression of the value of γ with increasing concentration, at fixed potential, which is observed even on part of the negative branch of the electrocapillary curve.

A case of specific adsorption of the cation is shown in Figure 9.7, for solutions of $TlNO_3$, in a supporting electrolyte consisting of $1.0\,M\,KNO_3$ and $0.01\,M\,HNO_3$. The potential of zero charge shifts in the positive direction, showing that specific

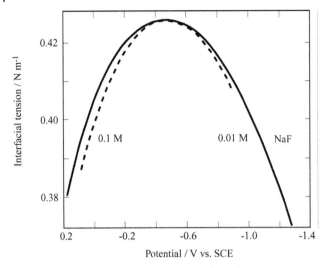

Figure 9.5 Electrocapillary curves for two concentrations of NaF. Data from D.C. Grahame and Soderberg, Tech. Rep. # 14, ONR, (1954). ($1\,\mathrm{N\,m^{-1}} = 10^3\,\mathrm{dyne\,cm^{-1}}$).

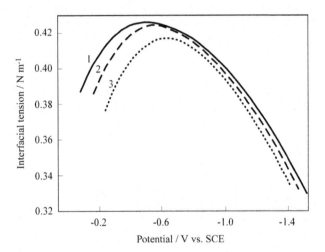

Figure 9.6 Electrocapillary curves for different concentrations of KBr. Lines 1, 2 and 3 for 0.01, 0.1 and 1 M KBr, respectively. Data from J. Lawrence, R. Parsons and R. Payne, *J. Electroanal. Chem.* **16** (1968). 193.

adsorption of the cation is predominant. The values of the surface tension are substantially lowered at all potentials, indicating high values of the surface excess of thallium, Γ_{Tl^+}

9.2.2.2 Adsorption of Neutral Molecules
The adsorption of neutral molecules is discussed in Chapter 13. Here we show electrocapillary curves obtained in the presence of n-butanol and compare them to an

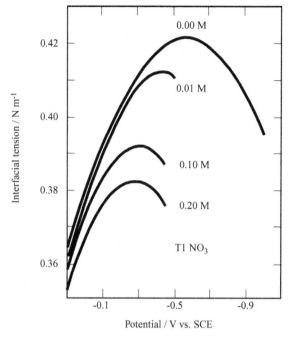

Figure 9.7 Electrocapillary curves for solutions of $TlNO_3$ in 1.0 M KNO_3, and 0.01 M HNO_3. Data from A.N. Frumkin, *Transactions of the Symposium on Electrode Kinetics*, Yeager, Ed. Wiley, pp. 1–12, (1961).

electrocapillary curve taken in a solution of NaF. Looking at Figure 9.8 we note that adsorption occurs mainly in the vicinity of E_{pzc}, and the organic substance seems to be "pushed out" of the interphase at both negative and positive values of the rational potential, $\bar{E} \equiv E - E_{pzc}$.

This behavior is not a property of the specific compound shown in Figure 9.8, nor can it be related to the structure of the adsorbed molecule or its dipole moment. Very similar behavior has been observed, for example, for benzene and phenol, although the former has no permanent dipole moment whereas the dipole moment of phenol is substantial. Results similar to those shown in Figure 9.8 were also obtained for n-butyl-cyanide (n-C_4H_9CN), although its dipole moment is quite different from that of n-butanol. It is logical, then, to conclude that the observed behavior is primarily a property of the solvent. Thus, the adsorption of an organic molecule requires the removal of a certain number of water molecules (depending on the size of the molecule being adsorbed), and it is the effect of potential on the energy required to remove these water molecules that determines, indirectly, the potential dependence of adsorption of neutral species. The adsorption of a neutral molecule and the simultaneous removal of an appropriate number of water molecules from the interphase is called *Electrosorption*. This is discussed in detail in Chapter 13.

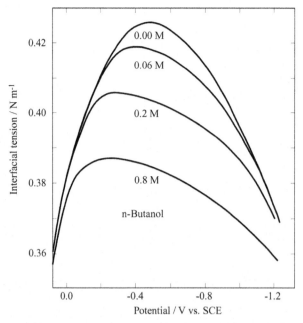

Figure 9.8 Electrocapillary curves in the presence of different concentrations of n-butanol in 0.1 M HCl. K. Muller, Ph.D. Dissertation, Fig. 37, University of Pennsylvania, 1965.

Another interesting feature of the curves plotted in Figure 9.8 is the relatively sudden convergence of the curves in the presence and in the absence of butanol at high positive and high negative rational potentials, corresponding to a sudden transition from strong adsorption to practically no adsorption. This leads to low values of the double-layer capacitance in the region of adsorption and to two sharp, *adsorption–desorption peaks* in the plot of C_{dl} versus E, which are shown in Figure 8.2.

10
Nanotechnology and Electrocatalysis

10.1
The Effect of Size on Phase Transformation

10.1.1
Introduction

Nanotechnology is a very active field of science and technology, which has major implications in electrochemistry and in many other fields. In this chapter we shall limit our discussion to some fundamental aspects of the physics of nanoparticles that are relevant to the preparation of electrode materials in electrochemical energy-conversion devices. The many ingenious methods of preparation and application of nanoparticles in electrochemistry are widely discuss in the literature and will not be treated here.

It was noted earlier in this book that the interface between two phases has unique properties that are different from the bulk properties of either phases in contact. This gives rise, among other things, to the double layer capacitance, discussed in Chapter 8. While atoms in the bulk are symmetrically coordinated, on the surface they lack that symmetry and, to a first approximation, have only half as many atoms within range of significant interaction. Hence, they are more active, have better ability to form bonds with a reactant or intermediate in solution and have a larger Gibbs energy of adsorption. Increasing the surface area requires added energy, given by

$$dG = \gamma \, dA \qquad (10.1)$$

where γ is the surface tension, defined in units of force per unit length [N m^{-1}] and dA represents the incremental increase in surface area, leading to a corresponding increase in the Gibbs energy dG. The surface tension can also be written in units of energy per unit surface area (J m^{-2}) according to Eq. (10.1).

Physical Electrochemistry: Fundamentals, Techniques and Applications. Eliezer Gileadi
Copyright © 2011 WILEY-VCH Verlag GmbH & Co. KGaA, Weinheim
ISBN: 978-3-527-31970-1

10.1.2

The Vapor Pressure of Small Droplets and the Melting Point of Solid Nanoparticles

The increase in the Gibbs energy of a small droplet of liquid with decreasing radius results in an increase in vapor pressure, expressed by the Kelvin equation

$$2.3 RT \log(p_r/p_0) = \frac{2\gamma}{r} \frac{M}{\rho} \qquad (10.2)$$

where M is the molecular weight [kg mol^{-1}], and ρ is the density [kg m^{-3}]. The ratio p_r/p_0 is the vapor pressure of a drop of radius r divided by that of a flat surface at the same temperature. A plot of p_r/p_0 versus the radius is shown in Figure 10.1a, for several liquids. Considering that the boiling point is defined as the temperature at which the vapor pressure reaches 1 atm, it follows that a small droplet of liquid will boil at a lower temperature than bulk liquid. In Figure 10.1 the curves for water and ethanol are almost the same, because the ratio $\gamma M/\rho$ happens to be the same, although the three terms in this ratio are quite different.

When a solid is involved, the Kelvin equation cannot be used, because the surface tension cannot be readily measured, the nanoparticles of metals are usually not spherical. Moreover, mechanical strain or stress that may occur in very small particles could influence the Gibbs energy of adsorption. Nevertheless, the physical rules are the same and the excess surface Gibbs energy has been observed as a lowering of the melting point of the metal.

We can calculate the percent of atoms on the surface, as a function of the size of the particle. To do this properly one should know the shape of the particle, which could be different for different metals. For simplicity we shall assume a spherical particle. The volume of a spherical shell is given by $4\pi\,r^2\Delta r$, where Δr is the thickness of the shell. Hence, the per cent of atoms on the surface of a sphere is given by

$$\frac{4\pi r^2 \Delta r}{(4\pi/3)r^3} \times 100 = \frac{3\Delta r}{r} \times 100 \qquad (10.3)$$

For a spherical nanoparticle of Pt the value of Δr was estimated on the basis of the number of atoms per cm^2 on the surface of Pt, known from the measured value of value of Γ_{max} for hydrogen adsorbed on Pt, which is 1.3×10^{15} atoms cm^{-2}, yielding a thickness of $\Delta r = 0.277$ nm. Note that this is the thickness of the spherical shell, corresponding to the diameter of a Pt atom. The result is shown in Figure 10.1b. The curves shown in Figure 10.1a and b are quite similar, as might be expected, since both are related to the special activity of the surface of a nanoparticle. The effect of size becomes significant for $r \leq 10$ nm, namely, where nanoparticles are concerned. In contrast, for a typical colloidal particle, ($r \approx 1$ μm) the extrapolated value of the curves in Figure 10.1a and b both yield an insignificant deviation of about 0.1% from bulk properties. It is interesting to note that, small as it may seem, the radius of 10 nm, is rather large on the atomic scale. For example, a water droplet of this radius contains about 1.4×10^5 molecules and a similar sized Pt sphere would contain about 2.8×10^5 atoms.

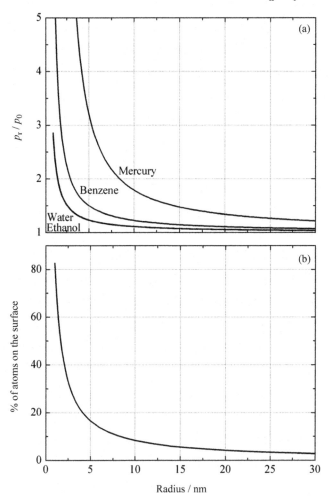

Figure 10.1 (a) The increase in vapor pressure of small drops of different liquids, with decreasing radius of the drop, at 25 °C. (b) The percent of atoms on the surface of a Pt nanoparticle, as a function of its radius.

The effect of particle size on the melting points of Al and Pb is shown in Figure 10.2a and b, respectively. The bulk melting point of Al (933.6 K) is reduced to about 838 K for a radius of 6.7 nm. It is noted also that the effect of size on the melting point becomes important only when the radius is somewhat below 15 nm, which is consistent with the calculations given in Figure 10.1. Similar (but by no means identical) behavior was observed for Pb, as shown in Figure 10.2b. In this case the melting point does not deviate from the bulk value until $r < 7.5$ nm, but then it is lowered sharply from 600.66 K to about 406 K, as the radius is reduced to 1.4 nm.

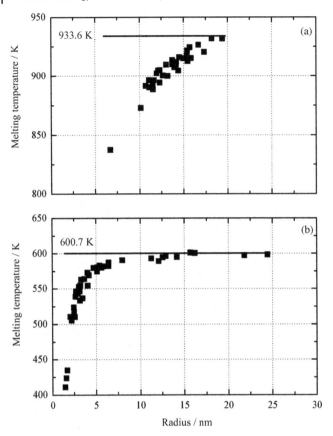

Figure 10.2 The melting point of (a) aluminum and (b) Pb, as a function of the radius of the nanoparticle. Based on data from D. Vollath in (2008) *Nanomaterials - An Introduction to Synthesis, Properties and Applications*, Wiley-VCH Verlag GmbH, Ch. 3, Figures 3.7 and 3.9.

10.1.3
The Thermodynamic Stability and Thermal Mobility of Nanoparticles

An isolated nanoparticle in the bulk of a solution or on the surface of a suitable catalyst can be stable, in spite of its high surface energy, expressed by the high vapor pressure shown in Figure 10.1a. or the lowering of the melting points shown in Figure 10.2. However, when two such particles collide and are fused together, their excess surface Gibbs energy per mole decreases. Assuming, for simplicity, that the particles have an equal radius, the radius of the new particle formed (assumed to be spherical) will increase by a factor of $\sqrt[3]{2} = 1.26$. The position of the new particle on the lines in Figure 10.1 will move to the right, lowering the vapor pressure or the melting point (and hence the excess surface Gibbs energy). Put differently, it is correct to say that a nanoparticle is thermodynamically unstable with respect to a process of merging it with one or more nanoparticles. This simple conclusion is very important in the area of catalysis or electrocatalysis, because it could lead to sintering of the catalytic nanoparticles into lumps of the same material, losing part of their active surface area, as well as any inherent catalytic activity they may have had as nanoparticles.

Being "thermodynamically unstable" is not an insurmountable impediment in technology. Indeed almost everything we use has this property: the wooden furniture in our houses is flammable, being unstable in respect to reaction with oxygen in the air. Nevertheless, furniture can survive for centuries without showing any sign of burning slowly The cutlery we use is usually made of stainless steel (an alloy of iron, chromium and nickel), all of which can combine with oxygen to form various oxides, lowering the Gibbs energy. Still, cutlery can be inherited and maintained for generations, showing little or no corrosion. This *apparent stability* is imparted by slow kinetics. Wood may dry out, but it cannot ignite and burn spontaneously. Cars are manufactured with protective coatings to minimize corrosion and metals such as aluminum, titanium and niobium create their own protective oxide coatings in air spontaneously. All of these lower the rate of corrosion, the reaction leading down the Gibbs energy slope towards equilibrium. The stability may be *apparent* from the thermodynamic point of view, but it can be very real from the engineering point of view.

How can this be applied to systems based on nanoparticles? The process we need to prevent, or at least slow down sufficiently, in order to impart a reasonable lifetime to the product, is sintering, that is aggregation of the nanoparticles to lumps, thereby losing some of their effective surface area. For this deterioration to happen, the particles need to have some mobility on the surface. Assuming that these particles are free to move in two dimensions, their average thermal velocity can be readily obtained by equating the kinetic energy with the average thermal energy. This leads to the simple relationship

$$0.5\,M\,v^2 = RT \tag{10.4}$$

where v is the average thermal velocity. Treating a Pt nanoparticle of $r = 10$ nm as a large molecule having 2.8×10^5 atoms (cf. Section 10.2.2), leads to an average thermal velocity of $0.30\,\mathrm{m\,s^{-1}}$, (while a similar calculation for a single water molecule yields a value of $v = 524\,\mathrm{m\,s^{-1}}$). Although the velocity of $0.30\,\mathrm{m\,s^{-1}}$ might look small, it is really very large on the scale of nanoparticles, for which it was calculated. Thus, if we write the velocity of a 10 nm particle of Pt, for which this calculation was made, as $3 \times 10^8\,\mathrm{nm\,s^{-1}}$, it can be seen that it will take 33 ns for a particle to move a distance equal to its own radius. This is the upper limit of the average velocity, ignoring all interactions among the particles. The actual movement may be restricted by different interactions, and this calculation is presented only in order to show that there is more than enough thermal energy (at room temperature) to allow the particles to move and collide with each other, forming aggregates. This mechanism of gradual decrease in the effective surface area is one of the possible modes of failure of electrocatalysts based on nanoparticles.

10.2
The Effect of Particle Size on Catalytic Activity

10.2.1
Does a Higher Energy of Adsorption Indicate Higher Catalytic Activity?

Considering that the surface atoms have a higher Gibbs energy than atoms in the bulk, the common wisdom is that the Gibbs energy of adsorption would be higher on the surface of nanoparticles than on large particles. While this makes sense, it is not self-evident that a higher energy of adsorption leads to higher catalytic activity. Considering the so-called *volcano plot*, it all depends on whether the specific rate constant, or the exchange current density that is proportional to it, is on the ascending or the descending branch of the plot of reaction rate versus the bond energy shown in Figure 7.4. If, as a result of the small size of the particle, the Gibbs energy of adsorption becomes too high, the intermediates formed in the reaction sequence could be too strongly adsorbed on the surface, reducing its catalytic activity. Indeed it was found in some studies of catalytic activity as a function of particle size, that there is a maximum in catalytic activity for particle sizes in the range $3 \leq r \leq 5$ nm, which is about where the fraction of the atoms on the surface starts rising sharply, and the melting points of metals start decreasing very significantly, as shown in Figures 10.1 and 10.2.

Another factor that is important in the context of the activity of nanoparticles in electrochemical systems is that adsorption in electrochemistry is a replacement reaction that can be described by an equation of the type

$$RH_{soln} + n(H_2O)_{ads} \rightarrow RH_{ads} + n(H_2O)_{soln} \tag{10.5}$$

where RH stands for an organic species that could be the reactant or one of the intermediates in a reaction sequence. This is referred to as *electrosorption*. It is characterized by the fact that the Gibbs energy of electrosorption is the difference between that of the energy of adsorption of RH and that of n molecules of water, as discussed in Section 13.1.

$$\Delta G_{ads} = \Delta G_{ads, RH} - n\Delta G_{ads, W} \tag{10.6}$$

This equation shows that the increase in the Gibbs energy of adsorption resulting from the decrease in size of the particle may be compensated for, or at least become much less significant, as a result of a similar increase of the Gibbs energy of adsorption of one or more water molecules, depending on the size of the species being adsorbed.

The purpose of the current section is not to claim that reduction of the size of the particles of a catalyst cannot lead to an increase in its catalytic activity, only to point out that such an increase is not universally correct, and the effect of size on activity should be tested experimentally for each system.

10.2.2

Nanoparticles Compared to Microelectrodes

The particular behavior of a single microelectrode or an ensemble of millions of microelectrodes is discussed in Section 14.3. Since a nanoparticle is the ultimate case of an ultramicroelectrode, it is appropriate to discuss some of the properties of nanoparticles employing the equations developed for microelectrodes, in order to calculate the increased rate of diffusion towards an isolated nanoparticle and the corresponding decrease in solution resistance.

For a single nanoparticle, assumed to be spherical, the limiting current is given by

$$j_L = \frac{nFDc_b}{\delta} = \frac{nFDc_b}{r} \tag{10.7}$$

where the radius of the particle plays the role of the Nernst diffusion-layer thickness, δ in the case of semi-infinite linear diffusion. The following numerical examples show the enhanced rate of diffusion and the associated decrease in the solution resistance for a nanoparticle of $r = 5\,nm$. Taking $n = 1$; $D = 6 \times 10^{-6}\,cm^2\,s^{-1}$ and $c_b = 1 \times 10^{-6}\,mol\,cm^{-3}$ yields, according to Eq. (10.7), a very large limiting current density of $j_L = 1.16\,A\,cm^{-2}$, for a rather dilute solution (1.0 mM) of the reactant.

It may be added here that stirring will have no influence on the limiting current density calculated above, because the Nernst-diffusion layer thickness is $\delta \geq 5\,\mu m$, even in vigorously stirred solution, which is three orders of magnitude higher than the radius of the nanoparticle. This can be seen from the equation for the limiting current density, taking into account both stirring and radius of the nanoparticle, (cf. Eq. (14.46)), which is given by

$$j_L = nFDc_b \left(\frac{1}{\delta} + \frac{1}{r} \right) \tag{10.8}$$

The solution resistance (cf. Eq. (3.8) and Section 14.3) for the same nanoparticle is given by

$$R_S = \frac{r}{\kappa} = \frac{5 \times 10^{-7}\,cm}{0.01\,S\,cm^{-1}} = 5 \times 10^{-5}\,\Omega\,cm^2 \tag{10.9}$$

where a moderate specific conductivity of $\kappa = 0.01\,S\,cm^{-1}$ has been assumed. At a current density of $j_L = 0.1\,A\,cm^{-2}$, the resulting potential drop is then given by

$$j\,R_S = 0.1\,A\,cm^{-2} \times 5 \times 10^{-5}\,\Omega\,cm^2 = 5\,\mu V \tag{10.10}$$

To be exact, one should point out that the above calculation is only approximate. Thus, microelectrodes are usually made as flat discs of the desired metal, embedded in an insulator and polished to be flush with the surface. Nanoparticles, on the other hand, are usually prepared separately and attached to the surface, so their interface with the electrolyte may be in the form of a hemisphere. In any case neither are really spherical. However, the calculation above is given to show orders of magnitude of the diffusion rate and the solution resistance, and an error by a factor of two or three does not matter.

This looks like an ideal situation for conducting electrochemical measurements at current densities far below the limiting current density, with an insignificant error caused by mass-transport limitation or by the potential drop across the solution resistance; but there is a problem. The surface area of a nanosphere of 5 nm radius is about 3×10^{-12} cm^2. Hence at 0.1 A cm^{-2}, the total current is only 3×10^{-13} A. This is measurable, but not useful for any device. In order to build a power source (e.g. a battery or a fuel cell) one would have to pack huge numbers of nanoparticles per unit geometrical surface area. On the other hand, when the nanoparticles are packed close together, mass transport by diffusion is reduced to the value found for planar electrodes, since the diffusion fields of all the particles completely overlap each other. The same applies to the solution resistance, which increases to values characteristic of planar electrodes.

10.2.3
The Need for High Surface Area

It has long been realized in the field of electrochemistry that increasing the surface area increases the *effective* catalytic activity. This is a simple-minded approach to electrocatalysis, ignoring the inherent properties of the catalyst, yet it is highly effective. In the time-honored method of preparing reversible hydrogen electrodes employing platinized-platinum, the real surface area of the electrode is increased by roughening it, without actually increasing the physical dimension of the electrode. In this way the effective exchange-current density can be increased by as much as two or three orders of magnitude, without changing the true catalytic activity of the surface. This approach is routinely applied successfully in the area of batteries and fuel cells. Improved performance is achieved using very high surface area materials

Methods of producing such electrodes have been developed over the years. The best substrate is based on carbon, which can produce electrodes having a specific area as high as 2.5×10^7 cm^2 g^{-1}). Such materials could be mixed with Pt or Pt–Rh alloys, for example, to prepare similarly high surface area catalysts.

Although there are several methods of measuring the surface area of porous materials, there seems to be some uncertainty regarding the true area where the electrochemical process can take place. Thus, employing the BET method, one uses the adsorption of an inert gas (usually argon or nitrogen), to calculate the total surface area, but some of the area interacting with a small gas molecule may not be accessible to the solvent or to the reactant. Another method has been to use the formation of a layer of oxygen or hydrogen atoms, followed by electrochemical stripping. This may have some advantage over methods based on adsorption of gases, because it measures the area of the surface that is in contact with the solution. In recent years the adsorption of CO, followed by its anodic stripping has been preferred, because CO is very strongly adsorbed and the results obtained are not sensitive to the presence of impurities. Nevertheless, it could be argued that this method measures the area accessible to a very small molecule, and some of the sites may not be available for adsorption of larger molecules, such as methanol.

There have been attempts to determine the catalytic activity of single nanoparticles, as a function of size. This is a delicate matter, since the area of such particles is of the order of 10^{-12} cm^2, and any error in the measurement (or better, the estimation) of the surface area could lead to a major error, when extrapolated to a macroscopic-sized electrode. The same applies to any background or stray current, bearing in mind that a background current of 1×10^{-13} A could lead to an error of 0.1 A cm^{-2} or more. In addition, the volume-to-surface area ratio in such measurements is extremely high, so that keeping the electrode clean enough, in order to allow a meaningful determination of its inherent catalytic activity, could be an insurmountable challenge, as discussed in Section 7.1.2).

This leaves the question of the dependence of catalytic activity on the size of the nanoparticle open to debate, at least from the point of view of the theory of electrocatalysis. On the other hand, there is no doubt that employing nanoparticles is a valid and highly effective method for producing high-surface area electrodes, thereby increasing the catalytic activity. Whether this should be attributed to an increase in the intrinsic catalytic activity associated with the small size, or just to the increase in electroactive surface area of the electrode may be of secondary importance, from the practical point of view, for example in the design of better anodes in fuel cells.

11
Intermediates in Electrode Reactions

11.1
Adsorption Isotherms for Intermediates Formed by Charge Transfer

11.1.1
General

An adsorption isotherm can be written in general form as

$$f(\theta) = g(c_b) \tag{11.1}$$

It represents the equilibrium between a species in solution and the same species adsorbed on the surface. The fractional surface coverage is defined as

$$\theta \equiv \Gamma/\Gamma_{max} \tag{11.2}$$

This definition implies that $0 \leq \theta \leq 1$, namely that adsorption is limited to a monolayer, which is valid if the bonding of the first layer of the adsorbate to the substrate is much stronger than that of subsequent layers to each other. Multilayer adsorption can occur under suitable conditions, but will not be discussed here.

11.1.2
The Langmuir Isotherm and Its Limitations

The *Langmuir isotherm* is written in chemistry in the form

$$\frac{\theta}{1-\theta} = Kc_b \tag{11.3}$$

where K is the equilibrium constant for adsorption. The standard Gibbs energy of adsorption is related to the equilibrium constant as in every chemical equilibrium, namely:

$$2.3\,RT\log K = -\Delta G^0_{ads} \tag{11.4}$$

Physical Electrochemistry: Fundamentals, Techniques and Applications. Eliezer Gileadi
Copyright © 2011 WILEY-VCH Verlag GmbH & Co. KGaA, Weinheim
ISBN: 978-3-527-31970-1

The Langmuir isotherm is only applicable if the following conditions are satisfied:

1) The Gibbs energy of adsorption is independent of coverage. This is a severe limitation! For this condition to apply it must be assumed that the surface is completely homogeneous and there are no lateral interactions among the adsorbed species. Solid surfaces are rarely homogeneous. There are highly active sites and others with low activity. As the fractional surface coverage increases, the most active sites are occupied first and the least active ones are occupied last. When this is the case, the absolute value of the standard Gibbs energy of adsorption decreases with increasing coverage

   The purists among us may object to the last statement. A *standard* Gibbs energy is defined for some standard state. From a formal point of view it cannot be said to depend on any variable. Yet, if we think of a heterogeneous surface as consisting of a very large number of minute homogeneous patches, we would have a different standard Gibbs energy for each little patch, and the experimentally observed standard Gibbs energy of adsorption would, in effect, be a function of coverage. To overcome this problem, it is customary to discuss the variation of *the apparent standard Gibbs energy of adsorption* with coverage. This distinction does not change the ensuing arguments, but it helps to soothe our scientific conscience. For brevity, we shall drop the word *"apparent"* in the following discussion.

   What about lateral interactions? These depend on the nature of the adsorbed species and on their average distance apart. Ion–ion interactions are long range, since they decay with the first power of the distance. Dipole–dipole interactions decay with r^{-3}, while chemical interactions decay with r^{-6}, and are expected to be felt only at higher values of the coverage. But what is considered a high coverage? In solution chemistry, a concentration of $c_b \leq 10$ mM is usually regarded to be a dilute solution, where interactions among solute molecules are small, while for $c_b \geq 1$ M the solution would be considered to be concentrated, and such interactions can usually not be ignored. In electrode kinetics values of $\theta \leq 0.1$ and $\theta \geq 0.9$ are considered to represent low and high coverage, respectively. This point will be discussed further in the next section.

2) The surface is assumed to have well-defined sites, with each molecule of the adsorbate occupying a single site, and each site able to accommodate only a single adsorbate molecule. For example, the adsorption of Ag on Au satisfies this condition, since the crystal radii of the two elements are almost equal (Ag 0.1445 nm; Au.0.1442 nm). On the other hand, the adsorption of Pb on Au cannot be fitted rigorously to the Langmuir isotherm, since the atomic radius of Pb is 0.175 nm.

3) Equilibrium between the species in the bulk and on the surface is assumed.

4) The isotherm is applicable only to monolayer adsorption. The last point may seem obvious when an isotherm is discussed in the context of adsorption of intermediates in electrode kinetics, but it should be emphasized here that this does not apply to the discussion of surface excess and the Gibbs adsorption isotherm, which were treated in Section 9.1.

11.1.3
Relating Bulk Concentration to Surface Coverage

How can we relate fractional surface coverage or surface concentration to bulk concentration, considering that the two have different units? This could be done, if we recognize that concentration can also be represented by the average distance between atoms or molecules, and calculate for any given value of fractional surface coverage, θ, the value of the bulk concentration, c_b, for which the average distance between the solute molecules is the same.

We have seen previously (cf. Section 4.1) that full coverage of hydrogen on Pt is $\Gamma_{max} = 1.3 \times 10^{15}$ atoms cm^{-2}. This corresponds to $3.6 \times 10^7 \times 3.6 \times 10^7$ atom cm^{-2} in a square array, leading to an average distance of 0.28 nm between atoms. The equivalent bulk concentration, for $\theta = 1$, is

$$\left(3.6 \times 10^7\right)^3 / 6 \times 10^{23} = 0.078 \text{ mol cm}^{-3} \tag{11.5}$$

where $N_A = 6 \times 10^{23}$ is Avogadro's number. Following this logic, the bulk concentration can be related to the fractional surface coverage by the equation

$$c_b \rightarrow \frac{\Gamma_{max}^{3/2}}{N_A} \theta^{3/2} \text{ mol cm}^{-3} \tag{11.6}$$

Note that we were careful not to use the "equal" sign here, because *volume* and *surface* concentrations have different units and therefore they cannot be equated with each other. The arrow is used instead to indicate that the two can be considered equivalent, if the criterion applied is that the average distance between the species on the surface and those in the bulk are the same. On the other hand, one should take into account that an atom in the bulk is surrounded spherically by all other atoms in the phase, while at the surface it enjoys only hemispherical symmetry. Thus, for a given average distance between molecules in the bulk and on the surface, the total Gibbs energy of interaction could be considered to be about twice as large in the bulk as on the surface.

11.1.4
Application of the Langmuir isotherm for Charge-Transfer Processes

Consider the application of the Langmuir isotherm to electrochemistry. For a simple charge-transfer process, such as the formation of an adsorbed chlorine atom, we write

$$Cl_{soln}^- + M \rightleftarrows Cl_{ads}^0 + e_M^- \tag{11.7}$$

The corresponding isotherm is

$$\left[\frac{\theta}{1-\theta}\right] = K\, c_b \exp\left(\frac{F}{RT} E\right) \tag{11.8}$$

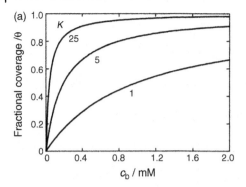

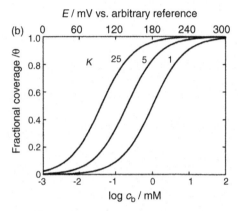

Figure 11.1 The Langmuir isotherm. The fractional surface coverage (a) θ versus bulk concentration and (b) θ versus E or $\log c_b$, for three values of the equilibrium constant $K = 1$, 5 and 25.

in which c_b is the bulk concentration of Cl^- ions. Plots of θ versus c_b and versus $\log c_b$ (or vs. E) are shown in Figure 11.1, for three values of the equilibrium constant K. Changing the equilibrium constant causes a change in the shape of the plot of θ versus c_b, whereas it causes only a parallel shift in the plot of θ versus $\log c_b$. Plots of θ versus E are similar to those of θ versus $\log c_b$. Changing the potential by $2.3RT/nF$, where n is the number of electrons transferred per atom adsorbed on the surface, has the same effect on θ as changing the concentration by a factor of ten. This is not surprising, in view of the dependence of the electrochemical potential on concentration and potential, given by

$$\bar{\mu}_i = \mu_i^0 + 2.3RT \log c_i + n_i \phi F \tag{11.9}$$

It may be recalled (cf. Chapter 6) that we used the Langmuir isotherm in electrode kinetics only under the special conditions of θ approaching either zero or unity. This relieves us of the most difficult assumption, namely, that the standard Gibbs energy of adsorption is independent of coverage. Thus, even if ΔG_{ads}^0 changes significantly with θ, this change can be considered to be negligible, when the coverage is small

$(\theta \leq 0.1)$, or when it is large $(\theta \geq 0.9)$. It is not surprising, then, that the Langmuir isotherm is often applicable in electrode kinetics, in spite of its many limitations. There are, however, cases in which it is not applicable, as discussed in the next section.

11.1.5
The Frumkin and Temkin Isotherms

The assumption that the standard Gibbs energy of adsorption is independent of coverage may be viewed as the first-order approximation. A better approximation will then be to assume a linear dependence of ΔG_{ads}^0 on θ:

$$\Delta G_\theta^0 = \Delta G_0^0 + r\theta \tag{11.10}$$

where ΔG_0^0 and ΔG_θ^0 are the standard Gibbs energies of adsorption for $\theta = 0$ and for a chosen value of θ, respectively. The parameter r is the rate of change of the standard Gibbs energy of adsorption with coverage. Considering that for the adsorption process both ΔG_θ^0 and ΔG_0^0 are negative, it may be better to write Eq. (11.10) as

$$\left|\Delta G_\theta^0\right| = \left|\Delta G_0^0\right| - r\theta \tag{11.11}$$

showing that the (absolute) value of the standard Gibbs energy of adsorption decreases with increasing coverage.

Keeping in mind the relationship between the standard Gibbs energy and the equilibrium constant, it is very easy to modify the Langmuir isotherm, taking into account the variation of ΔG_{ads}^0 with θ. To do this, we replace ΔG_{ads}^0 in Eq. (11.4) by ΔG_θ^0 and combine with Eq. (11.10), to obtain

$$2.3RT \log K = -\Delta G_\theta^0 = -(\Delta G_0^0 + r\theta) \tag{11.12}$$

which can also be written as:

$$K = \exp\left(-\frac{\Delta G_0^0 + r\theta}{RT}\right) = K_0 \exp\left(-\frac{r}{RT}\theta\right) = K_0 \exp(-f\theta) \tag{11.13}$$

where $K_0 = K \exp(-\Delta G_0^0/RT)$ and the Frumkin parameter, f, is defined as

$$f \equiv \frac{r}{RT}. \tag{11.14}$$

Substituting this value of K into Eq. (11.8) yields

$$\frac{\theta}{1-\theta} \exp\left(\frac{r}{RT}\theta\right) = K_0 \, c_b \exp\left(\frac{F}{RT}E\right) \tag{11.15}$$

This equation is known as the *Frumkin isotherm*. It should be clear that the Langmuir isotherm is a special case of the Frumkin isotherm, derived from it by setting $r = 0$. It

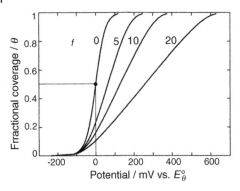

Figure 11.2 The dependence of θ on E for different values of the Frumkin parameter ($f \equiv r/RT$). The dotted lines on the curve for $f = 0$ meet at $\theta = 0.5$ and $E = E_\theta^0$.

can also be seen that, for reasonable values of the parameter r, (in the range of $(20-60)\ kJ\ mol^{-1}$), the exponential term on the left-hand side of this equation approaches unity for very small values of θ, and becomes constant when θ is close to unity. Thus, at extreme values of θ, the Frumkin and the Langmuir isotherms lead to the same dependence of coverage on potential, hence to the same rate equations in electrode kinetics.

The dependence of θ on potential is shown in Figure 11.2 for four values of the parameter $f = r/RT$.

What can be learned from Figure 11.2 and Eq. (11.15)? First we note that θ increases more slowly with potential as the value of the parameter f is increased. For example, it takes only 0.11 V to change the coverage from 0.1 to 0.9, if the Langmuir isotherm applies ($f = 0$). The same increase in θ occurs over about 0.52 V for the Frumkin isotherm, for $f = 20$. Secondly, at intermediate values of the coverage, the pre-exponential term $\theta/(1-\theta)$ varies little with θ compared to the variation of the exponential term. Thus, taking $\theta/(1-\theta) \approx 1$, the Frumkin isotherm can be written approximately as:

$$\exp\left(\frac{r}{RT}\theta\right) = K_0 c_b \exp\left(\frac{F}{RT}E\right) \tag{11.16}$$

or, in logarithmic form:

$$\theta = (2.3RT/r)\log(K_0 c_b) + \frac{F}{r}E \tag{11.17}$$

Equation (11.17) shows a linear dependence of θ on E for intermediate values of the fractional coverage, as seen in Figure 11.2. In the same intermediate region of coverage, Eq. (11.17) also shows a logarithmic dependence of the coverage on bulk concentration. A similar *logarithmic isotherm* was developed independently by Temkin. Consequently Eq. (11.17) is referred to in the field of electrode kinetics as the *Temkin isotherm*, characterized by a linear dependence of coverage on potential.

11.2

The Adsorption Pseudocapacitance C_ϕ

11.2.1

Formal Definition of C_ϕ **and Its Physical Significance**

The adsorption isotherms discussed in Section. 11.1 describe the potential dependence of the fractional surface coverage, θ. For intermediates formed in a charge-transfer process, (cf. Eq. (11.7)), the fractional coverage is associated with a Faradaic charge q_F that is given by

$$q_F = n\,F\,\Gamma \qquad (11.18)$$

where Γ is the surface concentration, corresponding to any given value of θ. If we denote the charge required to form a complete monolayer of a monovalent species by q_1, we have the simple relationship

$$q_F = q_1\theta \qquad (11.19)$$

Thus, the adsorption isotherm also yields the dependence of the Faradaic charge consumed in forming the adsorbed intermediate on potential. This allows us to define a new type of capacitance, which is called the *adsorption pseudocapacitance,* C_ϕ:

$$C_\phi \equiv (\partial q_F/\partial E)_{\mu_i} = q_1 (\partial\theta/\partial E)_{\mu_i} \qquad (11.20)$$

This rather important concept in interfacial electrochemistry warrants some clarification. Evidently, we are not dealing here with a pure capacitor, such as C_{dl}, because charge transfer is involved. Moreover, the very existence of the adsorption pseudocapacitance is linked to charge transfer. In contrast, the double-layer capacitance behaves as a pure capacitor: when charge is *brought* to one side (plate) of the capacitor, an equal but opposite charge is *induced* on the other side. An excess of electrons on the surface of the metal causes a rearrangement of the distribution of ions on the solution side of the interface, yielding an excess of positively charged ions, and vice versa. There is no transfer of charge across the interface.

Although not a pure capacitor, the adsorption pseudocapacitance exhibits the properties typical of capacitors. Whatever the type of isotherm applicable to the system, there is a singular relationship between the charge and the potential. Setting the potential determines the charge, and vice versa. When the potential is changed, a transient (Faradaic) current is observed. The current decays to zero when the charge passed is enough to bring the fractional coverage from its initial value to the value corresponding to the new potential. At a fixed potential, the steady-state current is zero. This is exactly the way a pure capacitor should behave. It allows the passage of transient currents but presents an infinite resistance to direct current.

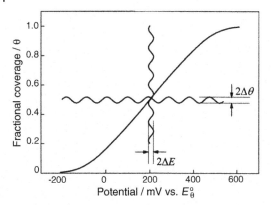

Figure 11.3 The variation of the coverage around $\theta = 0.5$, as a result of a small AC perturbation of the potential. The Frumkin isotherm with an interaction parameter $f = 20$ is used here, for illustration.

The capacitive nature of the adsorption pseudocapacitance can be further illustrated by considering its response to an alternating voltage (AC) perturbation. Let us assume that a low-amplitude sinusoidal voltage signal is applied to a system at equilibrium. The sinusoidal waveform can be expressed by the equation

$$E = E_{DC} + (\Delta E)\sin(\omega t) \tag{11.21}$$

Where E_{DC} is the DC potential, ΔE is the amplitude of the sine wave and ω is the angular velocity, related to the frequency as $\omega = 2\pi\nu$ where ν is the frequency. We assume here that the frequency is low enough, so that the coverage at any moment is equal to its equilibrium value, corresponding to the momentary value of the potential, as shown in Figure 11.3. The rate of change of coverage with time is proportional to the rate of change of potential with time:

$$\frac{d\theta}{dt} \propto \frac{dE}{dt} = (\omega\Delta E)\cos(\omega t) \tag{11.22}$$

The Faradaic current is given by

$$j_F = \frac{dq_F}{dt} = q_1 \frac{d\theta}{dt} \tag{11.23}$$

which can also be written as:

$$j_F = (\Delta j_F)\cos(\omega t) = (\Delta j_F)\sin(\omega t - \pi/2) \tag{11.24}$$

where Δj_F is the amplitude of the AC current. The phase retardation of $-\pi/2$ between potential and current is the expected response of a pure capacitor to a sinusoidal voltage perturbation.

11.2.2

The Equivalent Circuit Representation

What makes C_ϕ a *pseudo*capacitance, rather than a regular capacitance? Obviously it is the fact that it is intimately related to, and indeed dependent on, charge transfer across the interface. The equivalent circuit that represents the adsorption pseudo-capacitance itself is a resistor and a capacitor connected in series, unlike the double-layer capacitance, C_{dl}, which is represented as a pure capacitance, as shown in Figure 11.4.

Note that the resistance R_ϕ is an integral part of the physical phenomenon that gives rise to the formation of the adsorption pseudocapacitance. It is a Faradaic resistance, since C_ϕ is due to a charge-transfer process. The association of this charge-transfer process with the formation of an adsorbed intermediate, which can proceed only until the appropriate coverage has been reached, is manifested by placing the resistor in series with the capacitor. It should also be borne in mind that both C_ϕ and R_ϕ can depend on potential, as in other equivalent circuits representing the electrochemical interface

How is the adsorption pseudocapacitance affected by frequency? Returning to Figure 11.3, we recall that it was assumed that the change in θ can track the change in potential during the measurement. In other words, it was assumed that at every point in time, the value of θ is equal to its equilibrium value at that potential. But for this to happen, the charge-transfer reaction must occur fast enough, and the rate of this reaction is proportional to $1/R_\phi$. As the frequency is increased, the changes in partial coverage can no longer keep up with the changes in potential. Eventually, at a sufficiently high frequency, the adsorbed intermediate is effectively "frozen in" – the potential changes back and forth so fast that the coverage does not have a chance to follow. Since C_ϕ is proportional to $d\theta/dE$, the adsorption pseudocapacitance tends to zero when the time constant $\tau_\phi = C_\phi \times R_\phi$ of the circuit shown in Figure 11.4a, is much longer than the period of the AC signal.

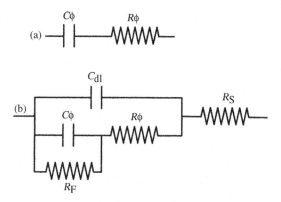

Figure 11.4 The equivalent circuits for (a) the adsorption pseudocapacitance C_ϕ and the corresponding resistance R_ϕ and (b) an interface containing an adsorption pseudocapacitance.

11.2.3
Calculation of C_ϕ as a function of θ and E

The adsorption pseudocapacitance can be readily calculated from the appropriate isotherm, with the use of its definition, given in Eq. (11.20). This is shown next for the Langmuir and Frumkin isotherms.

11.2.3.1 **The Langmuir Isotherm**
The Langmuir isotherm describing the formation of an adsorbed intermediate by charge transfer can be written in the form

$$\theta = \frac{K\, c_b \exp(EF/RT)}{1 + K\, c_b \exp(EF/RT)} \tag{11.25}$$

from which we obtain:

$$\frac{d\theta}{dE} = (F/RT)\left[\frac{K\, c_b \exp(EF/RT)}{[1 + K\, c_b \exp(EF/RT)]^2}\right] \tag{11.26}$$

Combining with Eq. (11.19) we have

$$C_L = q_1 \frac{d\theta}{dE} = \frac{q_1 F}{RT}\left[\frac{K\, c_b \exp(EF/RT)}{[1 + K\, c_b \exp(EF/RT)]^2}\right] \tag{11.27}$$

where C_L is the adsorption pseudocapacitance derived from the Langmuir isotherm. Taking the second derivative with respect to potential, it can be shown that the maximum value of the adsorption pseudocapacitance is reached when $K\, c_b \exp(F/RT)E = 1$, corresponding to $\theta = 0.5$. Thus we can write

$$C_L(\text{max}) = \frac{q_1 F}{4RT} \quad \text{and} \quad E_{\text{max}} = -\frac{2.3 RT}{F}\log(Kc_b) \tag{11.28}$$

were E_{max} is the potential where the pseudocapacitance reaches its maximum value.
The dependence of C_L on coverage can be obtained by rewriting the Langmuir isotherm in the form:

$$E = \frac{2.3 RT}{F}\log\left[\frac{\theta}{1-\theta}\right] - \frac{2.3 RT}{F}\log(Kc_b) \tag{11.29}$$

Differentiating with respect to charge one obtains

$$\frac{1}{C_L} = \left(\frac{1}{q_1}\right)\left(\frac{dE}{d\theta}\right) = \left(\frac{RT}{q_1 F}\right)\left[\frac{1}{\theta(1-\theta)}\right] \tag{11.30}$$

which can be rearranged to

$$C_L = \frac{q_1 F}{RT} \theta (1-\theta) \tag{11.31}$$

Clearly, C_L has its maximum value, given by Eq. (11.28), at $\theta = 0.5$. Setting the concentration in Eq. (11.28) equal to unity, we note that the potential at which C_L is a maximum can be regarded as the standard potential E_θ^0 for the adsorption process. It is given by

$$E_\theta^0 = -\frac{2.3RT}{F}\log K \tag{11.32}$$

The physical meaning of this choice is that the standard state of the system is chosen to be $\theta = 0.5$ and $c_b = 1.0 \, M$.

It is interesting to evaluate the numerical value of $C_{L, max}$. Taking a value of $q_1 = 0.23 \, mC \, cm^{-2}$ for a monolayer of single-charged species, we find

$$C_{L, max} = \frac{0.23 \times 10^{-3} \times 96.5 \times 10^3}{4 \times 8.31 \times 298} = 2.24 \, mF \, cm^{-2} \tag{11.33}$$

This is about two orders of magnitude higher than typical values observed for the double-layer capacitance. We can also evaluate the range of potential over which $C_L/C_{dl} \geq 1$. Substituting $C_L = 16 \, \mu F \, cm^{-2}$, which is a typical value of C_{dl}, in (Eq. 11.31), we find that the two values of θ that satisfy this relationship are 0.00187 and 0.99813. In other words, $C_L/C_{dl} \geq 1$ for values of θ between about 0.2% and 99.8%. The corresponding range of potential (found by introducing these values of θ into the Langmuir isotherm) is 0.16 V.

The important conclusion to be drawn from this numerical calculation is that great care must be exercised in interpreting double-layer capacitance measurements in systems in which an adsorbed intermediate can be formed. Even a minute fractional coverage can give rise to an adsorption pseudocapacitance comparable to or larger than the double-layer capacitance.

11.2.3.2 The Frumkin Isotherm

We proceed here to derive the expressions for the adsorption pseudocapacitance when the Frumkin isotherm applies, which we shall denote as C_F.

The isotherm (Eq. (11.15)) can be written in the form:

$$E = \frac{RT}{F} \ln\left[\frac{\theta}{1-\theta}\right] + \frac{r}{F}\theta - \frac{RT}{F}\ln(K_0 \, c_b) \tag{11.34}$$

We write the parameter r in dimensionless form as $f = r/RT$ and differentiate Eq. (11.34) to yield

$$\frac{dE}{d\theta} = \frac{RT}{F}\left[\frac{1}{\theta(1-\theta)}\right] + \frac{RT}{F}f = \frac{RT}{F}\left[\frac{1+f\theta(1-\theta)}{\theta(1-\theta)}\right] \tag{11.35}$$

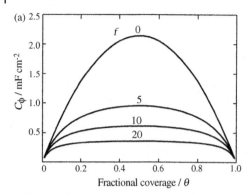

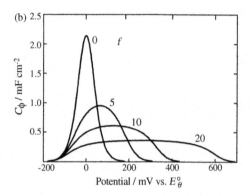

Figure 11.5 The dependence of C_ϕ on (a) coverage and on (b) potential, for different values of the parameter f, for $q_1 = 230 \, \mu C \, cm^{-2}$.

Thus, the adsorption pseudocapacitance under Frumkin conditions is given by:

$$C_F = \frac{q_1 F}{RT} \left[\frac{\theta(1-\theta)}{1+f \theta(1-\theta)} \right] = C_L \left[\frac{1}{1+f\theta(1-\theta)} \right] \tag{11.36}$$

This equation yields a maximum value of:

$$C_{F,\,max} = \frac{q_1 F}{4RT} \left[\frac{1}{1+f/4} \right] = C_{L,\,max} \left[\frac{1}{1+f/4} \right] \tag{11.37}$$

The dependence of C_ϕ on θ is shown in Figure 11.5a. We note that all curves are symmetrical around $\theta = 0.5$, irrespective of the value of the parameter f. The maximum declines rapidly with increasing values of f. The dependence of the adsorption pseudocapacitance on potential is shown in Figure 11.5b. This

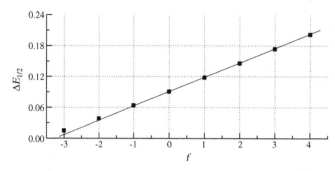

Figure 11.6 The width of the plot of C_F at half-height, $\Delta E_{1/2}$, as a function of the Frumkin parameter $f \equiv r/RT$.

function cannot be derived directly, since it is impossible to express θ as an explicit function of E from Eq. (11.15). However, since the dependence of both E and C_F on θ are known, the curves shown in Figure 11.5b can be readily calculated.

One may be tempted to calculate the parameter f from the value of $C_{F,\,max}$ as given by Eq. (11.37). The problem is that this quantity is calculated per unit of *real* surface area, whereas what one measures is $C_{F,\,max}$ per unit of *geometrical* surface area. The ratio between the real and the geometrical surface area is called the *roughness factor*, which is not known accurately in most cases. Moreover, even the value of the charge per monolayer, $q_1 = 230\,\mu C\ cm^{-2}$, which we used to calculate $C_{F,\,max}$, is only an estimate, which can vary between substrates and between different adsorbed intermediates. On the other hand, we note that the curves in Figure 11.5b become wider with increasing value of the parameter f. This provides a way to avoid the foregoing uncertainty, by determining the parameter f from measurement of the width of these curves at half height. Using the potential $\Delta E_{1/2}$ at half height to determine the value of the parameter f, we have made the result independent of the surface area. The variation of $\Delta E_{1/2}$ with the parameter f is shown in Figure 11.6. For positive values of the parameter f, this dependence is linear, with a slope of 27.3 mV per unit of the parameter f. For the Langmuir isotherm $(f = 0)$, a value of $\Delta E_{1/2} = 90.3$ mV is obtained. For $f < 0$ a slight deviation from linearity is observed.

11.2.4
The Case of Negative Values of the Parameter f

The parameter f can be negative only as a result of attractive lateral interaction between the adsorbed species. For values of $f \leq -4$, the plots of θ in Figure 11.7 show a maximum and a minimum. Between them the coverage *increased* with *decreasing* potential. This corresponds formally to a negative capacitance, which does not represent physical reality. Consequently, if the potential is swept in the positive direction, θ will jump from a very low to a very high value of θ. When the potential is

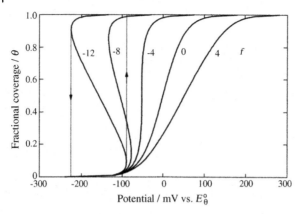

Figure 11.7 The Frumkin isotherm with both negative and positive values of the parameter f. The vertical lines show the hysteresis loop that is expected when the potential is swept first in the positive and then in the negative direction, for the case of $f = -12$.

swept back in the negative direction, it will decrease gradually and then change suddenly to a very low value, leading to a hysteresis loop shown by the vertical lines in Figure 11.7. The sudden transition from low to high coverage may be indicative of a two-dimensional phase formation.

12
Underpotential Deposition and Single-Crystal Electrochemistry

12.1
Underpotential Deposition (UPD)

12.1.1
Definition and Phenomenology

Underpotential deposition (UPD) is a process in which a monolayer of a metal is deposited on a different metal substrate, in the range of potential that is positive with respect to the reversible deposition of a metal in the same solution. Metal deposition is a reduction reaction that occurs at potentials negative with respect to the reversible potential. Thus it might appear that underpotential deposition defies the laws of thermodynamics, but careful analysis of the process shows that it does not.

In Figure 12.1 we show the deposition of Pb on a polycrystalline Ag surface. The experiment was conducted by the method of cyclic voltammetry, which will be discussed in detail in Section 15.4. In this method a triangular potential waveform is applied from a chosen initial potential (which is -0.10 V vs. Ag/AgCl for all curves in this figure), and the direction of the sweep is reversed at different potentials (which are between -0.50 V and -0.35 V in this figure). It can be seen that the cathodic limit of potential for curves 1–4 is positive with respect to the reversible potential for deposition of Pb, that is, it is in the UPD region. The cathodic sweep is where formation of a monolayer of Pb (or a fraction of it) takes place, while the anodic sweep is where the process is reversed, and anodic stripping of this layer takes place, according to the reaction

$$Pb^0_M \rightarrow Pb^{2+}_{soln} + 2e^-_M \tag{12.1}$$

The variation of potential with time is given by

$$E = E_{in} \pm vt \tag{12.2}$$

Hence the X-axis showing the potential could be replaced by one showing time, and the current under the anodic peak could be integrated to yield the corresponding

Physical Electrochemistry: Fundamentals, Techniques and Applications. Eliezer Gileadi
Copyright © 2011 WILEY-VCH Verlag GmbH & Co. KGaA, Weinheim
ISBN: 978-3-527-31970-1

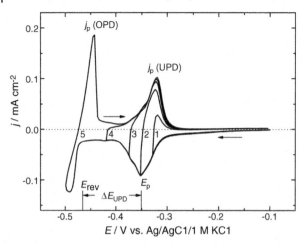

Figure 12.1 Underpotential deposition (UPD) of lead on polycrystalline silver, from a solution of 5.0 mM PbCl2 in 0.1 M HCl; $v = 25$ mV s^{-1}. The arrows show the direction of the potential sweep.

charge transferred, since

$$q = \int_{E_{in}}^{E} j_\phi dt \qquad (12.3)$$

where j_ϕ is the Faradaic current density associated with formation of a UPD layer. It is noted that lines 1 and 2 form only a partial monolayer while lines 3 and 4 essentially overlap, showing that at a potential of -0.375 V, about $+0.1$ V with respect to the reversible potential, a full monolayer is already formed. Moreover, it is noted that applying a negative overpotential, in the OPD region, where bulk deposition occurs, does not change the total charge in the UPD region – it is limited to one monolayer.

How can a metal be deposited at a potential positive with respect to its reversible potential? Deposition of Pb on an electrode made of the same metal can be represented by the equation

$$Pb_{soln}^{2+} + 2e_M^- \rightarrow Pb\text{-}Pb_M \qquad (12.4)$$

while deposition of Pb on an Ag substrate should be written as

$$Pb_M^{2+} + 2e_M^- + Ag_M \rightarrow Pb\text{-}Ag_M \qquad (12.5)$$

The two reactions are similar, but not identical. Equation (12.4) represents the formation of a bond between a Pb atom deposited and another Pb atom on the surface, while Eq. (12.5) represents the formation of the bond between a Pb atom deposited and an Ag atom on the surface. Hence, there is no reason a priori to assume that they should have the same reversible potential. Apparently the Pb–Ag bond is stronger than the Pb–Pb bond, and, hence, it can be formed at a less negative potential; the laws of thermodynamics have not been violated!

The difference between the potential of the peak current density for formation of a UPD layer and the reversible potential for bulk deposition of the same metal is shown as ΔE_{UPD}. It is given by

$$\Delta E_{UPD} \equiv E_p - E_{rev} \qquad (12.6)$$

If this is measured in a solution of standard concentration (1.0 M), and UPD formation is reversible, the potential of the peak current is usually chosen as the standard potential for formation of a UPD layer. It should be emphasized that ΔE_{UPD} depends on the substrate chosen. For example, its value for the Pb–Ag couple is not the same as that for the Pb–Au couple. This is to be expected, because this potential difference results from the difference between the bonding energy of Pb to Pb and that of Pb to the surface atom of the substrate. Hence, ΔE_{UPD} should be different for different substrates.

The charge corresponding to a monolayer can readily be determined by employing Eq. (12.3), but it may not be easy to calculate its value a priori. In Chapter 11 we stated that the charge associated with a monolayer of adsorbed intermediates on Pt is about 0.23 mC cm^{-2}, assuming that each adsorbed species transfers a single charge and occupies a single site on the surface[1]. Thus, formation of a full UPD layer of Pb should require a charge of about 0.46 mC cm^{-2}. But the crystal radii of Pb and Ag are 0.175 nm and 0.1445 nm, respectively, so that a Pb atom cannot be accommodated on a single site. Values of q_1 that are between one third and two thirds of 0.46 mC cm^{-2} have been reported in the literature for the Ag/Pb system. It should be noted, however, that the observed value of q_1 is calculated for the geometrical surface area, ignoring the effect of surface roughness. On the other hand, a monolayer of Cu on Ag, for example, could occupy all sites, because the crystal radius of copper is only 0.128 nm, smaller than that of Ag.

Figure 12.1 shows the essential features of underpotential deposition.

- UPD occurs, by definition, at potentials positive with respect to the reversible potential of deposition of the metal in the same solution.
- UPD is a self limiting process. Deposition is terminated as soon as a complete monolayer has been formed. This statement should be qualified. There are a few cases reported in the literature where more than a monolayer is formed, but rarely more than two atomic layers. However, even in such cases, the process is self-limiting
- Setting the potential anywhere in the UPD region determines the fraction of the surface covered, θ. Thus, an adsorption isotherm can be calculated, relating the fractional coverage to the applied potential
- An additional feature, not shown in Figure 12.1, has been observed experimentally. If the concentration of the metal ion in solution is very low (in the μM range), deposition becomes mass-transport limited. Under such conditions, even if a

---

[1] This is just a representative number. The exact number varies from metal to metal and even from one crystal face of the same metal to the other.

negative overpotential is applied, bulk deposition does not start until the formation of a full underpotential deposited layer of the metal has been completed.

- Formation of a uniform UPD layer is independent of the geometry of the part being deposited. This follows from the property of such layers of being self-limited. In the language of electroplating, this is referred to as an ideal *throwing power* (cf. Section 19.3). In electroplating this would indicate uniform thickness of the coating, independent of the shape of the part being coated. In the case of UPD, each occupied site on the surface of the metal becomes totally inactive (since the potential is positive with respect to the reversible potential) and a uniform UPD layer will be formed everywhere, on crevices as well as on protrusions, around the corner or even on the back side of an electrode. How can a single layer of atoms block the surface so efficiently? It is not because it forms a perfect barrier layer. It is because the potential is positive with respect to the reversible potential for bulk deposition of the metal. Formation of further layers is prevented by thermodynamic considerations, not by the ability of a single layer to act as an ideal barrier layer.
- Last but not least, the formation of a UPD layer is a prime example of an adsorption pseudocapacitance. This is discussed in Section 11.2 and the technique of cyclic voltammetry commonly used to study it is discussed in Section 15.4. When a potential is applied, a transient current will flow, which will decay to zero if the potential is held constant. This is exactly the behavior of a capacitor.

The equation controlling the current–potential relationships shown in Figure 12.1 is

$$ j_\phi = C_\phi \left(\frac{dE}{dt} \right) = C_\phi v \tag{12.7} $$

Thus, the current is proportional to the sweep rate, and its variation with potential represents the dependence of the adsorption pseudocapacitance on potential, as discussed in Section 11.2. Although the current observed increases linearly with sweep rate, the total charge needed to form a UPD layer is independent of it.

If the kinetics of deposition and stripping is fast and there is no mass transport limitation, the system behaves reversibly and the cathodic and anodic peaks are observed at the same potential, independent of the sweep rate and of the direction of the sweep. Increasing the sweep rate beyond a certain limit, which is characteristic to each system, the cathodic peak starts shifting in the negative direction and the anodic peak shifts in the positive direction. The system is no longer reversible and the difference in the peak potentials, ΔE_p, increases with increasing sweep rate. Thus, the curves shown in Figure 12.1 indicate that some degree of irreversibility exists in the case of formation of a UPD layer of Pb on Ag.

Underpotential deposition of metals is a commonly observed phenomenon, which has been found to occur for dozens of metal couples in both aqueous and non-aqueous solvents. The potential difference ΔE_{UPD} is independent of the concentration of the metal ion in solution, (as long as the UPD layer is formed reversibly and mass transport limitation is not involved,) since both E_p and ΔE_{rev} follow the same Nernst equation, albeit with different values of the standard potential.

The values of ΔE_{UPD} observed experimentally have been related to the difference in the electron work function of the two metals concerned. Much research has been devoted to the study of underpotential deposition on single-crystal metal substrates. As might be expected, ΔE_{UPD} is different for different crystal faces of the same metal, as are the work function and the potential of zero charge.

12.1.2
UPD on Single Crystals

The mercury/electrolyte interface played a major role in the early studies of the structure of metal/solution interfaces, and electrode kinetics in general. The surface of the liquid metal is highly reproducible and the low catalytic activity of Hg towards hydrogen evolution provided a rather wide range of potentials where the thermodynamic properties of the interface could be determined experimentally, allowing theories to be verified or discarded. However, mercury is of little industrial interest, and its use has been all but eliminated in recent decades because of its high toxicity and devastating influence on the environment.

The next logical step would have been to study single-crystal electrodes, which represent well-defined surfaces. This is not the way the field developed historically, because it took a long time before high-quality single-crystals of metals could be prepared in the laboratory or purchased commercially; but we need not be concerned with that here. The use of a single-crystal substrate has now become common in fundamental studies of UPD for three decades or more.

After a single crystal is prepared and its axis of symmetry is determined, it is cut at a chosen angle with respect to this axis, and embedded in an insulator, so that the solution will be in contact with only a single crystal face. An ideal crystal cut exactly at the correct angle would expose to the solution an atomically flat surface. An error in the angle of cutting will create terraces with atomic steps. With the introduction of scanning-tunneling microscopy (STM) to electrochemistry in the 1980s, it became possible to measure the size of such terraces, which were found to be large on the atomic scale, extending to hundreds of nanometers. As a rule, the larger the error in the angle of cutting, the shorter the terraces. Although single crystals may not be atomically flat on the macroscopic scale, the roughness caused by the terraces is not very significant. In any case, the shapes of the cyclic voltammograms, in particular the potential of the peak currents observed for UPD formation, are characteristic of each crystal face. Data for the formation of UPD layers of Pb on the three low-index faces of single-crystal gold are shown in Figure 12.2. On the (111) crystal face (Figure 12.2a), a very sharp peak is observed. Ideally this should be the only peak, but there are two smaller peaks, characteristic of two other crystal faces. In Figure 12.2b and 12.2c two major peaks are shown, one of which seems to be associated with the (111) crystal faces while the others are characteristic of the (100) and (110) faces, respectively. We shall not discuss the details of these peaks which are only shown here in order to emphasize that different crystal faces give rise to different current/potential curves, as discussed above.

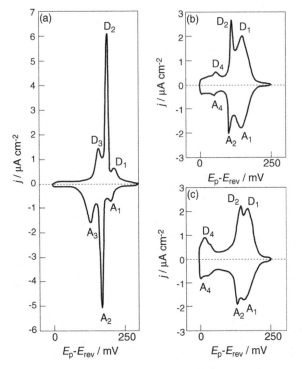

Figure 12.2 Cyclic voltammograms for UPD formation of Pb on different faces of single-crystals of Au. (a) (111), (b) (100), (c) (110). $v = 0.42\,mV\,s^{-1}$; 0.5 mMPb(ClO$_4$)$_2$ + 0.5 MNaClO$_4$ + 5 mMHClO$_4$. Reproduced with permission from E. Budevsky, G. Staikov and J. Lorenz, *Electrochemical Phase Formation and Growth*, VCH, Germany, 1966.

Often the peak current observed for formation of a UPD layer is found to be very narrow, as shown in Figure 12.2a, and this is generally assumed to represent two-dimensional phase formation. The formation of a two-dimensional phase is controlled by two opposing Gibbs energy terms. On the one hand we know that UPD must involve the formation of a chemical bond between atoms on the surface of the metal substrate and those in the UPD layer. This would lead to epitaxial deposition, which means that the structure of the deposited layer of atoms follows that of the particular crystal face exposed to the solution. However, this may not be the same as the most stable crystal structure of the metal deposited in the bulk form. Thus, lead on gold is much less likely to form a 2D phase than silver on gold, because the ratio of the crystal radii is only 1.002 for the Au/Ag couple, but 1.21 for the Pb/Au couple.

Formation of a 2D phase on polycrystalline metal surfaces is rarely if ever observed, because the metal surface is not atomically flat. Crystallites could be at different angles with respect to each other and different crystal faces may be exposed to the electrolyte at random positions, preventing long-range order. Nevertheless, in some cases the surface of a polycrystalline metal is found to behave as if only a single-crystal face were exposed to the solution. The extent of such behavior depends on the method of preparation and pretreatment of the substrate metal. For example, the crystals used for the electrochemical quartz-crystal microbalance (cf. Chapter 17) have a layer of gold sputtered on the surface, on top of a thin layer of titanium that serves as a binder between gold and the quartz. This gold surface has a preferred orientation in the (111) plane, although it is certainly not a single crystal in the usual sense.

12.1.3
Underpotential Deposition of Halogen Atoms

A phenomenon similar to the underpotential deposition of metals is also observed in the study of the anodic oxidation of halides. In Figure 12.3 we show the dependence of current on potential during an anodic potential sweep obtained on a platinum electrode in a non-aqueous medium. A peak current corresponding to the formation of atomic bromine on the surface is observed, about 0.4 V, before the potential for formation of molecular bromine has been reached. As in all other cases of UPD formation, the two reactions considered are different and are, therefore, expected to have different values of the reversible potential. For UPD of a layer of adsorbed bromine atoms the reaction is

$$Br^- + Pt \rightarrow Br\text{-}Pt + e_M^- \tag{12.8}$$

while bromine evolution is represented by the equation

$$2Br^- \rightarrow Br_2 + 2e_M^- \tag{12.9}$$

It is well known that molecular bromine is a much more stable species than atomic bromine. Thus, the formation of the latter at a less anodic potential (i.e., more easily), attests to the strong bond between atomic bromine and platinum atoms on the

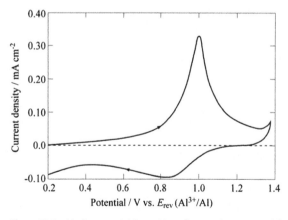

Figure 12.3 Underpotential deposition of atomic bromine and the beginning of bromine evolution on Pt in 1.0 M Al_2Br_6 and 0.8 M KBr in ethyl benzene. $v = 120\ mV\ s^{-1}$. Data from M. Elam and E. Gileadi, *J. Electrochem. Soc.* **126**, (1979) 1474.

surface. The interesting point to note is that bromine evolution does not occur on the bare platinum surface, but rather on a platinum surface covered by a monolayer of adsorbed bromine atoms. The same is true for the anodic formation of molecular chlorine and iodine. None of these halogen molecules is formed on the bare metal surface. This behavior may seem to be of little importance in, say, the industrial production of chlorine, but it could determine the catalytic activity and/or the stability of the electrodes used in this process. It is even more important to understand this behavior in the context of a fundamental study of such systems, and the interpretation of the transfer coefficient and other kinetic parameters observed experimentally.

12.1.4
Underpotential Deposition of Atomic Oxygen and Hydrogen

It was mentioned earlier, in the discussion of hydrogen evolution on Pt, that a layer of adsorbed hydrogen atoms is formed on the surface, even before the reversible potential for hydrogen evolution has been reached. This is similar to the case of the oxidation of halides, and could be considered as the formation of a UPD layer of hydrogen. The phenomenon is best seen in cyclic voltammetry, as peak currents appearing at characteristic potentials, positive with respect to the reversible hydrogen electrode in the same solution, as shown in Figure 12.4. The detailed behavior depends on the type of electrode used. Figure 12.4a is the typical curve obtained on a polycrystalline sample (e.g., a wire or a foil). Figure 12.4b was obtained on a spherical single crystal, (formed by melting the lower part of a Pt wire) on which different crystal orientations are exposed to the solution. The different peaks represent different energies of adsorption of atomic hydrogen. The area under all the peaks combined shows a charge of about 0.23 mC cm^{-2} of real surface area, corresponding to a monolayer of adsorbed hydrogen atoms. All the peaks are at potentials positive with respect to the reversible hydrogen electrode.

Cyclic voltammetry on noble metals has been studied extensively. The shape of the curve shown in Figure 12.4a, which is characteristic of Pt in sulfuric acid, is often used as a test for the purity of the system.

The peaks for adsorption and desorption of atomic oxygen are shown in Figure 12.4a. As with hydrogen, formation of adsorbed oxygen atoms precedes oxygen evolution. The area under the peak corresponds to deposition of a monolayer of oxygen atoms, requiring two electrons per atom deposited. Oxygen evolution does not take place on the bare platinum surface, but on a surface modified by a layer of adsorbed oxygen atoms. On the other hand, reduction of molecular oxygen takes place at much more negative potentials and could occur, at least in part, on a bare platinum surface.

The cyclic voltammogram on a gold electrode in the same solution is seen in Figure 12.5. The UPD layer of oxygen atoms is there, but hydrogen adsorption cannot be detected. Hydrogen evolution on gold seems to occur on the bare metal surface. The same is true for mercury, lead and other soft metals. It would seem that a low exchange current density is associated with the reaction taking place on the bare

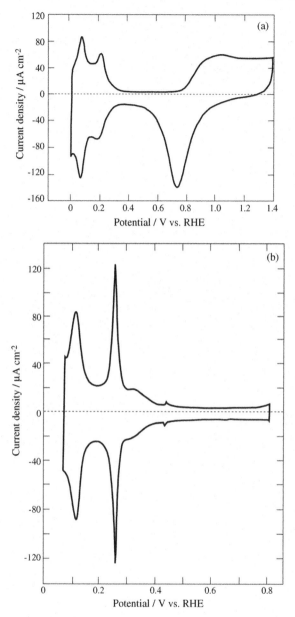

Figure 12.4 Cyclic voltammetry on Pt in 0.5 M H$_2$SO$_4$ (a) polycrystalline Pt; v = 15 mV s^{-1}; (b) spherical single crystal; v = 50 mV s^{-1}. The scale of potentials is different because (a) shows also the region of formation and removal of a layer of atomic oxygen, while (b) is limited to the region of adsorption and removal of hydrogen and the so-called double-layer region. Reprinted with permission from J. Clavilier and D. Armand, *J. Electroanal. Chem.*, **199** (1986) 187. Copyright Elsevier Sequoia.

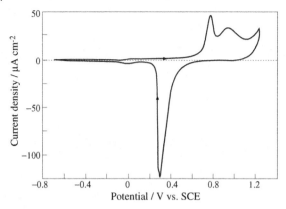

Figure 12.5 Cyclic voltammetry on a single-crystal (210) gold electrode in 10 mM NaF. $v = 20\,mV\,s^{-1}$. Based on data from A. Hamelin, *J. Electroanal. Chem.* **138** (1982) 395.

metal, whereas a high value of j_o is found when the surface is modified by a layer of adsorbed hydrogen atoms. On the other hand, oxygen evolution does not occur on the bare surface of any metal electrode.

We conclude this chapter by noting that underpotential deposition is a rather general phenomenon, occurring in both cathodic and anodic reactions. The surface is modified by the UPD layer and its catalytic activity is altered, usually for the better. The UPD layer is "transparent" to electrons (even when it consists of a layer of halogen atoms), and probably should be considered to be an extension of the metal, rather than a superficial layer of a foreign substance. A satisfactory understanding of the way in which the surface is modified by the UPD layer, which is essential for the understanding of electrocatalysis, is still lacking

13
Electrosorption

13.1
Phenomenology

13.1.1
What is Electrosorption?

Electrosorption is a replacement reaction. We have already discussed the role of the solvent at the interface, in the context of its effect on the double-layer capacitance. It is most important for our present discussion to know that the electrode is always solvated and the solvent molecules can be adsorbed on the surface both by electrostatic forces and by chemical bonds. Adsorption of a molecule on such a surface requires the removal of the appropriate number of solvent molecules to make place for the new occupant, so to speak. This is *electrosorption*. In this chapter we shall restrict our discussion to the electrosorption of neutral organic molecules from aqueous solutions, without explicit charge transfer[1]. Using the notation RH for an unspecified neutral organic molecule, we can then represent electrosorption in general by the reaction

$$RH_{soln} + n(H_2O)_{ads} \rightleftarrows RH_{ads} + n(H_2O)_{soln} \qquad (13.1)$$

Several important features of electrosorption follow from this simple equation. First, it becomes clear that the thermodynamics of electrosorption depends not only on the properties of the organic molecule and its interactions with the surface, but also on the properties of water. In other words, the Gibbs energy of electrosorption is the difference between the Gibbs energies of adsorption of RH and that of n water molecules:

$$\Delta G_{ads} = \left(\Delta G_{RH_{soln}} - \Delta G_{RH_{soln}} \right) - n \left(\Delta G_{W_{ads}} - \Delta G_{W_{soln}} \right) \qquad (13.2)$$

[1] It is recognized here that a species in contact with the surface, could interact with the delocalized electrons of the metal leading to partial charge transfer and imparting an effective charge to the adsorbed species. We distinguish here between this and an explicit charge transfer, which would be written as $RH_{soln} + n(H_2O) \rightarrow R_{ads}^- + H^+ + n(H_2O)_{soln}$

Physical Electrochemistry: Fundamentals, Techniques and Applications. Eliezer Gileadi
Copyright © 2011 WILEY-VCH Verlag GmbH & Co. KGaA, Weinheim
ISBN: 978-3-527-31970-1

The same relationships also apply to the enthalpy and the entropy of electrosorption. The enthalpy of electrosorption turns out to be less (in absolute value) than the enthalpy of chemisorption of the same molecule on the same surface from the gas phase. On the other hand, the entropy of chemisorption from the gas phase is, as a rule, negative, since the molecule RH is transferred from the gas phase to the surface, losing in the process three degrees of freedom of translation. This is also true for electrosorption, but in this case n molecules of water are transferred from the surface to the solution, leading to a net increase of $3(n-1)$ degrees of freedom. As a result, the entropy of electrosorption is usually positive. Remembering the well known thermodynamic relationship

$$\Delta G_{ads} = \Delta H_{ads} - T\Delta S_{ads} \tag{13.3}$$

we conclude that in chemisorption from the gas phase, a negative value of ΔG_{ads} is a result of a negative value of ΔH_{ads}, while the entropy term tends to drive the Gibbs energy in the positive direction. In electrosorption, the enthalpy term can be less negative, or even positive, and it is the positive value of the entropy of electrosorption that renders the Gibbs energy negative, in most cases. It can be said that *electrosorption is mostly entropy driven, whereas chemisorption is mostly enthalpy driven.*

While the above conclusion is intellectually intriguing, it may also have some important practical consequences, particularly in the area of fuel cells and organic synthesis. Thus, the common wisdom is that the extent of chemisorption *decreases* with *increasing* temperature. This follows formally from the well known equation:

$$\frac{d\log K_{ads}}{d(1/T)} = -\frac{\Delta H^0_{ads}}{2.3R} \tag{13.4}$$

with negative values of ΔH_{ads} for chemisorption. In contrast, studies of the electrosorption of ethylene on platinum electrodes from acid solutions yielded an enthalpy of adsorption close to zero. The equilibrium constant for the electrosorption of this compound was found to increase with increasing temperature. Thus, raising the operating temperature of a fuel cell employing ethylene as the fuel, in order to enhance the reaction rate, does not necessarily lower the extent of adsorption of the reactants or intermediates.

The second point to note is that electrosorption depends on the size of the molecule being adsorbed, *vis-à-vis* its dependence on the number of water molecules that have to be replaced for each RH molecule adsorbed. One may be led to think, on the basis of Eq. (13.2), that large molecules cannot be electrosorbed. This is not necessarily true, because both terms on the right-hand side of Eq. (13.2) may increase with increasing size of the molecule (i.e. with the parameter n), though not necessarily at the same rate.

An additional unique feature of electrosorption is that the coverage is a function of potential, at constant concentration in solution. Thus, we can discuss two types of isotherms: (i) those yielding θ as a function of c_b and those describing the dependence of θ on E. This *is not* a result of Faradaic charge transfer. Neither is it due to electrostatic interactions of the adsorbed species with the field inside the compact part of the double layer, considering that a potential dependence is observed even for neutral organic species having no permanent dipole moment. As we shall see, it turns out that the potential dependence of θ can be caused by the dependence of the Gibbs energy of adsorption of water molecules on potential.

13.1.2
Electrosorption of Neutral Organic Molecules

Electrosorption has been studied on mercury more than on any other metal, not because this is the most interesting system, either from the fundamental or the practical point of view, but because it is the easiest system to study, and because the results obtained are not complicated by uncertainties resulting from different features of the surface – a problem common to the study of solid surfaces. The dependence of θ on potential for the adsorption of butanol on mercury was shown in Figure 9.8. In Figure 13.1 we show plots of the fractional coverage, θ, for the electrosorption of phenol with methanol or water as a solvent. One should note that the dependence of θ on E is roughly bell-shaped in this case, and also for many other neutral molecules, whether they do or do not have a permanent dipole moment. Typically the maximum of adsorption occurs at a potential that is slightly negative with respect to the potential of zero charge.

If electrosorption is restricted to a monolayer, the fractional surface coverage can be related to the surface excess by simply writing:

$$\theta = \Gamma/\Gamma_{max} \tag{6.8}$$

This is a satisfactory approximation in the present and similar cases, since the interaction of a neutral organic molecule with the surface does not extend beyond a monolayer.

Comparing the data in water and in methanol, we note that a similar extent of coverage is reached in water in the range of concentrations of $(1-20)$mM, as in methanol in the range of $(0.05-1.5)$M. Considering that it is the same molecule being adsorbed on the same surface in both cases, the difference must be associated mainly with the difference in the solubility of phenol in the two solvents. This leads us to discuss the question of the appropriate scale of concentration to be used when comparing isotherms measured in different solvents. Usually chemists prefer to express concentrations in units of $mol\,l^{-1}$ (or $mol\,dm^{-3}$). This is fine for aqueous solutions (or at least for a fixed solvent), but it fails totally when different solvents are compared. Similarly, when comparing the adsorption of different solute molecules in the same solvent, the solubility of the adsorbed molecule also plays an important role. In such cases, it is best to use a dimensionless scale of c_b/c_{sat}. This follows directly from the fact that the chemical potential in a saturated solution is always equal to that of the solute in the pure solid phase, irrespective of either the solute or the solvent. Consequently, the choice of the above dimensionless scale of concentration permits us to compensate for the differences in the Gibbs energy of interaction between the solvent and the solute in the bulk of the solution, and the effects seen arise only from the different interactions of the solutes with the surface. This is a very good way of

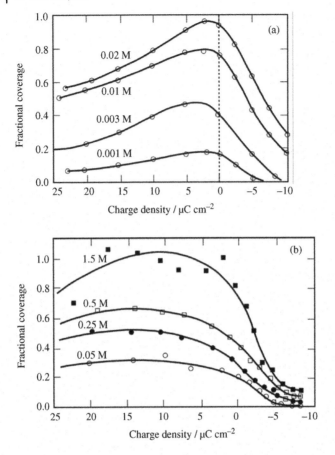

Figure 13.1 The electrosorption of phenol on mercury, from (a) water and (b) methanol as the solvent, as a function of the charge density. Supporting electrolyte: 0.1 M LiCl. The concentrations of phenol are marked on the curves. From Muller, Ph.D dissertation, University of Pennsylvania, (1965).

comparing the adsorption of *different solutes* from the *same solvent*. When the adsorption of the *same solute* from *different solvents* is studied, the choice of this dimensionless concentration still ignores the different interaction of the solvent molecules with the surface, as given by Eq. (13.2), so it is an approximation, albeit a much better one than using molar concentrations.

13.1.3
The Potential of Zero Charge, E_z, and Its Importance in Electrosorption

The concept of the potential of zero charge (PZC or E_z), has already been discussed in the context of electrocapillary thermodynamics, where we showed that, for an ideally polarizable interface, the PZC coincides with the electrocapillary maximum. In view of the very high accuracy attainable with the electrocapillary electrometer, it is possible to determine the value of E_z for liquid metals near room temperature to within about 1 mV.

Many methods have been used to determine the value of E_z on solid electrodes. The one that seems to be most reliable, and relatively easy to perform, is based on diffuse-double-layer theory. Measurement of the capacitance in dilute solutions ($c_b \leq 0.01$ M) should show a minimum at E_z, as seen in Eq. (8.10) and Figure 8.4. Lowering the concentration yields better defined minima. Modern instrumentation allows us to extend the measurement of capacitance to low concentrations of the electrolyte (cf. Sections 8.1.6 and 16.1) increasing the accuracy of the determination of E_z on solid electrodes. One should bear in mind, however, that the minimum in capacitance coincides with E_z only if a symmetrical electrolyte such as NaF is used. Asymmetric electrolytes of the type $A_m B_n$, (where $n \neq m$) also show minima in the plot of capacitance versus potential in dilute solutions, but the minima are shifted slightly from the PZC.

Although the instrumental aspects of measuring E_z can yield quite accurate results, chemistry is lagging behind. The most difficult problem, as always with solid electrodes, is the lack of reproducibility of surface preparation. The most reliable results have been obtained on single crystals of noble metals, where E_z depends on the particular crystal face exposed to the solution. An exceptional example of this is shown in Figure 13.2, where E_z is plotted for a large number of crystal faces of gold. We need not go into the details of these different crystal faces; Figure 13.2 is used just to illustrate that E_z can be measured rather accurately, even on solids, and that it is

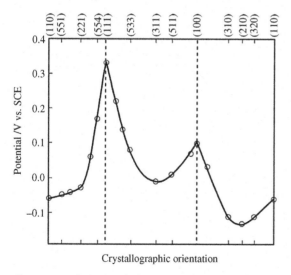

Figure 13.2 The potential of zero charge for single-crystal gold, plotted as a function of crystal orientation. Data from J. Lecoeur, J. Andro and R. Parsons, *Surf. Sci.* **114** (1982) 320.

clearly a function of the crystallographic orientation. It should also be noted that the difference in the values of E_z for different crystal faces can be quite large, amounting to about 0.4 V between the 110 and the 111 faces of gold.

On a polycrystalline sample (e.g., a wire or a foil) certain crystal faces may dominate, depending on the mechanical, thermal and electrochemical pretreatment of the sample, giving rise to different values of the PZC. It was shown, for example, that cycling the potential of a platinum electrode between oxygen and hydrogen evolution for a long time causes faceting, with the (111) crystal face becoming predominant.

Determination of E_z on base metals such as copper, nickel or iron can be complicated by the formation of oxide layers. The value of E_z measured may then correspond to an oxide-covered surface, rather than to the bare metal.

The occurrence of Faradaic reactions of any kind, and particularly those leading to the formation of adsorbed intermediates, can severely interfere with the determination of E_z, when based on measurement of the capacitance minimum. The high values of the adsorption pseudo-capacitance, C_ϕ, which extends over a significant range of potential, can distort the measurements of double-layer capacitance in dilute solutions, as discussed in Section 11.2.3.

Values of E_z are shown in Table 13.1, for a number of solid metals. The value for mercury is also included, for comparison. Electrosorption depends primarily on the excess charge density, q_M, which is related, albeit not quite linearly, to the potential measured with respect to the potential of zero charge. Coverage by neutral organic species decreases at both negative and positive values of the potential (or the charge), with a maximum of coverage at $q_M \approx -2\ \mu C\ cm^{-2}$. On the potential scale, the region of significant coverage extends over about 0.8 V, $\overline{E} \approx (-0.6\ to\ +0.2)V$. For aromatic

Table 13.1 The potential of zero charge, E_z/V, SHE for different metals.

Metal	E_z	Electrolyte	c_b/mM
Cd	−0.75	KCl	1
Tl	−0.71	NaF	1
Pb	−0.56	NaF	1
Zn	−0.63	HCl	1
Ga	−0.69	HCl	1
Bi	−0.39	HCl	1
Fe	−0.37	H_2SO_4	1
Sn	−0.38	K_2SO_4	2
Hg	−0.193	NaF	1
Ag(111)	−0.46	NaF	1
Ag(100)	−0.77	NaF	5
Au(111)	+0.19	NaF	5
Pt	+0.18	HF + KF	300 + 120
Pd	+0.10	$Na_2SO_4\ H_2SO_4$	5 + 1

compounds the coverage declines more slowly on the positive side, probably due to interaction of the π-electrons with the metal. For charged species the situation is more straightforward: positively charged ions are adsorbed mostly at negative values of $\overline{E}$, and vice versa. This effect is superimposed on other factors controlling electrosorption, so that a negatively charged molecule may be specifically adsorbed to some extent, even on a negatively charged surface, as a result of the chemical energy of interaction between the molecule and the surface. Naturally, the dependence of θ on potential will not be symmetrical in this case as it is, more or less, for neutral molecules.

It should be noted here that the adsorption of intermediates formed by charge transfer is not controlled by the potential of zero charge. When a process such as

$$Cl^-_{soln} + M \rightleftarrows M\text{-}Cl_{ads} + e^-_M \tag{11.7}$$

which leads to an adsorption isotherm of the form

$$\frac{\theta}{1-\theta} \exp\left(\frac{r}{RT}\theta\right) = K_0\, c_b \exp\left(\frac{F}{RT}E\right) \tag{11.15}$$

the region of potential over which θ is significant depends on the equilibrium constant K_0, which is related to the Gibbs energy of adsorption. The value of K_0 depends on the metal through its dependence on the Gibbs energy of the M–Cl bond, but it is not directly dependent on E_z. The case in which electrosorption occurs *with* charge transfer, so that both types of interactions have to be considered simultaneously, is discussed below.

13.1.4
The Work Function and the Potential of Zero Charge

The work function Φ is the energy required to remove an electron from a metal. It is an intrinsic property of the metal, which is measured under conditions of ultrahigh vacuum, on surfaces that have been meticulously cleaned. Even a small amount of impurity adsorbed on the surface (e.g. oxygen or an oxide, water, or carbonaceous molecule), could give rise to significant errors in the measured value of the work function. Moreover, the value measured on a polycrystalline metal surface is a weighted average of the contribution of the work function for different crystal faces. Thus, each metal has in fact several work function values, each characteristic of a different crystal face. The relevant scientific literature is replete with data with values of Φ for different metals and many of the common crystal faces for each metal.

The electrochemical properties of the meta/electrolyte interface follow a similar pattern, but involve some complication. Thus, the value of E_z is found to be different for different crystal faces, and so are the cyclic voltammograms characteristic for UPD formation, as shown in Chapter 12. It has been suggested that E_z and Φ should be linearly related to each other, and indeed such correlations have been reported in the literature. However, while Φ is a characteristic property of each metal and each crystal face, the PZC is a property of the metal/electrolyte interface, not just the metal. This is not surprising, in view of the fact that an electrode in contact with an electrolyte is never clean, on the level of the cleanliness of a surface in ultrahigh vacuum. In the best case it is covered by pure solvent molecules and influenced by the ions of the ubiquitously present solute needed to impart electrolytic conductivity. In addition, catalytic metals, which are of main interest in technology, may have adsorbed hydrogen, an oxide or some other oxygen-containing species.

An interesting approach to the determination of the PZC for active metals is to add a species that is strongly adsorbed at the metal surface, such as CO, replacing any other species, including the solvent, while the potential is held constant. Considering the simple relationship

$$q_M = C_{dl}(E - E_z) \tag{13.5}$$

it is clear that changing the value of C_{dl} at constant potential will give rise to a transient current. The sign of the current will depend on the position of the potential with respect to the PZC, namely on $(E - E_z)$. Repeating this kind of experiment at different fixed potentials will allow the determination of the value of E_z. This method applies strictly only when the interface is ideally polarizable, that is, when there is no Faradaic current flowing. Otherwise it can be treated as an approximate value of E_z (referred to in the literature as the potential of zero total charge, E_z^t), which represents the actual state of charge of the surface, including the effect of the Faradaic reaction taking place.

13.2
Methods of Measurement and Some Experimental Results

13.2.1
Electrosorption on Solid Electrodes

While studies of electrosorption on mercury were of great importance for the understanding of the structure of the interface, studies on solid electrodes are of much greater interest from the engineering point of view. It is a relevant factor in many fields, including electrocatalysis, electroplating, corrosion and bio-electro-chemistry, some of which are discussed below. Unfortunately, it is much more difficult to measure the surface coverage on solids, and the interpretation of data is complicated by limited reproducibility of the surface, by the competing formation of adsorbed layers of oxygen or hydrogen, and by the possibility of Faradaic reactions taking place during the electrosorption process.

In the early days of fundamental studies of fuel cells, great efforts were made to develop "direct-hydrocarbon fuel cells". Light paraffin molecules such as propane and ethane, as well as unsaturated compounds such as ethylene were studies in this context. The form in which hydrocarbons are adsorbed on platinum was the subject of controversy at the time, because it was assumed that it would influence the catalytic activity of the surface. In order to understand the difficulty involved, let us compare the electrosorption of ethane and ethylene.

The only way ethane could form chemical bonds with the surface is to break some C–H bonds, for example, according to the equation:

$$\underset{\substack{|\\ H}}{\overset{\substack{H \quad H\\ |\quad |}}{H-C-C-H}} + 4Pt \longrightarrow \underset{\substack{|\quad |\\ Pt \quad Pt}}{\overset{\substack{H \quad H\\ |\quad |}}{H-C-C-H}} + \underset{Pt}{\overset{H}{|}} + \underset{Pt}{\overset{H}{|}} \tag{13.6}$$

This was called *dissociative adsorption*. If the potential is sufficiently positive with respect to the reversible hydrogen electrode in the same solution, the adsorbed hydrogen atoms will be ionized, and Eq. (13.6) can be rewritten as:

$$\underset{\substack{|\\ H}}{\overset{\substack{H \quad H\\ |\quad |}}{H-C-C-H}} + 2Pt \longrightarrow \underset{\substack{|\quad |\\ Pt \quad Pt}}{\overset{\substack{H \quad H\\ |\quad |}}{H-C-C-H}} + 2H^+ + 2e^-_{crys} \tag{13.7}$$

Replacing ethane by ethylene, the same type of dissociative electrosorption could take place, namely:

$$\underset{\substack{|\quad |\\ H \quad H}}{\overset{}{H-C=C-H}} + 2Pt \longrightarrow \underset{\substack{|\quad |\\ Pt \quad Pt}}{\overset{}{H-C=C-H}} + 2H^+ + 2e^-_{crys} \tag{13.8}$$

But there is an alternative route, referred to as *non-dissociative electrosorption*, which can be represented by the equation

$$\underset{\substack{|\quad |\\ H \quad H}}{\overset{\substack{H \quad H\\ |\quad |}}{H-C=C-H}} + 2Pt \longrightarrow \underset{\substack{|\quad |\\ Pt \quad Pt}}{\overset{\substack{H \quad H\\ |\quad |}}{H-C-C-H}} \tag{13.9}$$

Some early results for adsorption of ethylene on Pt, obtained by the radiotracer method, are shown in Figure 13.3.

The important difference between dissociative and non-dissociative adsorption is that the former involves Faradaic charge transfer (at sufficiently positive potentials) while the latter does not. Consequently, it is easy to distinguish between these two modes of electrosorption. To do this, one injects a solution containing the hydrocarbon into a cell in which a platinum electrode is held at a potential of about $+0.5$ V versus RHE, where adsorbed hydrogen is rapidly ionized. If dissociative adsorption occurs, a transient current will be observed. The total charge during this transient should correspond to the ionization of two mol of hydrogen atoms for each mol of ethylene adsorbed. If adsorption is non-dissociative, only a small transient current is expected. The charge in this case is that associated with discharging the double-layer capacitor at constant potential, as a result of the decrease in its capacitance, caused by electrosorption. The results of such experiments led to the conclusion that saturated hydrocarbons are electrosorbed dissociatively, while unsaturated hydrocarbons tend to be adsorbed without dissociation. This may depend,

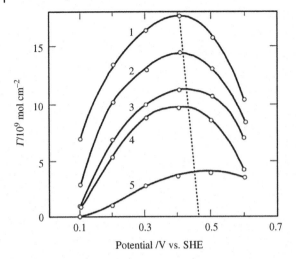

Figure 13.3 The surface concentration of ethylene, per unit of geometrical surface area, as a function of potential, on a platinized platinum electrode in 0.5 M H_2SO_4. (1) 17 μM; (2) 9.0 μM; (3) 4.6 μM; (4) 4.0 μM; (5) 2.1 μM. Reprinted with permission from E. Gileadi, B. Rubin and J.O'M. Bockris, *J. Phys. Chem.* **67**, 3335, (1965). Copyright 1965, the American Chemical Society.

however, on experimental conditions, and particularly on the temperature and the type of electrode used.

Methanol has been, and still is, a potentially attractive fuel for fuel cells, for two reasons: first, because it is relatively easily oxidized electrochemically, yielding only water and CO_2, and secondly because it can be cheaply manufactured and can, in effect, serve as a chemical means of storing hydrogen in the liquid form. It contains 12.5% hydrogen by weight, compared to less than 1.7% in Ti Fe $H_{1.8}$ and 1.3% in $LaNi_5 H_{5.8}$, which were both proposed as suitable media for chemical storage of hydrogen. Incidentally, ethanol is also rich in hydrogen (13 wt%) and can be readily manufactured from biomaterials, but a direct-ethanol fuel cell is not a viable alternative to methanol because its oxidation under conditions relevant for the operation of fuel cells is not complete, and yields acetic acid as the final product.

The electrosorption of methanol on platinum electrodes has, therefore, been studied extensively. Typical results obtained on bright platinum are shown in Figure 13.4. Although the dependence on potential is "bell-shaped," similar to that observed for the adsorption of organic molecules on mercury, one should be careful in interpreting these data. Thus, the decrease in coverage on the positive side of the peak may be due, at least in part, to the anodic oxidation of methanol in this range of potentials, while on the cathodic branch competition with adsorbed hydrogen may modify the form of the potential dependence observed. Moreover, spectroscopic evidence indicates that the electrosorption of methanol on platinum is probably a complex process, in which several different, partially oxidized, species may be formed, in relative concentrations that depend both on the potential and on the bulk concentration of methanol.

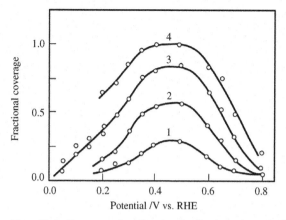

Figure 13.4 Electrosorption of methanol on bright Pt electrodes from 1.0 M H_2SO_4. Concentration of methanol: (1) 1×10^{-3} M; (2) 1×10^{-2} M; (3) 1×10^{-1} M.; (4) 1.0 M. Data from V.S. Bagotzky and Y. Vassiliev, *Electrochim. Acta*, **11** (1966) 1439.

13.2.2
The Radiotracer Methods

Radiotracer techniques are ideally suited for the detection of very small amounts of a chemical species, such as are found in a monolayer or a fraction of it, on the surface. When this method is employed, the electrode is placed on the window of a suitable radiation detector (in some cases the electrode constitutes the window) so that close to half the radiation emitted is directed toward the detector. The relative effect of the background from the environment can be suppressed by the use of a sufficiently high concentration of the radioisotope, but the background from solute molecules near the surface *which are not adsorbed* is increased proportionally. This effect depends on the penetrating length of the radiation through the solution. If an isotope emitting γ-radiation or high-energy β-radiation is used, the background from the solution is high, since it comes from a relatively large volume of the solution, and measurement of θ is difficult or impossible. Fortunately, the radio isotope of carbon, ^{14}C, emits low-energy β-particles, making it suitable for measurement of adsorption of organic molecules. The data shown in Figure 13.3 were obtained on a platinized Pt surface having a roughness factor of about 30. An additional advantage of the radiotracer method is that it can be used to follow adsorption as a function of time, namely to study the kinetics of adsorption.

13.2.3
Methods Based on the Change in Bulk Concentration

An obvious way to determine the amount of a substance adsorbed on the surface is to measure the resulting change in its bulk concentration. This is equivalent to measuring adsorption from the gas phase by determining the decrease in partial pressure of the relevant gas.

The sensitivity of such methods depends on the ratio between the volume of the solution and the surface area of the electrode. A typical electrochemical cell has about $10\,\text{cm}^3$ of solution per cm^2 of surface area. In view of the very small amounts of material needed to form a monolayer, there is a need to increase the sensitivity of this type of measurement, by using a high-surface-area electrode and low concentration in solution. The surface-to-volume ratio can also be increased by using porous electrodes or by rolling up an electrode in a minimum volume of solution. Care must be taken, however, to ensure that potential control and uniformity are maintained in this type of measurement, remembering that the extent of electrosorption is potential dependent.

13.2.4
The Lipkowski Method

In Chapter 9 we have shown the thermodynamic relationships among the surface tension, γ, (which is the same as the excess surface Gibbs energy), the excess charge density on the metal, q_M, the double layer capacitance, C_{dl}, and the surface excess of a particular species, Γ_i (cf. Eqs. (9.8) to (9.14)).

The surface tension is related to the double layer capacitance by the equation

$$-\left(\frac{\partial^2 \gamma}{\partial E^2}\right)_{\mu_i} = \left(\frac{\partial q_M}{\partial E}\right)_{\mu_i} = C_{dl} \tag{9.10}$$

The surface excess, Γ, which is a measure of the extent of adsorption, is calculated, using the equation

$$\Gamma_i = -\left(\frac{\partial \gamma}{\partial \mu_i}\right)_{E,\mu_{j \neq i}} \tag{9.11}$$

On liquid mercury, one can obtain the data necessary to use Eq. (9.11) in one of two ways: by measuring the surface tension directly, or by measuring the double-layer capacitance as a function of potential and integrating it twice.

$$\int_{E_z}^{E} C_{dl} dE = q_M \tag{13.10}$$

Followed by integration of the potential-dependent charge density on the surface

$$\int_{E_z}^{E} q_M \, dE = \gamma(E) - \gamma(E_z) \tag{13.11}$$

This strategy cannot be followed using solid electrodes, because the interfacial tension cannot be measured directly, and in particular the integration constant in Eq. (13.11) is not known. Nevertheless, one can obtain the relative values of $\gamma(E)$, as shown in Figure 13.5.

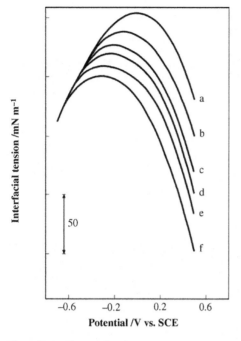

Figure 13.5 The interfacial tension as a function of potential, for different concentrations of pyridine on polycrystalline gold. Lines (a)–(f) correspond to concentrations of 0, 1×10^{-5} M; 4×10^{-5} M; 1×10^{-4} M; 5×10^{-4} M; 3×10^{-3} M, respectively. Data from L. Stolberg, J. Richter and J. Lipkowski, *J. Electroanal. Chem.* **207** (1986) 213.

To do this one needs to identify a potential at which there is no adsorption (i.e., $\Gamma = 0$). This is based on the idea that adsorption changes the value of the double layer capacitance significantly. Thus, if the capacitance is found to be the same in the base electrolyte, with and without the presence of the adsorbate, over a certain range of potential, it indicates that no adsorption occurs in that potential range. Stepping the potential from this to a series of values where adsorption does occur, and determining the charge associated with each step, makes it possible to determine the variation of the interfacial tension with potential and from it the surface excess as a function of potential can be obtained. The values of the interfacial tension shown in Figure 13.5 are only known to within a constant, which is represented by $\gamma(E_z)$ in Eq. (13.11).

It should be noted that this treatment is valid only if the interface is ideally polarizable in the range of potential studied. The details of the calculations involved in obtaining the surface excess by this method are outside the scope of this book. In Figure 13.5 we show the relative values of the interfacial tension obtained by this method, for the adsorption of pyridine on a polycrystalline gold surface.

13.3
Adsorption Isotherms for Neutral Species

13.3.1
General Comments

An isotherm describing electrosorption can be written in general form as:

$$f(\theta) = Kc_b \, g(E) \qquad (13.12)$$

where $f(\theta)$ and $g(E)$ represent some, as yet unspecified, functions of the fractional coverage and the potential, respectively. Up to now we have been concerned with the form of the function $f(\theta)$. The potential dependence of θ for an adsorbed species formed by charge transfer had the form of the Nernst equation.

$$E = -(2.3RT/F)\log K_1 + (2.3RT/F)\log\left[\left(\frac{1}{c_b}\right)\left(\frac{\theta}{1-\theta}\right)\right] \qquad (6.21)$$

Here, we shall discuss the dependence of coverage on potential in the absence of charge transfer, paying particular attention to the size of the adsorbed molecules, expressed in terms of the number of sites, n, on the surface occupied by each molecule, which is equal to the number of water molecules replaced from the surface per molecule of the adsorbed species.

One of the early triumphs of the Langmuir isotherm was in distinguishing between physically adsorbed molecular hydrogen and chemically adsorbed atomic hydrogen. For the former one writes

$$H_2 + M \rightleftharpoons M\text{-}H_2 \qquad (13.13)$$

which should follow the Langmuir isotherm,

$$\frac{\theta}{1-\theta} = K \, p_{H_2} \qquad (13.14)$$

On the other hand, for the adsorption of atomic hydrogen the appropriate equilibrium is

$$H_2 + M \rightleftharpoons 2\,M\text{-}H \qquad (13.15)$$

In this case the rate of adsorption is proportional to $(1-\theta)^2$, and the rate of desorption is proportional to θ^2, since we consider empty and occupied sites to be reactants for the adsorption and the desorption reactions, respectively. The corresponding isotherm is:

$$\left[\frac{\theta}{1-\theta}\right]^2 = K \, p_{H_2} \qquad (13.16)$$

Thus, at low coverage, dissociative adsorption leads to

$$\theta = \sqrt{K \, p_{H_2}} \qquad (13.17)$$

while adsorption of molecular hydrogen leads to a linear dependence of θ on the pressure. The adsorption of molecular hydrogen was studied in this manner on catalytic metals, such as platinum, palladium and nickel and found to be dissociative, following Eq. (13.17).

When a larger molecule taking up n sites on the surface is being adsorbed, we could follow the same line of reasoning and write:

$$\frac{\theta}{(1-\theta)^n} = K p \, (\text{or} = K c_b) \tag{13.18}$$

This is not a rigorously correct equation, but it is a better approximation than the Langmuir isotherm, suitable to describe the adsorption of larger molecules.

For the electrosorption of a large organic molecule, represented by the equilibrium

$$RH_{soln} + n(H_2O)_{ads} \rightleftarrows RH_{ads} + n(H_2O)_{soln} \tag{13.1}$$

A better approximation can be derived for the function $f(\theta)$ in Eq. (13.12) by expressing the equilibrium constant in terms of the mole fractions of the species involved.

$$K = \frac{\left(X_{RH,ads}\right)\left(X_{W,soln}\right)^n}{\left(X_{RH,soln}\right)\left(X_{W,ads}\right)^n} \tag{13.19}$$

Substituting the appropriate values for the mole fraction of the adsorbed molecule in the solution and on the surface, we obtain the expression

$$K c_{b,RH} = \left(\frac{\theta}{(1-\theta)^n}\right) \left(\frac{[\theta + n(1-\theta)]^{n-1}}{n^n}\right) \tag{13.20}$$

It is easy to see that if we set $n = 1$ in Eq. (13.20), it reverts to the Langmuir isotherm. This should not be surprising, because no coverage-dependent Gibbs energy of electrosorption has been introduced in deriving Eq. (13.20).

13.3.2
The Parallel-Plate Model of Frumkin

We now turn to the potential dependence of electrosorption of neutral molecules, considering first the model developed by Frumkin. This is a phenomenological model, which depends on considerations of the changes in the electrostatic energy of the interface caused by adsorption. Assuming that measurements are taken in concentrated solutions of a supporting electrolyte, we can neglect diffuse-double-layer effects and focus our attention on the Helmholtz part of the double layer, considered as a parallel-plate capacitor. In the pure electrolyte, the capacitance C_0 and the corresponding surface charge density, q_0, are determined by the properties of water at the interface, mainly its effective dielectric constant and its dimensions (which determine the thickness of the capacitor). We have already seen that adsorption of an organic molecule tends to decrease the capacitance. This property can be associated with the combined effects of a lower dielectric constant and an increase in thickness of the parallel-plate capacitor. Writing q_0 and q_1 for the charge densities at $\theta = 0$ and $\theta = 1$, respectively, Frumkin proposed the equation:

$$q_0 = C_0 \overline{E} \text{ and } q_1 = C_1 \left(\overline{E} - E_N \right) \tag{13.21}$$

where C_0 and C_1 are the values of the double-layer capacitance at $\theta = 0$ and $\theta = 1$, respectively, and $\overline{E} \equiv E - E_z$ is the potential with respect to the potential of zero charge. The parameter E_N represents the shift in the potential of zero charge, caused by a full monolayer of adsorbed species.

The Frumkin isotherm is based on the assumption that at any value of the coverage, the interface can be viewed as two capacitors connected in parallel. It follows immediately from this assumption that the charge q_θ corresponding to a given value of θ can be written as:

$$q_\theta = q_0(1-\theta) + q_1\theta = C_0 \overline{E}(1-\theta) + C_1(\overline{E} - E_N)\theta \tag{13.22}$$

The derivation of the Frumkin isotherm is rather involved and is not given here. We note only that it is based on calculating the difference in electrostatic energy of charging the double-layer capacitor with and without the adsorbed species. The final result is as follows:

$$K_{\overline{E}} = K_{E_z} \exp\left(-\frac{0.5\,(C_0 - C_1)\overline{E}^2 + C_1 \overline{E}\,E_N}{RT\,\Gamma_{\max}} \right) \tag{13.23}$$

where $K_{\overline{E}=0}$ is the value of the equilibrium constant at E_z. If we ignore for a moment the term containing E_N, we find that this isotherm predicts a maximum of adsorption at E_z, with θ declining symmetrically on either side. The term $C_1 \overline{E}E_N$ accounts for the failure of the potential of maximum adsorption observed experimentally to coincide with E_z.

All the quantities that appear in the exponent of Eq. (13.23) can be measured, at least in principle, thus permitting the isotherm to be tested by comparison with experiment. This is a great advantage of any theory, inasmuch as it does not contain any adjustable parameters. The disadvantage of the Frumkin model is that it makes no attempt to explain the observed phenomena on the molecular level. Thus, the values of $(C_0 - C_1)$ and E_N are taken as such; it is not explained why they have their observed values or how these values depend on molecular size, on the orientation of the molecules at the interface and on the nature of the interactions with the surface.

The assumption underlying the derivation of the Frumkin isotherm is tantamount to assuming that the surface charge density is a linear function of coverage at constant potential, as seen in Eq. (13.22). This is by no means generally correct, although it may constitute a good approximation in many cases.

Equation (13.22) can be rewritten in the form:

$$\theta = \frac{q_\theta - q_0}{q_1 - q_0} \tag{13.24}$$

which may be used to determine the coverage from differential capacitance measurements, if the potential of zero charge is known with sufficient accuracy to allow the determination of the charge by integration of the capacitance as a function of potential.

An approximate form of Eq. (13.24) is sometimes used to determine coverage on solid electrodes. Differentiating the charge in Eq. (13.22) with respect to potential, one has

$$\frac{dq}{dE} = \frac{dq_0}{dE}(1-\theta) + \frac{dq_1}{dE}\theta - (q_0-q_1)\frac{d\theta}{dE} \tag{13.25}$$

If one neglects the potential dependence of θ, it follows from this equation that

$$\theta \cong \frac{C_\theta - C_0}{C_1 - C_0} \tag{13.26}$$

This may be a reasonable approximation in certain cases, but it must be considered inherently inconsistent, when the purpose of the experiment is to determine the potential dependence of θ.

13.3.3
The Water-Replacement Model of Bockris, Devanathan and Muller

The isotherm derived by Bockris, Devanathan, and Muller (BDM) is centered on the role of water at the interface.

Electrosorption is a replacement reaction, as pointed out above, and its standard Gibbs energy of adsorption is the difference between the standard Gibbs energies of adsorption of the organic species and of the water molecules it replaces from the surface (cf. Eqs. (13.1) and (13.2)). Thus, the dependence of θ on potential is associated with the variation with potential of the standard Gibbs energy of adsorption of water. The electrostatic energy of interaction of the water dipole $\vec{\mu}$ with the field $\vec{F}$ in the Helmholtz part of the double layer is given by:

$$U_\mu = \vec{\mu}\,\vec{F}\cos\varphi \tag{13.27}$$

A fundamental premise of the BDM isotherm is that water molecules can take up only one of two positions at the interface: with the dipole vector either in the direction of the field or in the opposite direction. In other words, the angle φ between the direction of the field and the orientation of the dipole can either be zero or π, and the energy of interaction between the water dipoles and the field is either $\vec{\mu}\vec{F}$ or $\overleftarrow{\mu}\vec{F}$, (which equals $-\vec{\mu}\vec{F}$) depending on its orientation.

Considering only the energy of interaction of the water dipole with the surface charge density, all the water molecules should have their positive ends facing the surface when it is negative and vice versa, with a sharp transition at the E_z. There are, however, several factors that modify this transition, making it gradual.

1) First, there is the entropy term. A situation in which all dipoles are oriented in one direction represents the minimum entropy for this system. Completely random orientation corresponds to maximum entropy. Thus, enthalpy and entropy affect orientation in opposite directions, as is often found in chemistry,

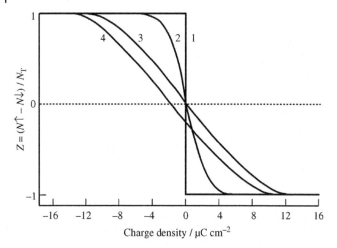

Figure 13.6 Schematic representation of the variation of water orientation at the interface with charge density. (1) Only the energy of interaction between the dipole and the field in the double layer is considered, (2) an entropy term is added, (3) lateral dipole–dipole interactions are also considered, (4) the difference in chemical energy of adsorption of water in the two orientations is taken into account.

and a compromise, determined by the condition of lowest Gibbs energy, is reached. The resulting orientation of water molecules at the interface depends both on temperature and on potential (or charge).

Let us define a parameter Z such that

$$ Z = \frac{\vec{N} - \overleftarrow{N}}{N_T} \tag{13.28} $$

where $\vec{N}$ and $\overleftarrow{N}$ are the numbers of water molecules per square centimeter in the two allowed orientations, depending on charge density, as shown schematically in Figure 13.6.

2) Next there is a lateral interaction term. If left alone (i.e., in the absence of an external field), dipoles would pair up to minimize the total energy of the system, yielding a value of $Z = 0$. This opposes the tendency of the field to orient all the dipoles in one direction, making the transition from one orientation to the other even slower.

3) Finally, there is the chemical interaction of water with the surface. A water molecule cannot be regarded as a structureless electric dipole. It is more strongly adsorbed when oriented with the negative end of the dipole (i.e., the oxygen atom) facing the surface. Thus, at $q_M = 0$ the parameter Z is negative. In order to force Z to be zero, one must have a small negative charge density in order to overcome the difference in the chemical energy of adsorption of water in the two orientations. In the case of adsorption of neutral organic molecules on mercury, this charge is found to be about $q_M = -2\ \mu C\ cm^{-2}$. We shall again

skip the tedious part of deriving the isotherm and write the final result in simplified form as follows:

$$K(E) = K(E_z)\exp\left[-nZ\left(\frac{\vec{\mu}\,\vec{F}-Z\,\sigma}{kT}\right)\right] \tag{13.29}$$

It is important to consider the physical origin of the various terms in this equation. The term nZ is the number of water molecules removed from the surface for each organic molecule adsorbed, multiplied by the fractional excess of water molecules oriented one way or the other. The parameter Z is defined in such a way that the product $\vec{\mu}\times\vec{F}$ is always positive, leading to a symmetrical decrease in the equilibrium constant (hence in the coverage) on both sides of $Z=0$. The parameter σ is the lateral interaction term. It is the total electrical energy of interaction of one water dipole with all other dipoles surrounding it. It is also multiplied by Z, since interactions with oppositely oriented dipoles cancel each other.

The complete isotherm has the form:

$$\left[\frac{\theta}{(1-\theta)^n}\right]\left[\frac{[\theta+n(1-\theta)]^{n-1}}{(n^n)}\right] = K(E_z)c_{b,RH}\exp\left(-nZ\frac{\vec{\mu}\,\vec{F}-Z\,\sigma}{kT}\right) \tag{13.30}$$

Although this may look rather complex, it is a simplified form of the isotherm. Thus, the function $f(\theta)$ on the left-hand side of Eq. (13.30) does not take into account changes in the standard Gibbs energy of adsorption due to surface heterogeneity or to lateral interactions between the adsorbed molecules[2]. Since the unique feature of this isotherm is the potential dependence of the coverage, we prefer to write it in the following form:

$$f(\theta) = K(E_z)c_{b,RH}\exp\left(-nZ\frac{\vec{\mu}\,\vec{F}-Z\,\sigma}{kT}\right) \tag{13.31}$$

This is the BDM isotherm, not taking into account the small shift between the potential of zero charge and the potential of maximum adsorption, which arises from the different chemical energies of interaction of water with the surface in its two allowed orientations.

Equation (13.31) is not in a form that can be conveniently used by electrochemists. We would like to express the surface coverage in terms of the charge or the potential, rather than the field. The field is related to the charge through the Gauss theorem, namely

$$\vec{F} = \left(\frac{\partial\phi_x}{\partial x}\right)_{x=0} = -\frac{q_M}{\varepsilon_0\varepsilon} \tag{13.32}$$

---

2) The parameter σ on the right-hand side relates to lateral interactions between water dipoles, not between molecules of the adsorbate.

If it is preferred to write the BDM isotherm in terms of potential, we can use the relationship

$$\vec{F} = \frac{E - E_z}{\delta} = \frac{\bar{E}}{\delta} \tag{13.33}$$

in which δ is the thickness of the Helmholtz double layer.

This leads to the BDM isotherm in a form that relates the fractional surface coverage, θ, to the potential, namely:

$$f(\theta) = K(\bar{E} = 0)\, c_{b,\,RH} \exp\left[-nZ\left(\frac{\vec{\mu}\bar{E}/\delta - Z\sigma}{k_B T} \right) \right] \tag{13.34}$$

Equation (13.34) is not quite correct, since it implies that the absolute metal–solution potential difference at the PZC is zero. In other words, it implies that we can measure the absolute metal–solution potential difference, which is not the case, as we showed in detail at the outset of this book. The error is, however, only a constant, which can be lumped into the equilibrium constant and has no effect on the potential dependence of θ.

The Frumkin isotherm (cf. Eq. (13.23)), can be written in the form:

$$f(\theta) = K(\bar{E} = 0)\, c_{b,\,RH} \exp\left[-\frac{0.5\,(C_0 - C_1)\bar{E}^2 + C_1 \bar{E}\, E_N}{RT\,\Gamma_{max}} \right] \tag{13.35}$$

Both isotherms can predict the results obtained experimentally. A detailed study that would make it possible to determine which fits the experimental results more closely and for a larger number of systems has yet to be performed.

14
Experimental Techniques

14.1
Fast Transients

14.1.1
The Need for Fast Transients

In this chapter we shall focus our attention on the use of transients for the separation between activation and mass-transport-controlled processes.

The need to enhance the rate of mass transport has already been discussed briefly in Section 4.2.2. It follows from the fact that mass transport and charge transfer are consecutive processes, and therefore, the slower of the two will be rate determining. This is expressed mathematically by equation Eq. (14.1).

$$\frac{1}{j} = \frac{1}{j_{ac}} + \frac{1}{j_L} \tag{14.1}$$

For values of $j/j_L \ll 1$, the measured current density is close to the activation controlled current density, and thus the latter can be determined accurately. When the measured current density approaches its limiting value (where j/j_L is in the range 0.1–0.9), the value of j_{ac} can still be calculated, employing Eq. (14.1), providing that the limiting current density is known, but the accuracy is reduced as the measured current density approaches mass transport limitation. This follows immediately from Eq. (14.1), which can be rewritten as

$$\frac{1}{j_{ac}} = \frac{j_L - j}{j_L \times j} \tag{14.2}$$

For values of the measured current density approaching the limiting current density, the value of j_{ac} calculated from the above equation depends on a small difference between two large numbers, hence its accuracy is diminished. To illustrate this point, assume that the values of j_L and j are 100 and 90 mA cm^{-2}, respectively, and the accuracy in measuring each of these quantities is ± 1 mA cm^{-2}. In this case the difference $j_L - j = 10$ mA cm^{-2} will be determined with an accuracy of about

Physical Electrochemistry: Fundamentals, Techniques and Applications. Eliezer Gileadi
Copyright © 2011 WILEY-VCH Verlag GmbH & Co. KGaA, Weinheim
ISBN: 978-3-527-31970-1

$\pm 1.4 \, \text{mA cm}^{-2}$. Consequently the error in determining j_{ac} will be about 14%, although the raw measurement of current is accurate to within 1%.

Two points of interest should be noted in relation to the above calculation:

1) Using the above values of j_L and j, we find that $j_{ac} = 900 \, \text{mA cm}^{-2}$. This may appear to be odd. The limiting current, by its very definition, represents the highest rate at which the reaction can proceed, so how can the activation controlled current be nine times higher? The answer is given by examining Eq. (14.1) carefully. The activation controlled current is not the rate at which the reaction occurs. It represents *the rate at which is would have occurred had there been no mass transport limitation*. When it is much higher than the limiting current, it has very little effect on the observed rate of the electrochemical reaction. In other words, the measured current will be close to the mass-transport limited current, which is exactly what we have assumed in the above calculation. So, one can determine the rate of an electrochemical charge-transfer process, (represented by j_{ac}) even when this rate exceeds the rate of mass transport – albeit with significant loss of accuracy.

2) Is the random error of 1%, assumed above, reasonable? A current of 100 mA could be measured with an accuracy of $1 \, \mu\text{A}$, employing equipment available in most electrochemistry research laboratories. However, in our *digital era*, the sensitivity and accuracy of such measurements are no longer limited by the quality of the electronic instrumentation – they are determined by the stability of the chemical system. Many factors determine the latter, and we shall not dwell on them here. However, careful experimental practice (such as high purity of the solution, careful elimination of oxygen, good control of temperature, and proper correction for residual background currents) can reduce the random error by a factor of 10 at least. Figures 4.6 and 7.2 were obtained under such conditions, allowing determination of the activation controlled current density up to about 20 times j_L, corresponding to values of $j \approx 0.95 j_L$.

It can be concluded that increasing the rate of mass transport is important in studying electrode kinetics, because it increases the upper limit of current density where the kinetic parameters can be determined.

The limiting current density, j_L can generally be written in the form

$$j_L = nFDc_b/\delta \tag{1.6}$$

where the Nernst diffusion-layer thickness δ is determined by the mode of mass transfer. In quiescent solutions (where convective mass transport is absent) having an excess of supporting electrolyte, (where mass transport by migration in the electric field in solution is avoided) mass transport is dominated by diffusion. Under these conditions

$$\delta = (\pi D t)^{1/2} \tag{4.12}$$

and the diffusion-limited current density is given by

$$j_d = \frac{nFDc_b}{\sqrt{\pi D t}} \tag{14.3}$$

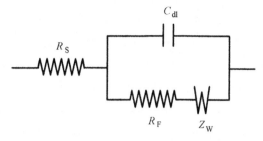

Figure 14.1 The equivalent circuit for partial mass-transport limitation by diffusion. The Warburg impedance, Z_W, is in series with the Faradaic resistance, R_F, and in parallel with the double-layer capacitance.

Thus, the diffusion-limited current density is proportional to $t^{-1/2}$, therefore, employing fast transients can extend the range over which the activation controlled current density can be determined.

The equivalent circuit for a system in which diffusion plays a significant role is shown in Figure 14.1. The symbol W stands for the *Warburg impedance*, which accounts for mass-transport limitation by diffusion. This is an odd element in the equivalent circuit, since it behaves neither as a capacitor nor as a resistor, as will be discussed in Section. 16.2.

The strength of fast-transient methods in the study of electrode reactions is shown in Table 14.1. The fastest methods of stirring the solution can be matched by transients of a few ms.

The limitations of fast transient measurements, both from the experimental and the theoretical points of view, are discussed below.

14.1.2
Small-Amplitude Transients

Transient measurements can be of two types: small-amplitude transients, which give rise to a linear response, and large-amplitude transients, which result in a nonlinear,

Table 14.1 Comparison of values of δ obtained by different methods. $n = 2$; $D = 6 \times 10^{-6}$ cm^2 s^{-1}.

Type of stirring	$\delta/\mu m$	Corresponding time for transient/s $t = \delta^2/\pi D$	$j_d/mA\,cm^{-2}$ ($c_b = 10$ mM)
Natural convection	150–250	12–33	0.80–0.48
Magnetic stirrer	50–100	1.3–5.3	2.4–1.2
RDE at 400 rpm	30	0.48	4.0
RDE at 10^4 rpm	6.0	19×10^{-3}	20
Fast impinging jet	2.0	2×10^{-3}	60
Fast pulse	0.14–1.4	$1 \times 10^{-5} - 1 \times 10^{-3}$	830–83

often exponential, response. We have already seen (cf. Section 5.2.4) that a system at equilibrium responds linearly to a small perturbation in potential or in current, according to the equation[1]

$$\frac{j}{j_0} = \frac{n\eta F}{v\,RT} \tag{14.4}$$

A "small" perturbation in this context is one for which $n\eta F/v\,RT \ll 1$ (corresponding to $j/j_0 \ll 1$). The linearity of the response allows easier and more rigorous mathematical treatment and is, therefore, often preferred.

It is interesting to note that a linear response is also obtained when a small perturbation is applied to a system far away from equilibrium. To prove this, we write the usual rate equation for an activation controlled process in the linear Tafel region

$$j = j_0 \exp\left(\frac{\alpha\eta F}{RT}\right) \tag{14.5}$$

Equation (14.5) represents the current–potential relationship at steady state. If the system is perturbed by a small signal, $\Delta\eta$, the resulting current can be expressed by the relation:

$$j + \Delta j = j_0 \exp\left(\frac{\alpha(\eta + \Delta\eta)F}{RT}\right) \tag{14.6}$$

Subtracting Eq. (14.5) from Eq. (14.6) one has:

$$\Delta j = j_0\,\exp(\alpha\,\eta F/RT)[\exp(\alpha\,\Delta\eta F/RT)-1] = j[\exp(\alpha\,\Delta\eta F/RT)-1] \tag{14.7}$$

For a sufficiently small perturbation, $\alpha\,\Delta\eta\,F/RT \ll 1$, the above equation can be linearized to yield

$$\left(\frac{\Delta j}{j}\right) = \frac{\alpha F}{RT}\Delta\eta \tag{14.8}$$

The relative change in current density, $\Delta j/j$, is linearly related to the perturbation $\Delta\eta$, and is independent of the steady-state current density or the overpotential[2].

The result shown in Eq. (14.8) should not be surprising. The physical meaning of this equation is that even when the j/η relationship is exponential, a small interval of this curve, near any given steady state value, can be linearized.

A small current or voltage perturbation also implies that the changes in concentration of the reactants and products near the electrode surface are small, and the

---

1) The parameter v in this equation is the stoichiometric number, defined as the number of times the rate determining step has to occur in a multistep reaction, for the overall reaction to occur once.

2) Note that η and j in Eq. (5.6) have exactly the same meaning as $\Delta\eta$ and Δj in Eq. (14.8), except that in the former the overpotential and the current density before application of the perturbation in overpotential are zero.

associated equations of mass transport can also be linearized, to simplify the mathematical treatment.

Large perturbations of the potential or current are treated quite differently. The most common example of a large perturbation signal is the linear potential sweep or cyclic voltammetry, which is discussed in Chapter 15.

14.1.3
The Sluggish Response of the Electrochemical Interface

Considering Table 14.1 above we note that the use of fast transients in the range of $0.01-1.0$ ms yields limiting currents that are higher than those achieved by the fastest methods of stirring. It would be nice to extend this method by use of very short pulses, in the range of nanoseconds, to study fast electrode reactions. This is, unfortunately, not possible, because of the sluggishness of the interface, which is due to the need to charge the double-layer capacitor. If we wish to change the potential by 10 mV, for example, we need to add a charge of

$$\Delta q = C_{dl} \times \Delta \eta = 20 \, \mu F \, cm^{-2} \times 10 \, mV = 200 \, nC \, cm^{-2} \qquad (14.9)$$

Supplying this amount of charge in 1 ns requires an average current density of $200 \, A \, cm^{-2}$, which is clearly impractical. Extending the pulse duration to $t = 0.01$ ms, an average current density of only $0.02 \, A \, cm^{-2}$ is required, which is well below the diffusion limited current density shown in Table 14.1 for this time of measurement. In practice, studies with pulses of 0.01 ms duration are relatively easy. Very careful design of the electrochemical cell and its connection to the measuring system can allow measurements at $10^{-6}-10^{-7}$ s, but not at shorter times. This has nothing to do with the limitations of the electronic instruments. It is an inherent physical limitation of the behavior of the metal/solution interface.

14.1.4
How can the Slow Response of the Interface be Overcome?

14.1.4.1 Galvanostatic Transient
Imagine that we wish to determine the current–potential relationship for a given reaction, over a wide range. We can apply a series of galvanostatic steps and observe the steady-state potential corresponding to each current density. The trick is to do the measurement rapidly, before diffusion limitation starts to play a role. On the other hand we must wait long enough for the double layer to be charged up to its steady-state value, when the potential across it is given by $j \, R_F$. The results of this type of experiment, for different kinetic parameters, are shown in Figure 14.2. There are two parameters in this figure that are characteristic to the interface – the time-constants of the parallel combination of a capacitor and a resistor, τ_c, given by

$$\tau_c \equiv C_{dl} \times R_F = \frac{v}{n} \frac{RT}{F} \frac{C_{dl}}{j_0} \qquad (14.10)$$

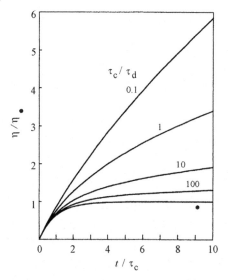

Figure 14.2 Variation of the overpotential with time during a galvanostatic transient, for different values of the parameter τ_c/τ_d. When $\tau_c/\tau_d \geq 10^3$, the reaction is under purely activation control, and η stops changing with time when $t/\tau_c \geq 3$.

and the characteristic time constant for the diffusion process τ_d given by

$$\tau_d^{1/2} \equiv \frac{RT}{(nF)^2} \frac{2C_{dl}}{c_b D} \tag{14.11}$$

It may look unusual that the product of capacitance and resistance would have the units of time, but this can be readily proven, by substituting the appropriate units in Eq. (14.10). Thus

$$C_{dl} \times R_F \rightarrow \frac{F}{cm^2} \times \Omega\, cm^2 \rightarrow \frac{Q}{V} \times \Omega \rightarrow \frac{A \times s \times \Omega}{V} \rightarrow s \tag{14.12}$$

A similar analysis shows that τ_d also has the units of time, thus the ratio of τ_c/τ_d used in Figure 14.2 is dimensionless. Since τ_c is inversely proportional to the exchange current density, while τ_d is independent of it, a large value of τ_c/τ_d indicates that the rate of charge transfer is low, and vice versa.

14.1.4.2 The Double-Pulse Galvanostatic Method

Consider the equivalent circuit shown in Figure 14.3, ignoring for the moment the Warburg impedance associated with mass-transport limitation. When a galvanostatic pulse is applied to such a circuit, the response is that shown in Figure 14.3. The equation describing the change of overpotential with time during the transient is

$$\eta = \eta_\infty [1 - \exp(-t/\tau_c)] + j\, R_S \tag{14.13}$$

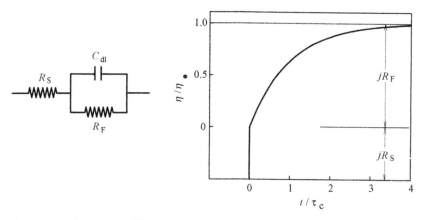

Figure 14.3 The response of the overpotential to a galvanostatic transient under purely activation controlled conditions. $\tau_c \equiv R_F C_{dl}$. Presented in dimensionless coordinates, as η/η_∞ versus t/τ_c.

where the term jR_S is the ohmic overpotential (i.e., the current density multiplied by uncompensated solution resistance between the working and the reference electrodes).

There is nothing to be gained by increasing the applied current density since, as we have seen, the relaxation time is independent of it. In fact, all transients taken at any current density are described by the single line shown in Figure 14.3, because it is plotted in dimensionless form. It is possible, however, to charge the interface faster, by applying two consecutive pulses, the first substantially shorter but larger than the second, as shown in Figure 14.4a.

The idea behind this method is to charge the double-layer capacitance rapidly with a large current pulse j_p, and interrupt this pulse just as the overpotential has reached the correct value, corresponding to steady state at the lower current density, j, namely when $\eta = \eta_\infty = jR_F$ and $j \ll j_p$. One does not know this "correct" value of the overpotential, of course, since it is the quantity being measured. This disadvantage can be overcome by trial and error, as shown in Figure 14.4b. An overshoot or an undershoot can be detected and the correct value of the ratio j_p/j can be found.

The reader should be warned here that performing this type of experiment is not as easy as might be inferred, upon viewing the simulated transients shown in Figure 14.4b. Factors such as electronic noise, poor impedance matching between the instruments and the electrochemical cell and saturation of the input amplifier as a result of a large jR_S potential drop, may distort the pulse and make the determination more difficult, and sometimes impossible. Chemical factors, such as modification of the surface during the initial pulse, may also make it hard to choose the correct value of j_p/j. Most of these problems can, however, be overcome by following correct experimental procedures, and the double pulse method can yield very useful kinetic data for fast electrode reactions.

14.1.4.3 The Coulostatic (Charge-Injection) Method

Consider the application of a very short current pulse to the interface. The charge $q_p = j \times t_p$ injected during the pulse changes the potential across the double-layer

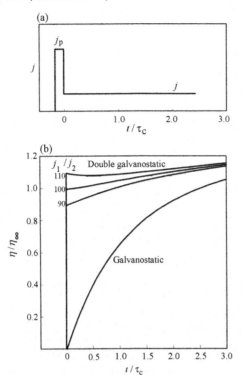

Figure 14.4 Simulated representation of the double-pulse galvanostatic method. (a) the pulse shape (b) the response, calculated for $\tau_c/\tau_d = 100$. A single-pulse galvanostatic transient for the same value of τ_c/τ_d is also shown for comparison. The optimal ratio of currents in this particular case turns out to be in the range $j_p/j = 100-110$.

capacitance by an amount $\Delta E_{pulse} = q_p/C_{dl}$. Starting from the open circuit potential, this will be equal to the overpotential η_0, as seen in Figure 14.5. We use here the subscript *zero*, because this is the initial overpotential (at $t = 0$) for the open circuit decay transient studied in a coulostatic experiment.

We shall treat here the simplest case, in which the pulse duration is very short compared to the time constant for charging the double-layer capacitance, $(t_p/\tau_c \ll 1)$ and diffusion limitation can be ignored. Under such conditions, no Faradaic reaction takes place during the charging pulse. Once on open circuit, the capacitor will be discharged through the Faradaic resistor, R_F. It is easy to derive the form of the decay transient. On the one hand, the current is given by.

$$j = -C_{dl}(d\eta/dt) \tag{14.14}$$

On the other hand, since we are dealing with small perturbations, it is also given by

$$j = \frac{j_0 \, nF\eta}{\nu \, RT} = \frac{\eta}{R_F} \tag{14.15}$$

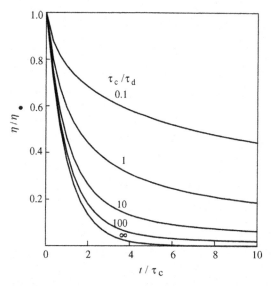

Figure 14.5 The coulostatic method (charge injection is followed by open-circuit decay). At low values of τ_c/τ_d, the decay transient is slowed down by partial diffusion control, as expected from the equivalent circuit shown in Figure 14.1.

Remember that the current density in Eq. (14.15) is an *internal* current, since the decay of overpotential is followed at open circuit. It is also interesting to note here that during the charging pulse, C_{dl} and R_F are effectively connected in parallel, while during open-circuit decay the same two circuit elements must be considered to be connected in series, and the same (internal) current is flowing through both.

Combining Eqs. (14.14) and (14.15) and rearranging, we can write

$$\int_{\eta_0}^{\eta} d\ln \eta = -\frac{1}{\tau_c} \int_0^t dt \tag{14.16}$$

which shows that the overpotential decays exponentially with time, following the equation:

$$\eta = \eta_0 \exp(-t/\tau_c) \tag{14.17}$$

The relaxation time for charge transfer, τ_c, can be obtained from the slope of a plot of $\ln \eta$ versus t. The intercept at $t = 0$ yields $\eta_0 = q_p/C_{dl}$, from which C_{dl} is obtained, since q_p is the experimentally controlled parameter. With C_{dl} known, R_F and j_0 can readily be obtained from τ_c.

If diffusion limitation is considered, the overpotential decays more slowly, as shown in Figure 14.5. This should be evident, since the Warburg impedance shown in Figure 14.1 is added in series with the Faradaic resistance, R_F. In this case the plot of $\ln \eta$ versus t is not linear and a more complex mathematical treatment, taking into account the diffusion equations, must be applied to calculate the kinetic parameters.

The unique feature of the coulostatic method is that measurements are made at open circuit. This leads to two important consequences:

1) Since the charge is injected in a very short time (preferably 1 μs or less), measurement can often be completed before diffusion limitation has become significant. In this respect the charge-injection (coulostatic) method is similar to the double-pulse galvanostatic method, except that one has more freedom in the choice of the parameters of the pulse, since there is no need to match it to the second pulse.
2) Electrode reactions can be studied in poorly conducting solutions, since there is no error due to the jR_S potential drop (also referred to as the *resistance overpotential*) in the course of an open-circuit measurement. This feature may be particularly useful for studies in non-aqueous solutions and at low temperatures. Although this is fundamentally correct, there are practical limitations to its applicability. To show an extreme example, one cannot follow the open-circuit decay of potential over a range of 10 mV, if the jR_S potential change during the pulse is, say, 10 V. Thus, even though measurements are taken at open circuit, it is clearly advantageous to use highly conducting solutions whenever possible, and this becomes essential if we wish to study the rate of fast reactions. These considerations become even more critical when the effects of diffusion limitation are included.

14.2
The Time-Dependent Diffusion Equation

14.2.1
The Boundary Conditions of the Diffusion Equation

In the study of the diffusion of species to and from the electrode surface, we use the notation $c_i(t, x)$ to describe the concentration of the ith species as a function of time and distance from the electrode surface. The time-dependent diffusion equation in its general form is written as:

$$[\partial c(x, t)/\partial t] = \nabla^2[Dc(x, t)] \tag{14.18}$$

where ∇^2 is the *Laplace operator*, corresponding to the second derivative of the concentration with respect to distance, in the appropriate coordinates. Two assumptions are commonly made in electrochemistry, in order to simplify Eq. (14.18): (i) The diffusion coefficient, D, is assumed to be constant, independent of concentration, and (ii) the equation is solved in one dimension (perpendicular to the surface), assuming the so-called semi-infinite linear diffusion. With these assumptions Eq. (14.18) is simplified to

$$[\partial c(x, t)/\partial t] = D[\partial^2 c(x, t)/\partial x^2] \tag{14.19}$$

How severe are these assumptions? The concentration of the electroactive species in the Nernst diffusion layer can vary from zero (at $x = 0$ and $j = j_L$) to the bulk concentration, (for $x = 0$ and $j = 0$). The bulk concentration is typically a few mM. Since measurements are conducted in the presence of a large excess of supporting electrolyte, this represents a very small change in the *total* concentration, and the variation in D as a function of the distance from the surface can be considered to be negligible.

What is *semi-infinite* in this case? How far is *infinity*, in the context of diffusion of a species in an aqueous solution? It should be far enough from the surface for the concentration to have reached its value in the bulk. This applies when $(x \geq 5\,\delta)$. Employing Eq. (4.12) we find that $\delta \approx 6 \times 10^{-2}$ cm at 100 s, hence *infinity* lies less than 0.3 cm away from the surface. One does not even have to use a planar electrode in order to achieve one-dimensional (planar) diffusion. A cylindrical electrode (i.e., a wire) or a spherical electrode will also look "planar" as long as the Nernst diffusion layer thickness is small compared to the radius of curvature of the electrode. On the other hand, a microelectrode, typically having a radius of $r \leq 10\,\mu m$ will not follow the equations for semi-infinite linear diffusion, as discussed in Section 14.3 below.

If the conditions for reaching an "infinite" distance from the surface are so easy to implement, why is it called "semi-infinite"? Because we deal with the direction perpendicular to the surface into the solution, not into the electrode, namely, only *half* of the space is taken into consideration.

There are two aspects to solving the diffusion equation. One must first set the initial and boundary conditions, then find a mathematical procedure for solving the equations. Here, we shall concentrate on the former aspect. Mathematical techniques for solving the diffusion equation are discussed in many texts, since this problem is not unique to electrochemistry.

The initial and boundary conditions under which the diffusion equation is solved define, in mathematical language, the kind of experiment being performed and the initial conditions of the experiment. It is important to realize that the resulting equations hold true only if these initial and boundary conditions have been maintained. This can be explained with the use of a few examples.

The equation to be solved (Eq. 14.19) is a second-order differential equation in two variables. It requires, therefore, three initial and boundary conditions. Consider a simple reaction of the type

$$Ox + n\,e_M^- \rightleftharpoons Red \tag{14.20}$$

In which both the oxidized and the reduced form are in solution. One actually must solve two similar diffusion equations simultaneously, one for the reactants and one for the products. Thus, there are six initial and boundary conditions that must be defined, which are discussed below.

14.2.1.1 Potential Step, Reversible Case (Chrono-amperometry)
The initial conditions for this and all other cases to be discussed below are as follows:

$$c_{Ox}(x, 0) = c_{b,Ox} \quad \text{and} \quad c_{Red}(x, 0) = c_{b,Red} \tag{14.21}$$

In words, these equations state that the concentrations of both reactant and product are uniform everywhere in solution at $t = 0$ (i.e., before we have applied the potential pulse). For simplicity we shall assume here that there is no product initially in solution, but this is not essential. These initial conditions may seem self-evident, but they really are not. In particular, when the experiment is conducted by applying a series of pulses (e.g., each at a different potential) care must be taken to make the solution homogeneous before each pulse – for instance, by stirring for a short time and then allowing the solution to become completely quiescent.

The next two equations arise from the condition of *infinity*. They are written as follows:

$$c_{Ox}(\infty, t) = c_{b,Ox} \quad \text{and} \quad c_{Red}(\infty, t) = c_{b,Red} = 0 \tag{14.22}$$

Since, as we have shown, "infinity" is less than 0.3 cm away during diffusion-controlled mass-transport processes, these conditions generally apply, except when a conscious effort is made to place the working and counter electrodes very close to each other, as in thin-layer cells.

Next we have an equation of mass balance, written as follows:

$$D_{Ox}[\partial c_{Ox}(0, t)/\partial x] + D_{Red}[\partial c_{Red}(0, t)/\partial x] = 0 \tag{14.23}$$

For the reaction assumed here, one molecule of the reduced form, Red, is produced for each molecule of the oxidized form, Ox, which has been consumed; hence the flux of the oxidized species reaching the surface must be equal to the flux of the reduced species leaving it. Since the two fluxes are in opposite directions, their sum must be zero. This boundary condition leads to some restrictions on the use of the resulting diffusion equation. Equation (14.23) is valid only if both species are soluble. It does not apply, for example, to an electroplating process, because the product stays on the surface. Even for a reaction such as bromine evolution, where both reactants and products are soluble, Eq. (14.23) would have to be slightly modified, to take into account the fact that two reacting species combine to form a single molecule of the product. The diffusion equation can, of course, be solved for these and other cases as well. The point we want to emphasize is that, before applying a diffusion equation found in the literature, it is important to know the boundary conditions under which that particular equation was solved, to ensure that it applies to the experiment being analyzed.

The sixth and last boundary condition follows from the assumption of reversibility. If the reaction is very fast, the reactants and products *at the electrode surface* are in equilibrium at all times, and their concentrations will conform to the Nernst equation. This boundary condition can be written as:

$$E = E^0 + (RT/nF)\ln\left[\frac{c_{Ox}(0, t)}{c_{Red}(0, t)}\right] \tag{14.24}$$

Often this equation is written in a different form as

$$\Theta \equiv \frac{c_{Ox}(0, t)}{c_{Red}(0, t)} = \exp\left[\frac{nF}{RT}(E - E^0)\right] \tag{14.25}$$

Note that we have ignored the ratio of activity coefficients in the last two equations. This is a good approximation, since a large excess of supporting electrolyte is used.

Having established the physical conditions and the six initial and boundary conditions, one can proceed to solve the diffusion equation. We shall skip the mathematical derivation and proceed directly to the solution, applicable to a potential step under reversible conditions, which is

$$j = \left[\frac{n F D c_b}{\sqrt{\pi D t}} \right] \left[\frac{1}{1 + \Theta} \right] \tag{14.26}$$

In this equation we have made the simplifying assumption that $D_{Ox}/D_{Red} \approx 1$.

As the diffusion-limited current density is reached, the concentration of reactant at the surface is reduced to zero, and, therefore, $\Theta = 0$. Substituting in Eq. (14.26) we have

$$j_d = \frac{n F D c_b}{\sqrt{\pi D t}} \tag{14.27}$$

Combining Eqs. (14.26) and (14.27), we can relate the current density to potential and time by the simple equation

$$j(t) = \frac{1}{1 + \Theta} j_d(t) \tag{14.28}$$

The potential dependence, which is "hidden" in Θ, can be introduced explicitly, employing Eq. (14.25), to yield

$$E = E^0 + \frac{RT}{nF} \ln \left(\frac{j_d}{j} - 1 \right) \tag{14.29}$$

We have discussed the assumptions under which Eq. (14.26) is valid, but we have assumed implicitly that the potential changes from its reversible value to a preset value, E, *instantaneously*. This is never the case in practice. The assumption is justified if the rise-time of the pulse is short compared to the duration of the experiment. This assumption can be a major source of error if the uncompensated potential drop in solution, jR_S, is not negligible in comparison with the size of the potential step applied, as shown in Figure 14.6.

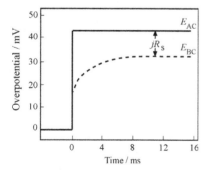

Figure 14.6 Schematic presentation of the change of potential across the double layer E_{BC} following a potential step, E_{AC}, at the output terminals of the potentiostat. A significant residual uncompensated solution resistance is assumed (see the equivalent circuit in Figure 4.3).

But where exactly has this implicit assumption been introduced? If we had known exactly how the potential changes with time during the transient, where might we have introduced this variation in the boundary conditions of the differential equation? Examining our six initial and boundary conditions we find that the one that would have been affected is Eq. (14.24), since the potential applied to the interface would not be constant. This problem has been dealt with in the literature: the result is significantly more complicated than the equations given here and is not of general interest. We raised the point only to show that one should always be aware of *implicit* assumptions, which could lead to erroneous results under certain experimental conditions.

Solving the diffusion equation under the foregoing initial and boundary conditions yields the concentration profile near the electrode surface, as a function of time.

$$c\,(x,t) = c_b\,\mathrm{erf}\left[\frac{x}{\sqrt{4Dt}}\right] \tag{14.30}$$

Plots of the dimensionless concentration c/c_b as a function of distance at different times are shown in Figure 14.7. The gradual development of the Nernst diffusion layer with time can be clearly seen. The concentration profile very close to the surface is linear, but a deviation from linearity is observed farther away, as the concentration approaches its bulk value.

14.2.1.2 Potential Step, High Overpotential Region (Chrono-amperometry)

Our second example is also a potential step experiment. Here, however, it is assumed that the kinetics of the reaction is slow, and equilibrium is not maintained at the interface. Moreover, we assume that the potential range studied is far from equilibrium, so that only the forward reaction need be considered. In the original literature this condition was referred to as the *totally irreversible case*, a term which we consider to be rather misleading, because the course of the reaction can be reversed, if a potential step in the opposite direction is applied.

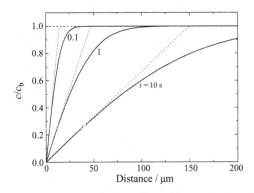

Figure 14.7 Evolution of the concentration profile with time near the electrode surface, just after the potential has been stepped to the limiting current region. $D = 7 \times 10^{-6}\,\mathrm{cm^2\,s^{-1}}$

The first five initial and boundary conditions of the differential equation listed above remain unchanged, only the sixth (Eq. (14.24)) is different. The new boundary condition is obtained by relating the current density to the heterogeneous rate constant of the forward reaction k_h and to the flux of reactant at the electrode surface:

$$\frac{j}{nF} = k_h\, c(0, t) = D\left(\frac{\partial c(0, t)}{\partial x}\right) \qquad (14.31)$$

The solution of the diffusion equation yields the current density as a function of time and potential, as:

$$j(t) = nFk_h c_b \exp\left(\lambda^2\right)\mathrm{erfc}(\lambda) \qquad (14.32)$$

in which the dimensionless parameter λ is given by

$$\lambda \equiv k_h\sqrt{t/D} \qquad (14.33)$$

The heterogeneous rate constant k_h depends on potential exponentially, following a Tafel-like relationship:

$$k_h = k_h^0\left[-\frac{\alpha F(E-E^0)}{RT}\right] \qquad (14.34)$$

Combining Eq. (14.32) with the expression for the diffusion-limited current density, which we have obtained earlier (cf. Eq. (14.27)), and with Eq. (14.33) we have:

$$\frac{j}{j_d} = F_1(\lambda) = \pi^{1/2}\lambda \exp\left(\lambda^2\right)\mathrm{erfc}(\lambda) \qquad (14.35)$$

The function $F_1(\lambda)$, which has been tabulated in detail in the literature, can be used to obtain λ for any given ratio of j/j_d. In this way k_h, which is proportional to the *activation-controlled* current density, can be evaluated as a function of potential. A plot of $\ln k_h$ (or of $\ln \lambda$) vs. E is equivalent to the traditional Tafel plot, in which $\ln j_{ac}$ is plotted versus E or versus η.

What are the limitations imposed on the validity of Eq. (14.32) and (14.33)? In writing the sixth boundary condition we made the assumption that the reaction is first order with respect to the reactant. This is a serious limitation, because in a study of the mechanism of electrode reactions, the reaction order is one of the quantities we wish to determine experimentally. Obviously the values of k_h obtained from Eq. (14.35) in electrolytes containing different concentrations of the reactant cannot be used to evaluate the reaction order, since this equation is valid only if the reaction order is unity.

14.2.1.3 Current Step (Chronopotentiometry)

Our third example, in which a current step is applied, could be called a galvanostatic experiment, in the sense that the current, rather than the potential, is the externally controlled parameter.

The first five initial and boundary conditions of the diffusion equation remain unaltered, and it is again the sixth that must be changed, to make the result applicable to this particular experimental technique. Since the current is externally controlled, one controls, in effect, the flux at the electrode surface. This is expressed mathematically by the equation

$$\frac{j}{nF} = -D\left[\frac{\partial c(0,\ t)}{\partial x}\right] \tag{14.36}$$

Note that the heterogeneous rate constant k_h is not part of this equation, because the reaction is *forced* to proceed at a rate determined by the applied current.

Solving the diffusion equation, one obtains the concentration as a function of time

$$\frac{c}{c_b} = 1-(t/\tau)^{1/2}\left[\exp(-\chi^2)-\pi^{1/2}\chi\ \mathrm{erfc}(\chi)\right] \tag{14.37}$$

where the dimensionless distance parameter, χ, is given by

$$\chi = x/(4Dt)^{1/2} \tag{14.38}$$

and τ is the *transition time*, defined by

$$\tau^{1/2} \equiv \frac{n\,F\sqrt{\pi\,D}}{2j}c_b \tag{14.39}$$

This is the *Sand equation*, derived in 1901. The transition time τ is the time taken for the surface concentration of the reacting species to be reduced to zero.

Concentration profiles calculated from Eq. (14.37) for different times are shown in Figure 14.8. The important thing to note in this figure is that in a galvanostatic experiment the flux at the surface is constant[3] but the concentration decreases with time. In contrast, in a potentiostatic experiment, the surface concentration is held constant, but the gradient of concentration at the surface, $-D(\partial c/\partial x)_{x=0}$ decreases with time, as seen in Figure 14.7.

A chronopotentiometric experiment must be performed in quiescent solutions. Thus, while the concentration at the surface declines to zero, the bulk concentration is unchanged, as prescribed by the boundary conditions (Eq. (14.22)). The meaning of the transition time, τ, may be clarified by considering Figure 14.9, which displays a transient obtained for a reversible reaction. The corresponding equation describing the variation of potential with time is

$$E = E_{1/4} + \ln\left(\frac{\tau_{1/2}}{t_{1/2}}-1\right) \tag{14.40}$$

For a slow reaction, the shape of the transient is different, but the transition time τ is independent of the kinetics of the reaction, because application of a constant

---

3) The flux is proportional to the gradient of concentration at the electrode surface, namely to $(\partial c/\partial x)_{x=0}$, which is constant in this type of experiment.

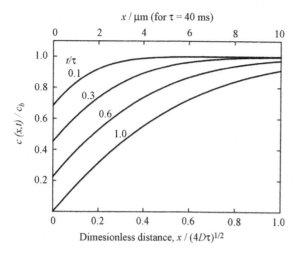

Figure 14.8 Development of the concentration profile with time during a constant-current (chronopotentiometric) transient. $\tau = 40$ ms, $D = 6 \times 10^{-6}$ cm^2 s^{-1}. The distance, x, is given in dimensionless form at the bottom and in micrometers at the top.

current forces the reaction to proceed at the same rate, irrespective of its heterogeneous rate constant.

The quarter-wave potential $E_{1/4}$ used in Eq. (14.40) is equal to the polarographic half-wave potential and is therefore characteristic of the electroactive species in solution.

14.2.2
Open-Circuit-Decay Transients

Whereas the charge-injection method is a small-amplitude perturbation method, in which measurement is conducted during open-circuit decay, we now discuss a

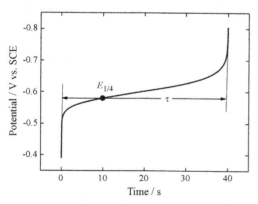

Figure 14.9 Chronopotentiometric transient ($j = $ const.) for a reversible reaction, showing the meaning of τ and of $E_{1/4}$. ($\tau = 40$ ms; $j = 0.16$ mA cm^{-2}; $c_b = 2$ mM Cd^{2+} in 1.0 M KNO$_3$).

different open-circuit measurement, in which the initial overpotential is high, in the linear Tafel region. The equations we need to solve are similar to Eqs.(14.14) and (14.15), except that the value of the current density in Eq. (14.15) is that corresponding to the linear Tafel region, namely

$$j = j_0 \exp{(\eta/b)} \tag{14.41}$$

Combining with Eq. (14.14) this yields

$$j = j_0 \exp(\eta/b) = -C_{dl}(d\eta/dt) \tag{14.42}$$

Assuming that C_{dl} is independent of potential in the range of interest, we can write

$$\int_{\eta_0}^{\eta} \exp\left(-\frac{\eta}{b}\right) d\eta = -\frac{j_0}{C_{dl}} \int_0^t dt \tag{14.43}$$

The justification for writing these equations is that one assumes that the over-potential depends on the current density in the same manner during external polarization and on open circuit. This must be so, because all one is really saying is that the potential developed across the faradaic resistance depends only on the current flowing through it, not on the source driving this current. During open circuit the current flowing is a *real* current, associated with the discharge of the capacitor, although it cannot be detected in the external circuit[4].

Integrating Eq. (14 43) we arrive at an expression of the form

$$\eta = a - b\ln(t + \tau) \tag{14.44}$$

in which

$$a = -b\ln\left(\frac{j_0}{b\,C_{dl}}\right) \quad \text{and} \quad \tau = \frac{b\,C_{dl}}{j_0}\exp\left(-\frac{\eta_0}{b}\right) \tag{14.45}$$

Thus, the Tafel slope can be determined from the slope of the open-circuit decay curve, once the parameter τ is known. The latter is found by trial and error, as the number that yields the best straight line on a semi-logarithmic plot of η vs. $\ln(t + \tau)$. This line must merge with the plot of η versus $\ln t$ at long times, when $t \gg \tau$.

If the current–potential relationship can be determined experimentally and compared to the open-circuit-decay behavior, the validity of the assumption that the capacitance is independent of potential can be tested. Indeed, under these conditions the capacitance can readily be found from the open-circuit-decay curve, with the use of Eq. (14.14), by determination of the slope $d\eta/dt$, as a function of η, and employing the value of j corresponding to each overpotential.

We might ask ourselves why is it that the results in the two cases of open-circuit decay are so different? For small transients we found that $\ln \eta$ is proportional to t

---

4) This is similar to the corrosion current at open circuit, associated with the mixed potential

(cf. Eq. (14.17)), whereas for large transients we find η proportional to $\ln(t + \tau)$. The answer is very simple. In the former case, the Faradaic resistance is taken to be a constant, independent of overpotential, while in the latter it is an exponential function of potential. The situation can become even more complicated if there is a strong dependence of the capacitance on potential. This is the case when the coverage by adsorbed intermediates is high and a large adsorption pseudocapacitance is involved. A large pseudocapacitance gives rise to a slow decay of potential with time at open circuit. In fact, such a plateau on the open-circuit-decay plot is a good indication of significant coverage by adsorbed intermediates.

14.3
Microelectrodes

14.3.1
The Unique Features of Microelectrodes

So far we have restricted our discussion of diffusion-controlled processes to the case of *semi-infinite linear diffusion*, which corresponds to a planar electrode in a cell where the solution extends to infinity. It has already been pointed out that the word "infinity" should not frighten us, since the dimension should be "infinitely large" only compared to the Nernst diffusion-layer thickness, δ, which is typically less than 0.1 cm. There are always edge effects at the periphery of the electrode, but their influence on the observed current–potential relationship is usually negligible, as long as the characteristic length of the electrode (the radius of a circle or the shorter side of a rectangle) is large compared to δ. As one decreases the size of the electrode, edge effects become more pronounced. Eventually, when a microelectrode is considered, edge effects become predominant. A microelectrode can be considered to be "all edge". This depends, of course, on the actual size of the microelectrode and on the time scale used. For $r = 10\ \mu m$ the Nernst diffusion layer thickness will be equal to the radius after about 30 ms. For an ultra-microelectrode of $r = 0.25\ \mu m$, the same is true in less than 20 μs.

The situation at a miniature disc microelectrode embedded in a flat insulator surface (such as an RDE of very small size) can be approximated by spherical symmetry, obtained for a small sphere situated at the center of a much larger (infinitely large, in the present context) spherical counter electrode. How will the change of geometry influence the diffusion-limited current density? This is shown qualitatively in Figure 14.10.

For a planar electrode the diffusion limited current density decreases with $t^{-1/2}$. In the spherical configuration the electroactive material diffuses to each segment on the surface from a cone of given solid angle. Thus, while δ increases with time, the cross-section for diffusion increases with δ^2 As a consequence, the bottleneck for the rate of diffusion is very close to the surface, (at approximately $\delta/r \leq 5$, leading to a diffusion limited current density that becomes independent of time, as we shall see below.

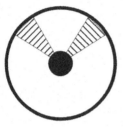

Figure 14.10 Schematic representation of planar and spherical geometry for diffusion (the latter is shown in two dimensions, for simplicity). Hatched areas represent lamina and cones from which electroactive material can diffuse to the surface.

The effect of geometry on the resistivity can be understood in a similar way. The real bottleneck is very close to the surface, where the resistance is high and the cross-section for conductance (which is the equivalent of the flux in the case of diffusion), is small. Farther out, the cross-section increases with the square of the distance. Hence the contribution of this region to the total resistance soon becomes negligible, as seen in. Figures 3.3 and 3.4.

14.3.2
Enhancement of Diffusion at a Microelectrode

The response of a spherical electrode to a potential-step function in the limiting current region is given by

$$j_d = nFDc_b \left[\frac{1}{(\pi Dt)^{1/2}} + \frac{1}{r} \right] \tag{14.46}$$

At long times, spherical diffusion is predominant, the current becomes independent of time, and Eq. (14.46) takes the form

$$j_d = \frac{n\,FDc_b}{r} \tag{14.47}$$

Thus, in a spherical field of diffusion (which is achieved for a microelectrode after a time determined by its radius), one obtains an equation similar to that given for semi-infinite linear diffusion, except that the radius of the electrode plays the role of the Nernst diffusion-layer thickness and the limiting current density is independent of time. The validity of Eq. (14.47) is one of the incentives for fabricating ultra-microelectrodes. Thus, for example, the limiting current obtained on an RDE operated at 10^4 rpm can be equalled by the current at a microelectrode of $r = 5\,\mu m$ in a quiescent solution. Such a device is relatively easy to fabricate and would not be considered to be an *ultra*-microelectrode. Smaller electrodes, having a radius of $0.25\,\mu m$ have been prepared in several laboratories. Using Eq. (14.47) we note that the limiting current density at such an electrode is about $0.8\,A\,cm^{-2}$ when $n = 2$ and $c_b = 10\,mM$. Such limiting current densities, which cannot be reached at steady state by any other method, substantially increase the range over which the current–potential relationship can be observed under activation-controlled conditions, as discussed in Section 14.1.1.

Another advantage of having a very large limiting current density at steady state is in the analysis of trace elements. Using Eq. (14.47) for the same-sized electrode, we obtain a current density of about $8\,\mu A\,cm^{-2}$ for a concentration of $0.01\,ppm$, (assuming a molecular weight of 100 for the purpose of this calculation). Thus, measurements in the ppb (part per billion) range should be possible with ultra-microelectrodes.

When an ultra-microelectrode is used as an electroanalytical tool, the diffusion-limited current density is not affected by the rate of flow, which typically creates a diffusion layer thickness of the order of $50-100\,\mu m$. This can be a great asset for on-line monitoring in industrial applications, where the flow rate may fluctuate and would otherwise have to be measured and corrected for.

14.3.3
Reduction of Solution Resistance

The jR_S potential drop due the uncompensated solution resistance associated with different geometries was discussed in Section 3.2.3. For a spherical electrode, which is of interest here, we can write

$$E_{jR_S} = j\frac{d}{\kappa}\left[\frac{r}{r+d}\right] \tag{14.48}$$

where κ is the specific conductivity and d is the distance of the probe from the electrode surface. Of interest to us here is the limiting form of this equation, for large values of d ($d/r \gg 1$), namely, far away from the electrode. In this case Eq. (14.48) is simplified to

$$jR_S = j\frac{r}{\kappa} \tag{14.49}$$

showing that the resistance (in units of $\Omega\,cm^2$ i.e., normalized for unit surface area), is proportional to the radius of the ultra-microelectrode.

As usual, a numerical example might help to illustrate the advantage of ultra-microelectrodes, from the point of view of solution resistance. In Section 14.3.2 we obtained a limiting current density of $0.8\,A\,cm^{-2}$ for an electrode having a radius of $0.25\,\mu m$, in a $10\,mM$ solution of the reactant. If we assume a specific conductivity of $\kappa = 25\,mS\,cm^{-1}$, the solution resistance R_S, according to Eq. (14.49), is $1 \times 10^{-3}\,\Omega\,cm^2$. We assumed here a solution of medium specific conductivity, and yet arrived at an ohmic potential drop of less than 1 mV at a very large current density of $0.8\,A\,cm^{-2}$. Thus, using an ultra-microelectrode extends the range of measurable current densities because (i) the limiting current density is inversely proportional to the radius and (ii) the uncompensated resistivity is proportional to the radius. More concisely, we could say that both the *diffusion-limited current density* and the *conductivity* are inversely proportional to the radius.

14.3.4
The Choice Between Single Microelectrodes or Ensembles of Thousands of Microelectrodes

Microelectrodes have many advantages over regular-sized electrodes, but they also have two major disadvantages. First, since the electrode is very small, the total current flowing in the circuit is minute and may be difficult to measure accurately. The ultra-microelectrode just discussed has a total surface area of about $2 \times 10^{-9}\,cm^2$. For a current density of $1.0\,mA\,cm^{-2}$, the total current observed is hence only $2 \times 10^{-12}\,A$. Although this is measurable in the research laboratory environment, the measurement is by no means easy, and accuracy is limited. This makes the use of ultra-microelectrodes for routine applications, particularly in industrial environments, impractical. The second disadvantage entails the extremely high volume-to-surface ratio, which makes it impossible to purify solutions to a level that would ensure that impurities could not accumulate on the surface during measurement, as discussed in Section 7.1. Thus, for a regular electrode, one may have a V/A ratio of $10\,cm^3/cm^2$. For a thin-layer cell this ratio could be as low as $10^{-3}\,cm^3/cm^2$, while for an ultra-microelectrode the ratio is of the order of $10^8\,cm^3/cm^2$. This is an inherent difficulty, which cannot be overcome by improved instrumentation. These numbers mean that, given a desired level of purity *of the surface*, the allowed level of impurities in solution would have to be nine orders of magnitude lower for the ultra-microelectrode than for a regular electrode. Bearing in mind that proper electrochemical measurements must be conducted in highly purified solutions, even when electrodes of macroscopic dimensions are employed, the level of purity needed for work with a single ultra-microelectrode is clearly not achievable.

One way around this problem, which retains most of the advantages of ultra-microelectrodes while largely overcoming their disadvantages, is to use ensembles (sometimes also referred to as arrays) of ultra-microelectrodes. Suppose that the surface of an insulator is dotted with a regular array of conducting spots, serving as the ultra-microelectrodes, which are all connected at the back to a common current collector, as shown in Figure 14.11.

The fraction of the surface that is active is given by $(d/L)^2$, where d is the diameter of each electrode and L is the distance between their centers. Designing such an ensemble of microelectrodes, one must compromise between the desire to make the ratio d/L as small as possible (to decrease the overlap between the diffusion fields of the individual electrodes), and the desire to make d/L as large as possible (to increase the total active area). Values of d/L in the range 0.03–0.1, corresponding to 0.1%–1% of active area, seem to be a reasonable choice, as we shall see.

Side view   Top view

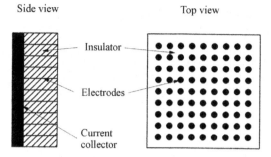

Figure 14.11 An ensemble of ultra-microelectrodes. Note that the distance between the electrodes is large compared to their diameter

Interestingly, the problem just described was first solved for an entirely different physical situation, referred to as a *partially blocked electrode*. In this case one assumes that the surface of a regular macroscopic electrode is partially covered by some non-conducting material, which could be an oxide formed electrochemically or an impurity sticking to the surface. The physical situation is the same, whether a large part of the surface has become inactive accidentally, by the accumulation of some impurity, or whether it was made inactive by design.

The solution of the diffusion equation is best obtained by digital simulation. It should be noted here that Eq. (14.46) is applicable for a spherical electrode, while in most cases microelectrodes are imbedded in a matrix of isolating material, exposing only a flat surface to the solution. Thus, a hemispherical diffusion field is probably a better approximation, but this would only introduce a small numerical factor, and has no bearing on the general behavior of single or multiple microelectrodes. The numerical calculations take this shape effect into consideration, of course.

Fortunately, one can understand the behavior of such ensembles qualitatively, and the conclusions reached in this way are in good agreement with the results of numerical calculations.

We can discuss this problem in terms of the ratio between the Nernst diffusion-layer thickness given by $\delta = (\pi D t)^{1/2}$, and the radius of the electrode on the one hand, and that between δ and the distance between two electrodes, on the other hand. To do this, we shall list the various possibilities, and derive the corresponding behavior qualitatively.

1) For $\delta/r \leq 3$ the system is in the range of semi-infinite linear diffusion. The current, per unit of total surface area, including the non-conduction matrix, is given by:

$$I_d = \left[nFDc_b / (\pi\, Dt)^{1/2} \right] (r/L)^2 \tag{14.50}$$

Where the term $(r/L)^2$ represents the fraction of the geometrical surface area that is electrochemically active (the rest being the insulator into which the microelectrodes are imbedded).

2) For $\delta/r \geq 3$ but $\delta/L \leq 0.3$, the diffusion field around each electrode is nearly spherical and the overlap between the diffusion fields of neighboring electrodes is still negligible. The diffusion-limited current, per unit of total surface area, is given approximately by

$$I_d = (nFDc_b/r)(r/L)^2 \tag{14.51}$$

3) For $\delta/L \geq 3$, complete overlap between the diffusion fields of the individual microelectrodes can be assumed. The total current is given in this case by

$$I_d = \frac{nFD}{\sqrt{\pi Dt}} c_b \tag{14.52}$$

Note that in the last three equations I_d represents the total current per cm^2 of the ensemble of microelectrodes. The current density at the active sites for the case of total overlap is larger by a factor of $(L/r)^2$ than at short times, when Eq. (14.50) applies. The situation is similar to the case of rough electrodes, where the geometrical surface area should be multiplied by the roughness factor, to obtain the real surface area, except that in the present case the roughness factor is expressed by $(r/L)^2$ and is much less than unity.

The development of the diffusion field at an ensemble of microelectrodes is shown schematically in Figure 14.12.

Equation (14.51) corresponds to spherical diffusion to each ultra-microelectrode, with negligible overlap between the individual electrodes. This is the region in which the total limiting current is nearly independent of time. Whether such a region is or is not observed in practice depends on the design of the ensemble, that is, on the ratio (δ/L), since for this equation to hold it is required that $r \ll \delta \ll L$.

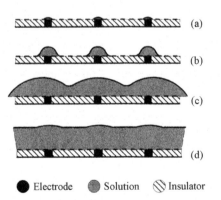

● Electrode ● Solution ◇ Insulator

Figure 14.12 Development of the diffusion field near the surface of an ensemble of microelectrodes. (a) semi-infinite linear diffusion; (b) spherical diffusion with no overlap; (c) spherical diffusion with substantial overlap; (d) total overlap, equivalent to planar diffusion to the whole surface.

14.3.5
Shapes of Microelectrodes and Ensembles

We have considered so far only disc-shaped microelectrodes, for which spherical diffusion can be applied, to a good approximation. Other forms have been used, mainly because they might be easier to fabricate. Most noted among these is the linear or strip microelectrode, which is *macro*scopic in length but *micro*scopic in width. The diffusion field at such electrodes can be approximated satisfactorily by diffusion to a cylinder. The enhancement of diffusion is less than that for a disc microelectrode, of course, but the increase in surface area alleviates, to some extent, the problems of very low total currents to be measured and the exceedingly large volume-to-surface ratio.

An entirely different class of microelectrodes consists of electrodes used in biological and medical research, mostly for application *in vivo*. In this case the small size enables the researcher to introduce the electrode into the living organism with minimum damage to the tissue being studied, and to study local effects on the scale of living cells. The electrochemical properties of microelectrodes, namely enhancement in the rate of diffusion and decrease in the resistance, are not the main issue in such uses, although they should by no means be ignored in the design of the microelectrode and in the evaluation of its response.

15
Experimental Techniques (2)

15.1
Linear Potential Sweep and Cyclic Voltammetry

15.1.1
Three Types of Linear Potential Sweep

Linear potential sweep is a potentiostatic technique, in the sense that the potential is the externally controlled parameter. The potential is changed at a constant rate

$$v = \frac{dE}{dt} \tag{15.1}$$

and the resulting current is followed as a function of time.

In most cases the potential is swept forward and backward between two fixed values, a technique referred to as *cyclic voltammetry* (CV). In this way the current measured at a particular potential on the anodic sweep (going from negative to positive potentials) can readily be compared with that measured at the same potential on the cathodic sweep, (going from positive to negative potentials). A typical cyclic voltammogram is shown in Figure 15.1.

Linear potential sweep measurements are generally of three types:

1) Very slow sweeps

When the sweep rate is very low, in the range of $v = (0.1-5)\,mV\,s^{-1}$, measurement is conducted under quasi-steady-state conditions. The sweep rate plays no role in this case, except that it must be slow enough to ensure that the reaction is effectively at steady state along the course of the sweep. This type of measurement is widely used in corrosion and passivation studies, as we shall see, and also in the study of some fuel cell reactions in stirred solutions. Reversing the direction of the sweep should have no effect on the current–potential relationship, if the sweep is slow enough. Deviations occur sometimes as a result of slow formation and/or reduction of surface oxides or passive layers. Because the sweep rate is slow, the potential is often swept only in one direction, and the experiment is then referred to as *linear sweep voltammetry* (LSV).

Physical Electrochemistry: Fundamentals, Techniques and Applications. Eliezer Gileadi
Copyright © 2011 WILEY-VCH Verlag GmbH & Co. KGaA, Weinheim
ISBN: 978-3-527-31970-1

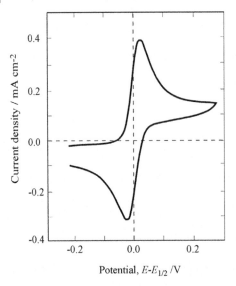

Figure 15.1 A simple cyclic voltammogram, $v = 0.01 \, V \, s^{-1}$; 3 mM FeSO$_4$ and 1.0 M H$_2$SO$_4$.

2) Studies of oxidation or reduction of species in the bulk of the solution

In the second case, the sweep rate is usually in the range of $(0.01-10) \, V \, s^{-1}$. The lower limit is determined by the need to maintain the total time of the experiment below 10–50 s (before mass transport by natural convection becomes significant). The upper limit is determined by the double-layer charging current and by the uncompensated solution resistance, as discussed in Section 15.2 below.

3) Studies of oxidation or reduction of species adsorbed on the surface

The redox behavior of species that are adsorbed on the surface is usually activation controlled and influenced by the remaining number of free sites on the surface. Hence, this may be conducted in stirred solutions. The typical sweep rates are also in the range $(0.01-10) \, V \, s^{-1}$, but here the lower limit is determined by background currents from residual impurities in solution (and perhaps by the desire of the experimenter to collect more data in a given time) while the upper limit is determined by the uncompensated solution resistance and by instrumentation. The Faradaic current associated with this process is proportional to the sweep rate, and so is the double-layer charging current, so that the relative effect of j_{dl} on the measured current is independent of sweep rate.

Although the amount of material adsorbed in a monolayer is very small (of the order of $(1-2)$ nmol cm^{-2}, mass transport can be a limiting factor under certain circumstances, when the bulk concentration of the adsorbate is low (typically below a few mM and/or when the sweep rate is high.

In this chapter we shall discuss only the second and third cases, since very slow sweeps are just a convenient method for scanning the potential automatically, under conditions in which the sweep rate or its direction have practically no effect on the current observed.

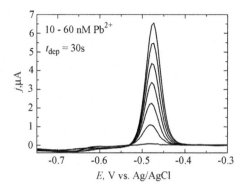

Figure 15.2 Subtractive anodic stripping voltammetry (SASV) following deposition for 30 s on a rotating disc Ag electrode, from solutions containing (10–60) nM Pb.

Supporting electrolyte: 10 mM HNO₃ + 10 mM HCl. Data from Y. Bonfil and E. Kirowa-Eisner, *Anal. Chim. Acta* **457** (2002) 285.

Although this book does not deal with electroanalytical chemistry it is interesting to mention here the method of anodic striping voltammetry, (ASV) employed for the study of trace concentrations of metallic elements, such as Pb, Cd and Cu.

This is a pre-concentration technique, in which the metal is deposited under conditions of mass-transport control on a rotating-disk electrode for a relatively long time, (10–100) s, followed by an anodic linear sweep at a relatively high sweep rate, during which it is removed. An anodic peak is observed, from which the total charge and hence the concentration of the trace element can be evaluated. This method is most useful when deposition occurs in the UPD region and accurate results can be obtained even when the fractional surface coverage reached during the deposition stage is less than 0.1%, allowing determination of the concentration in solution in the nanomolar region. An example of data in anodic stripping of Pb is shown in Figure 15.2.

The anodic sweeps observed in this figure for solutions containing (10–60) × 10^{-9} M solution of Pb^{2+} were actually acquired by using a slight modification of ASV, referred to as subtractive anodic sweep voltammetry (SASV), in which the background current is measured immediately after the anodic sweep, and is subtracted from the current during the anodic sweep itself.

15.1.2
Double-Layer-Charging Currents

We cannot discuss the CV and LSV methods properly without appreciating the importance of the double-layer-charging current in this type of experiment. We use the simple equation:

$$j_{dl} = C_{dl}(dE/dt) = C_{dl}\,v \tag{15.2}$$

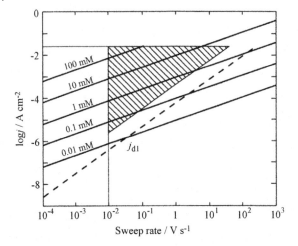

Figure 15.3 The optimum range of concentration and sweep rate (hatched area) for measurements in cyclic voltammetry on a smooth electrode. $n = 1$. The double-layer-charging current j_{dl} was calculated for $C_{dl} = 20\,\mu F\,cm^{-2}$.

If double-layer charging is the only process taking place in a given potential region, the only current observed will be j_{dl}. Such measurements can be used to determine the value of C_{dl}, as discussed in Section 8.1.6 and shown in Figures 8.7 and 8.8. In addition to the determination of C_{dl}, it is also important to determine the numerical value of j_{dl}, since this represents an error term, that must be corrected for in the analysis of the current–potential relation during CV. For a very slow sweep experiment using, for example, $v = 1\,mV\,s^{-1}$ and $C_{dl} = 20\,\mu F\,cm^{-2}$ a value of $j_{dl} = 20\,nA\,cm^{-2}$ is obtained, which is negligible, even with respect to the small currents observed on passivated electrodes, and can, hence, be ignored. However, cyclic voltammetry is often conducted at higher sweep rates, where j_{dl} can no longer be ignored.

The interplay between double-layer charging and oxidation or reduction of an electroactive material in the bulk of the solution is illustrated in Figure 15.3. The Faradaic currents shown by the solid lines are the peak currents j_p calculated according to Eq. (15.9), which are discussed in Section 15.2 below.

It is important to note that the Faradaic currents change with $v^{1/2}$, while j_{dl} is proportional to v. As a result, double-layer charging becomes more important with increasing sweep rate. The hatched area in Figure 15.3 represents the region of sweep rates and concentrations in which accurate measurements can be made, on the basis of the following assumptions:

1) The peak current of the Faradaic process being studied should be at least ten times larger than the double-layer-charging current, in order to allow reliable correction for the latter.
2) The sweep rate should be $v \geq 10\,mV\,s^{-1}$, chosen so as to limit errors due to natural convective mass transport;

3) The peak Faradaic current should be $j_{peak} \leq 20$ mA cm^{-2}, in order to limit the error due to the uncompensated solution resistance.
4) The bulk concentration should be $c_b \leq 100$ mM, to ensure that it will always be possible to have an excess of supporting electrolyte, to suppress mass transport by migration.

The numerical values used to construct Figure 15.3 are somewhat arbitrary. They do represent, however, the right order of magnitude, and are probably correct within a factor of two or three. The important thing to learn from Figure 15.3 is the type of factors which must be considered in setting up an experiment. One could readily construct a similar diagram using somewhat different assumptions, but the conclusions would not be fundamentally different. Thus, one notes that it is difficult to make measurements at concentrations below 0.1 mM or at sweep rates above 100 V s^{-1}. The best concentration to use is in the range (1–10) mM. Interestingly, this conclusion does not depend strongly on the four assumptions made above, since it refers to the "middle of the field" of applicability of the method.

An interesting point to consider is the effect of surface roughness on the range of applicability of the linear potential sweep method. The double-layer-charging current j_{dl} in Figure 15.3 is calculated for $C_{dl} = 20$ µF cm^{-2}, which is a relatively low value for solid electrodes, implying a highly polished surface with a roughness factor of 1.5 or less. If the roughness factor is increased, for example, by using platinized platinum instead of bright platinum, the charging current could increase by a factor of typically 20–100, while the diffusion current remains essentially unchanged. Using corresponding values of j_{dl} in Figure 15.3, we note that the range of applicability is seriously limited – to concentrations above 1 mM and to sweep rates below a few volts per second. Thus, linear-sweep voltammetry should be conducted, whenever possible, on smooth electrodes. The application of this method to the study of porous electrodes, of the type used in fuel cells and in metal–air batteries, is rather limited.

We should perhaps conclude this section by noting that the above considerations are valid for large electrodes, operating under semi-infinite linear diffusion conditions. If ultra-microelectrodes are employed, the diffusion rate is higher and the uncompensated solution resistance is significantly lower, increasing the range of measurement by an extent which depends on the radius of the microelectrodes, as seen in Section 14.3. Indeed, one can find examples in the literature where sweep rates as high as $v = 1 \times 10^6$ V s^{-1} were applied. Although this may be valid, from the points of view of mass transport and solution resistance, one should keep in mind that the double-layer charging current density may become very large at such high sweep rates.

15.1.3
The Form of the Current–Potential Relationship

We have seen a typical j/E curve obtained in a CV experiment in Figure 15.1. This particular curve was observed in a solution containing 3 mM Fe^{2+} ions in 1 M H$_2$SO$_4$. Thus, the anodic peak corresponds to the simple reaction

$$Fe^{2+} \rightarrow Fe^{3+} + e_M^-$$

(15.3)

in which both reactant and product are stable species and both are soluble. The initial concentration of Fe^{3+} is zero, but at the peak current its *surface concentration* is very close to the bulk concentration of Fe^{2+}. As a result, the cathodic reduction peak is nearly equal to the anodic oxidation peak, as long as the product of the cathodic reduction is stable in solution.

Why is a peak observed in this type of measurement? In the experiment shown in Figure 15.1 the sweep is started at a potential of -0.2 V versus the standard potential for the Fe^{2+}/Fe^{3+} couple, where no Faradaic reaction takes place.[1] At about -0.05 V, the anodic current starts to increase with potential. Initially, the current is activation controlled, but as the potential becomes more positive, diffusion limitation sets in. The observed current is the inverse sum of the activation and the diffusion limited currents, (cf. Eq. (1.5)). As time goes on, the activation-controlled current increases (due to the increase in potential with time) but the diffusion-controlled current decreases. As a result, the observed current increases first, passes through a maximum and then decreases. At higher sweep rates each potential is reached in a shorter time, when the effect of diffusion limitation is less, hence the peak current is found to increase with sweep rate.

In CV the potential is made to change linearly with time between two set values. Often the current during the first cycle is quite different from that in the second cycle, but after 5–10 cycles the system settles down, and the current traces the same line as a function of potential, independent of the number of cycles. This is often referred to as a "steady-state voltammogram", which is an odd name to use, considering that the current keeps changing periodically with potential and time. The term is used here in the sense that *the voltammogram as a whole* is independent of time. This is a very nice experiment to perform, because it tends to be highly reproducible. Moreover, the voltammograms are stable over long periods of time, sometimes in the range of hours. Interpretation of the data is another matter. Using the results for *qualitative* detection of reactions taking place in a given range of potential is fine, but quantitative treatment, using the equations developed for a single linear potential sweep is wrong because (i) the initial and boundary conditions of the diffusion equation have change and (ii) convective mass transport may already play an important role.

15.2
Solution of the Diffusion Equations

The boundary conditions for solving the diffusion equation for linear potential sweep are really the same as those written for the potential step experiment, as discussed in Section 14.2. because in both cases the potential is the externally controlled parameter. As before, we can distinguish between the reversible case, in which it is assumed that the concentrations at the surface are determined by the potential via

---

1) In fact, there is a small cathodic current flowing, which must be attributed to the reduction of some impurity, but this is of no interest here.

the Nernst equation (cf. Eq. (14.24)) and the high-overpotential region, where the specific rate constant is related to the surface concentration $c(0,t)$ and to the flux at the surface, as given by Eq. (14.31). There is one important difference, however. During a potential-step experiment, the potential is assumed to be constant throughout the transient, while in CV it is assumed to change linearly with time, following the simple equation

$$E = E_{in} \pm vt \tag{12.2}$$

where E_{in} is the initial potential at $t = 0$, which is chosen in a range where no Faradaic reaction takes place. Incorporating this equation into the appropriate boundary condition will then produce a result which is applicable to the CV experiment.

The change in the boundary conditions complicates the mathematics, to the point that an explicit algebraic solution for the whole j/E curve has not been found. Numerical solutions are, however, available for the different cases. The discussion here is limited to the coordinates of the peak, namely to the values of the peak current j_p and the peak potential E_p, as a function of sweep rate and the kinetic parameters of the reaction involved.

15.2.1
Reversible Region

The peak potential in the case of a reversible linear potential sweep is given by the equation

$$E_{p,\ rev} = E_{1/2} \pm 1.1 \frac{RT}{n\,F} \tag{15.4}$$

where $E_{1/2}$ is the polarographic half-wave potential, which is very close to the standard potential E^0

$$E_{1/2} = E^0 + \frac{2.3RT}{n\,F} \log \left[\frac{\gamma_{Ox}}{\gamma_{Red}} \left(\frac{D_{Red}}{D_{Ox}} \right)^{1/2} \right] \tag{15.5}$$

The positive sign in Eq. (15.4) is applicable to an anodic sweep and the negative sign applies to a cathodic peak. In either case, the peak should appear at $(28/n)$ mV *after* $E_{1/2}$, in the direction of the sweep. The peak potential is independent of the sweep rate in the reversible case. This characteristic can, in fact, be used as a criterion for reversibility. It is also independent of concentration. This, however, is correct only if one considers a simple reaction, such as represented by Eq. (15.3). For a slightly more complex stoichiometry, such as

$$2\,Cl^- \rightarrow Cl_2 + 2\,e_M^- \tag{15.6}$$

the half-wave potential, and with it $E_{p,\ rev}$ depend logarithmically on concentration

$$E_{p,\ rev} = E^0 + \frac{2.3RT}{n\,F} \log \left[\frac{\gamma_{Ox}}{\gamma_{Red}} \left(\frac{D_{Red}}{D_{Ox}} \right)^{1/2} \right] + \frac{2.3RT}{nF} \log c_{b,Cl^-} \tag{15.7}$$

The peak current density for a reversible linear potential sweep is given by

$$j_{p, \text{rev}} = \left[0.44n\, F \left(\frac{n\, F}{RT} \right)^{1/2} D^{1/2} \right] c_b v^{1/2} \tag{15.8}$$

which can be written, for room temperature, as

$$j_{p, \text{rev}} = \left[2.72 \times 10^5 n^{3/2} D^{1/2} \right] c_b v^{1/2} \tag{15.9}$$

In this equation $j_{p, \text{rev}}$ and c_b are given in units of $[\text{A cm}^{-2}]$ and $[\text{mol cm}^{-3}]$ respectively.

15.2.2
High-Overpotential Region

In the high-overpotential region the peak potential depends logarithmically on sweep rate, following the equation

$$E_{p, \text{irrev}} = E_{1/2} + \frac{b}{2} \left[1.04 - \log \frac{b}{D} + \log \frac{v_c}{k_f^2} \right] \tag{15.10}$$

From a plot of $E_{p, \text{irrev}}$ versus $\log v$ one can obtain the Tafel slope, b, and from the intercept at $E_{p, \text{irrev}} = E_{p, \text{rev}}$, the specific rate constant can be calculated. Note that the value of k_f obtained in this way is that corresponding to $E = E_{\text{rev}}$ only; but since the Tafel slope is known, k_f at any other potential can readily be calculated.

The peak current density in the high-over potential region is given by

$$j_{p, \text{irrev}} = \left[3.01 \times 10^5 n\, \alpha^{1/2}\, D^{1/2} \right] c_b v^{1/2} \tag{15.11}$$

In this equation α is the transfer coefficient, which is obtained directly from the Tafel slope.

The dependence of E_p on sweep rate over a wide range is shown in Figure 15.4.

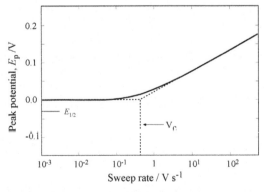

Figure 15.4 Variation of the peak potential E_p over a wide range of sweep rates, covering both the reversible and the high-overpotential regions. $E_{p,\text{rev}}$ is arbitrarily taken as zero. The "critical" sweep rate v_c is the one to be used in Eq. (15.11) to calculate k_f.

The ratio of the peak current densities in the two regions is given by

$$\frac{j_{p, \text{irrev}}}{j_{p, \text{rev}}} = 1.07 \left(\frac{\alpha}{n}\right)^{1/2} \tag{15.12}$$

Since (α/n) is usually smaller than unity, the peak current in the high-overpotential region is, as a rule, smaller than that in the reversible region. The difference is not very large, however. For $\alpha = 0.5$ and $n = 1$ the above ratio is equal to 0.78. For $\alpha = 0.5$ and $n = 2$, it is reduced to 0.55.

15.3
Uses and Limitations of LPS and CV

Linear potential sweep (LPS) and cyclic voltammetry are at their best for *qualitative* studies of the reactions occurring in a certain range of potential. In Figure 15.5, for example, we see the cyclic voltammogram obtained on a mercury-drop electrode in a solution of *p*-nitrosophenol in acetate buffer.

Starting at a potential of $+0.3$ V vs. SCE and sweeping in the negative direction, one observes the first reduction peak at about -0.1 V vs. SCE. This potential corresponds to the reduction of *p*-nitrosophenol to *p*-phenol-hydroxylamine, as shown in Eq. (15.13)

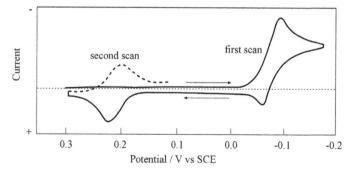

$$\qquad \text{ON} \qquad + 2H^+ + 2e_M^- \rightleftarrows \qquad \text{HNOH} \qquad \text{first sweep} \tag{15.13}$$

On the return sweep the reverse reaction is clearly seen as a peak at -0.05 V vs. SCE, but another anodic peak is observed at about $+0.22$ V vs. SCE. We did not observe a corresponding cathodic peak *in the first sweep*, but such a peak is clearly seen in the

Figure 15.5 Cyclic voltammogram in a solution of *p*-nitrosophenol in acetate buffer, on a mercury-drop electrode. Reprinted with permission from "*Instrumental Methods in Electrochemistry*" p. 199. The Southampton Group, copyright 1985, Ellis H.

second and subsequent sweeps. A chemical reaction following charge transfer must have taken place, producing a new redox couple. This has been identified as the decomposition of *p*-phenol-hydroxylamine to *p*-imide-quinone and water

$$
\text{(structure: benzene ring with OH top, HNOH bottom)} \rightarrow \text{(structure: ring with O top, NH bottom)} + H_2O \qquad \text{Hydrolysis} \qquad (15.14)
$$

which forms a redox couple with *p*-aminophenol, as shown in Eq. (15.15).

$$
\text{(structure: ring with O top, NH bottom)} + 2H^+ + 2e^-_M \rightleftharpoons \text{(structure: ring with OH top, } H_2N \text{ bottom)} \qquad \text{second sweep} \qquad (15.15)
$$

Since the two redox couples have widely different standard potentials, they can be easily detected on the cyclic voltammogram. This is an example of a reaction sequence commonly referred to as an *ece mechanism*, indicating that an electrochemical step is followed by a chemical step; which is in turn followed by an electrochemical step.

It should be obvious that the relative peak heights in Figure 15.5 depend on the sweep rate and on the homogeneous rate constant for hydrolysis. Increasing the sweep rate causes an increase in the first oxidation peak and a decrease in the second, and vice versa. Methods of evaluating the rate constants of reactions preceding or following charge transfer, from the dependence of the peak currents on sweep rate, have been described in the literature in detail and will not be discussed here.

Determining the peak current density in cyclic voltammetry can sometimes be problematic, particularly for the reverse sweep, or when there are several peaks, which are not sufficiently separated on the scale of potential. The usual way to determine the peak currents is shown in Figure 15.6. For the forward peak, the correction for the baseline is small and does not substantially affect the result. For the two reverse peaks, however, the baseline correction is quite large and may introduce a substantial uncertainty in the calculated value of the peak current density. In fact, there is no theory behind the linear extrapolation of the baselines shown in Figure 15.6 and this leaves room for some degree of "imaginative extrapolation". This is one of the weaknesses of cyclic voltammetry, when used as a *quantitative tool*, in the determination of rate constants and reaction mechanisms.

However, the greatest drawback in the *quantitative* use of linear potential sweep is due to the uncompensated solution resistance. We have already discussed this point in some detail with respect to galvanostatic and potentiostatic measurements. It was shown that if this uncompensated resistance cannot be reduced to a negligible level, (either by proper cell design or by electronic means, or preferably by both), galvanostatic measurements are, as a rule, more reliable. The reason, as

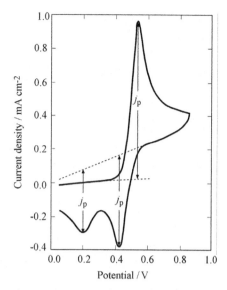

Figure 15.6 Commonly used graphical method of extrapolating the baseline, to measure the peak current densities in cyclic voltammetry.

we have demonstrated, is that in a galvanostatic measurement the experiment is conducted correctly, even if jR_S is substantial. One is then left with the challenge of measuring a small activation overpotential on top of a large (but constant) signal due to solution resistance, which is a matter of using high quality measuring instruments. In a potentiostatic experiment, the uncompensated solution resistance distorts the shape of the pulse, so that the experiment itself is not conducted under the presumed conditions, since the potential step is no longer a sharp step, as seen in Figure 14.6, a problem that cannot be corrected *post factum* by increasing the sensitivity and accuracy of the measuring equipment. Linear potential sweep, being a potentiostatic technique, suffers from the same drawback, but the problem in this case is even worse, as we shall see in a moment.

In Figures 4.3–4.5 we compared the potential applied to the interface during linear potential sweep with and without an uncompensated solution resistance. Clearly, the error is a maximum at the peak, where the current has its highest value. Just before and during the peak, the effective sweep rate imposed on the interface is much less than that applied by the instrument. The assumption that the sweep rate is constant, which was used as one of the boundary conditions for solving the diffusion equation, does not apply in this region. In this sense the experiment is no longer conducted "correctly".

The problem is aggravated by the common practice of extracting the kinetic information from the coordinates of the peak, namely from the values of j_p and E_p and their dependence on sweep rate. In other words, *the information is obtained from the point at which the error is the highest!* Moreover, this is by no means a constant error. As the sweep rate is increased, the peak current increases and the error due to jR_S

becomes worse. The effect can be quite dramatic, as seen by comparing the two cyclic voltammograms shown in Figure 4.4, which were obtained in the same solution, with and without electronic jR_S compensation. Admittedly, we have used rather extreme conditions, in which the voltammogram is highly distorted, but it should be evident that even much smaller values of jR_S might make the results of a *quantitative* analysis questionable.

We conclude this section by noting that linear potential sweep and cyclic voltammetry are excellent *qualitative* tools in the study of electrode reactions. However, their value for obtaining *quantitative* information is rather limited. The best advice to the novice in the field is *that cyclic voltammetry should always be the first experiment method used to study a new system, but never the last.*

15.4
Cyclic Voltammetry for Monolayer Adsorption

In Section 11.2 we discussed the concept of the adsorption pseudocapacitance and its dependence on potential and the fractional coverage. The phenomenon of underpotential deposition is discussed in Chapter 12, noting that UPD is a prime example in which the adsorption pseudocapacitance plays a role. Both phenomena are studied in most cases (but not exclusively) by applying cyclic voltammetry. Here we discuss the theory behind cyclic voltammetry associated with the above two phenomena

15.4.1
Reversible region

Consider a simple charge-transfer process, leading to the formation of an adsorbed intermediate, such as

$$H_3O^+ + e_M^- \rightleftarrows H_{ads} + H_2O \tag{15.16}$$

As we sweep the potential from an initial value where $\theta = 0$ to a final value where θ is essentially unity and back to the initial value, we observe a Faradaic current, associated with the formation and removal of a monolayer of adsorbed species. If we hold the potential constant at any value within this range, the current decays to zero, since the coverage is a function of potential and does not continue to change with time at constant potential. This is the behavior characteristic of a capacitor and, in fact, the current measured during the sweep, which we denote j_ϕ, is that required to charge and discharge the adsorption pseudocapacitance C_ϕ, discussed in Section 11.2. Assuming that bulk Faradaic processes (such as discussed in Section 15.2) and double-layer charging are negligible, we can write

$$j_\phi = q_1 \frac{d\theta}{dt} = q_1 \frac{d\theta}{dE} \frac{dE}{dt} = C_\phi \times v \tag{15.17}$$

This makes life easy, because we already know how C_ϕ depends on potential for different isotherms. If adsorption follows the Langmuir isotherm, the current during cyclic voltammetry is given by

$$j_\phi = \frac{q_1 F}{RT}\left[\frac{K_0\,c_b\exp(-E\,F/RT)}{[1+K_0c_b\exp(-E\,F/RT)]^2}\right] \times v \qquad (15.18)$$

which we obtained by substituting the value of C_ϕ from Eq. (11.27) into Eq. (15.17). We cannot derive an explicit form of the dependence of C_ϕ on potential for the Frumkin isotherm, but the shape of the curve can readily be obtained numerically from the dependence of C_ϕ on θ, on the one hand, and from the dependence of θ on E on the other hand. The peak current density is obtained, by combining Eq. (11.37) with Eq. (15.18).

$$j_{F,p} = \left[\frac{q_1 F}{4RT}\right] \times \left[\frac{1}{1+(f/4)}\right] \times v \qquad (15.19)$$

in which $f \equiv r/RT$ is the dimensionless parameter defining the rate of change of the standard Gibbs energy of adsorption with coverage. It is important to note that the peak current density, and indeed the current density at any value of the potential, is proportional to the sweep rate v. This makes it relatively easy to distinguish between surface-controlled processes and diffusion-controlled bulk processes, for which the peak current density is proportional to $v^{1/2}$.

The peak potential can readily be obtained from the Frumkin isotherm

$$\left[\frac{\theta}{1-\theta}\right]\exp\left(\frac{r}{RT}\theta\right) = K_0\,c_b\exp\left(-\frac{F}{RT}E\right) \qquad (11.15)$$

Bearing in mind that the maximum value of the pseudocapacitance always occurs at $\theta = 0.50$, irrespective of the value of the parameter f (cf. Figure 11.5). This yields

$$E_p = \frac{2.3RT}{F}[\log K_0c_b - f/2] \qquad (15.20)$$

Setting $c_b = 1$ and $f = 0$, we can define the standard potential for adsorption as

$$E_\theta^0 \equiv \frac{2.3RT}{F}\log K_0 \qquad (15.21)$$

Substituting this into Eq. (15.20) we have

$$E_p = E_\theta^0 + \frac{2.3RT}{F}[\log c_b - f/2] \qquad (15.22)$$

which shows that the peak potential is shifted from its value for the Langmuir isotherm by an extent which is proportional to the Frumkin interaction parameter $f = r/RT$.

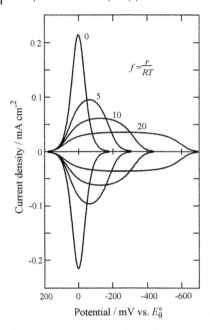

Figure 15.7 simulated cyclic voltammogram for monolayer adsorption and desorption of species formed with charge transfer. $v = 0.10 \, V \, s^{-1}$, $q_1 = 0.23 \, mC \, cm^{-2}$. The standard state is chosen, as $\theta/(1-\theta) = 1$, $c_b = 1.0 \, M$ and $f = 0$).

Cyclic voltammograms calculated for different values of the parameter f are shown in Figure 15.7. One should note that in the case of adsorbed intermediates discussed here, the peak current density is independent of concentration, while the peak potential depends on it through a Nernst-type equation. This is in contrast with the similar equations for reaction of a bulk species, where j_p is proportional to the bulk concentration (Eq. (15.9) and Eq. (15.11)), while E_p is independent of it, for a simple stoichiometry (Eq. (15.4) and Eq. (15.10)). Also, the anodic and cathodic peaks for formation and removal of an adsorbed species occur at exactly the same potential, unlike the case of reaction of a bulk species.

Let us calculate the peak current density, for comparison with the current densities resulting from other processes, which may take place simultaneously. For the Langmuir case, $C_{\phi, \, max} = 2.3 \, mF \, cm^{-2}$, which gives rise to a peak current density of $j_{\phi,p} = 2.3 \, mA \, cm^{-2}$ at $v = 1.0 \, V \, s^{-1}$, about 100 times the typical values of j_{dl} at the same sweep rate. Consequently, double-layer charging does not interfere seriously with measurement of the cyclic voltammogram. Moreover, since both j_{dl} and $j_{\phi p}$ depend linearly on sweep rate, their ratio is independent of it. Last but not least, increasing the roughness factor is advantageous in this case, for the same reason that it is a disadvantage in the case shown in Figure 15.3, since $j_{\phi,p}$ is proportional to the *real* surface area, while the diffusion limited peak current density is proportional to the *geometrical* surface area (i.e., it is essentially independent of the roughness factor).

15.4.2
High-Overpotential Region

In the reversible case the rates of adsorption and desorption are so fast that the value of θ at any moment during the transient is equal to its equilibrium value. Having the fractional surface coverage controlled totally by the potential, through the appropriate adsorption isotherm, is equivalent to having the surface concentrations of reactants and products totally controlled by the Nernst equation, which is the assumption made for the reversible cyclic voltammetry. For the case of formation of atomic hydrogen on the surface of a platinum electrode, this assumption holds up to a sweep rate of about 1.0 V s^{-1}. For the formation of OH_{ads} on the same surface, it does not hold even for a slow sweep rate of 1 mV s^{-1}. The transition from reversible to irreversible conditions is, of course, gradual. Here we shall present the equations for the *high-overpotential region*, namely the case in which only the forward reaction needs to be considered.

In this region the current associated with formation of an adsorbed species can be written as

$$j_\phi = F k\, c_b (1-\theta) \exp\left(-\frac{\beta F}{R T} E \right) \tag{15.23}$$

which is simply the expression for the forward rate of the reaction shown in Eq. (15.16). It was shown that in this case the peak current is given by

$$j_{\phi,p} = \frac{q_1 F}{5.44 R T} \times v \tag{15.24}$$

This value of $j_{\phi,p}$ is applicable to the Langmuir isotherm, and should therefore be compared with Eq. (15.19) with $f=0$. The peak in the high-overpotential region is thus somewhat lower than in the reversible region, as shown in Figure 15.8.

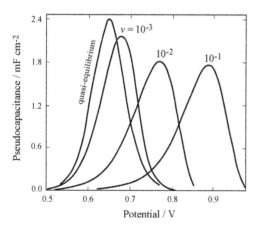

Figure 15.8 Calculated j/E plots for different sweep rates in the reversible and the high-overpotential region. $q_1 = 0.23 \text{ mC cm}^{-2}$, $f = 0$. The vertical axis is given in units of pseudo-capacitance, $C_\phi = j/v$. From S. Srinivasan and E. Gileadi, *Electrochim. Acta*, **11**, 321, (1966). Reprinted with permission from Pergamon Press.

The peak potential is shifted with sweep rate, following the equation

$$E_p = -\frac{2.3RT}{\beta F} \log\left(\frac{v}{k}\right) + \text{constant} \tag{15.25}$$

Note that the ratio between the reversible and the irreversible peak currents is $4.0/5.44 = 0.74$ which is close to the value of $1.07(\alpha/n) = 0.76$ for $\alpha = 0.5$ and $n = 1$, calculated for a redox reaction in the bulk, as given in Eq. (15.12).

16
Experimental Techniques (3)

16.1
Electrochemical Impedance Spectroscopy (EIS)

16.1.1
Introduction

The use of a phase-sensitive voltmeter for the study of the electrical response of the interface can provide highly accurate measurements of the double-layer capacitance. But this instrument has far more important uses in electrochemistry than just the measurement of capacitance. By combining a phase-sensitive voltmeter (also called a lock-in amplifier) with a variable frequency sine-wave generator, one obtains an electrochemical impedance spectrometer (EIS). This makes it possible to probe the interface over a wide range of frequencies, and to record and analyze the data. Modern instrumentation can cover a frequency range of about 12 orders of magnitude, from 10^{-5} to 10^7 Hz. This is a very wide range of frequencies indeed, when compared to other fields of spectroscopy. We recall, for instance, that visible light extends over a factor of just under two in frequency. In fact, the range of frequencies that can be used in EIS measurements is limited by the electrochemical aspects of the system, not by instrumentation. Thus, measurements at very low frequencies take a long time, during which the interface may change chemically. While it is technically possible to make measurements at, say, 10^{-5} Hz, this would take longer than a day, and the changes in the interface during the measurement at a single frequency could make the result meaningless. At the high frequency end, stray capacitances and inductances combine with possible non-uniformity of current distribution to make the results unreliable. For these reasons, EIS experiments can be conducted in the range $(10^{-3}-10^5)\,\mathrm{Hz}$, but one would be well advised to limit measurements to a more narrow range of $(1 \times 10^{-2}-5 \times 10^4)\,\mathrm{Hz}$.

For a circuit containing both capacitors and resistors, the ratio between the applied voltage signal and the resulting current signal is the impedance $Z(\omega)$, which is a

Physical Electrochemistry: Fundamentals, Techniques and Applications. Eliezer Gileadi
Copyright © 2011 WILEY-VCH Verlag GmbH & Co. KGaA, Weinheim
ISBN: 978-3-527-31970-1

function of frequency. The impedance of a pure resistor is simply its resistance R, while the impedance of a pure capacitor is given by

$$Z_C = -\frac{j}{\omega C} \tag{16.1}$$

The total impedance can be written in concise form as

$$Z(\omega) = \mathrm{Re}Z - j \times \mathrm{Im}Z \tag{16.2}$$

where $\mathrm{Re}Z$ and $\mathrm{Im}Z$ stand for the *real* and the *imaginary* parts of the impedance, respectively, and $j \equiv \sqrt{-1}$. It follows that the absolute value of the impedance vector is given by

$$|Z(\omega)| = \left[(\mathrm{Re}Z)^2 + (\mathrm{Im}Z)^2 \right]^{1/2} \tag{16.3}$$

When a sinusoidal voltage signal

$$E = \Delta E \sin(\omega t) \tag{16.4}$$

is applied across the interface, the result is a sinusoidal current signal of the same frequency but displaced somewhat in time,

$$j = \Delta j \sin(\omega t + \varphi) \tag{16.5}$$

namely, having a *phase shift* φ.

For an ideally polarizable interface with negligible solution resistance, the phase angle is $-90°$. For an ideally non-polarizable interface the phase angle is zero. Real systems do not behave ideally, of course. The actual phase angle will, therefore, be somewhere in between, and it will depend on frequency in most cases. The phase-sensitive voltmeter can measure the absolute value of the impedance vector $|Z|$ and the phase angle simultaneously.

For a capacitor and a resistor in series one has

$$Z(\omega) = R_S - j/\omega C_{dl} \tag{16.6}$$

and for a capacitor and resistor in parallel the impedance is given by

$$\frac{1}{Z(\omega)} = \frac{1}{R_F} - \frac{\omega C_{dl}}{j} \tag{16.7}$$

It is convenient to display the results of EIS in the *complex-plane impedance* representation.[1] The X-axis on this plot is $\mathrm{Re}Z$, which is the Ohmic resistance, and the Y-axis is $-\mathrm{Im}Z$, which, in the present case, is the capacitive impedance $-j/\omega C$.

---

1) We shall refer to it as the complex-plane impedance plot. The terms Cole–Cole plot and Nyquist plot are also found in the literature.

The absolute values of the impedance vector $|Z|$ and the phase angle φ are given by

$$|Z(\omega)| = \left[(\mathrm{Re}Z)^2 + (\mathrm{Im}Z)^2 \right]^{1/2} \tag{16.3}$$

and

$$\tan\varphi = \frac{\mathrm{Im}Z}{\mathrm{Re}Z} \tag{16.8}$$

This is shown in Figure 16.1, which is a vector representation of $|Z|$ for both the series and parallel combinations.

As the frequency is increased, the capacitive impedance decreases, while the resistive impedance is unchanged. In the series combination this makes the circuit behave more and more like a pure resistor, causing a decrease in phase angle, as seen in Figure 16.1a. In the parallel combination, it makes the circuit behave more and more like a capacitor, causing an increase in phase angle, as shown in Figure 16.1b.

Consider now a more realistic situation, in which both the series and the parallel resistance must be taken into account. The equivalent circuit and the corresponding complex-plane impedance plot are shown in Figure 16.2.

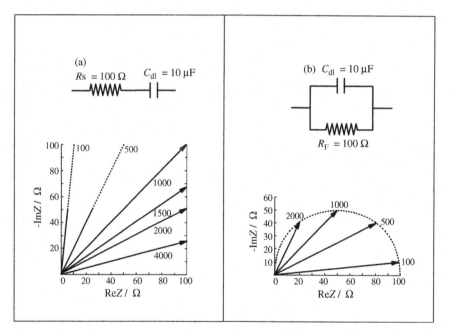

Figure 16.1 Vector representation of the impedance of (a) a series and (b) a parallel combination of a capacitor and a resistor, showing the variation of φ with ω (expressed in rad s^{-1}). The ends of the arrows show the absolute values of the vectors, except at the two lowest frequencies in the series combination.

Let us derive the mathematical expression for the impedance applicable to this circuit, as a function of the angular velocity ω:

$$Z(\omega) = R_S + \frac{1}{1/R_F - \omega\, C_{dl}/j} = R_S + \frac{R_F}{1 + j\, \omega\, C_{dl}\, R_F} \qquad (16.9)$$

The following simple manipulation allows us to separate the real from the imaginary part of the impedance:

$$Z(\omega) = R_S + \left[\frac{R_F}{1 + j\, \omega\, C_{dl}\, R_F}\right] \times \left[\frac{1 - j\, \omega\, C_{dl}\, R_F}{1 - j\, \omega\, C_{dl}\, R_F}\right] \qquad (16.10)$$

which leads to the expression

$$Z(\omega) = R_S + \frac{R_F}{1 + (\omega\, C_{dl}\, R_F)^2} - j \times \frac{\omega\, C_{dl}\, R_F}{1 + (\omega\, C_{dl}\, R_F)^2} \qquad (16.11)$$

The result is a semicircle having a diameter equal to R_F, with its center on the $\mathrm{Re}\,Z$ axis and displaced from the origin of coordinates by $R_S + R_F/2$. Each point on the semicircle in Figure 16.2 represents a measurement at a given frequency. At the limit of high frequencies, the Faradaic resistance is effectively shorted out by the double-layer capacitance, leaving the solution resistance R_S as the only measured quantity. At

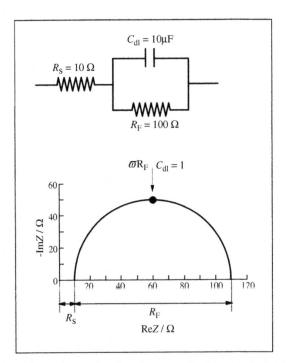

Figure 16.2 Complex-plane representation of the impedance of an interface. ReZ and ImZ are the real and imaginary components of the impedance, respectively.

the limit of low frequency the opposite occurs: the capacitive impedance becomes very high and one measures only the sum of the two resistors, $R_S + R_F$, in series.

The double-layer capacitance can be obtained from this plot, since the maximum on the semicircle satisfies the equation

$$R_F \times C_{dl} \times \omega_{max} = 1 \tag{16.12}$$

When one studies an (almost) ideally polarizable interface, such as the mercury electrode in pure deaerated acids, the equivalent circuit is a resistor R_S and a capacitor C_{dl} in series. The high accuracy and resolution offered by modern instrumentation allows measurement in such cases in very dilute solutions or in poorly conducting, non-aqueous media, which could not have been performed in the early days of studying the mercury/electrolyte interface.

16.1.2
Graphical Representations

The results of electrochemical impedance spectroscopy (EIS) can be displayed in a number of different forms. In Figures 16.1 and 16.2 we used the complex-plane impedance representation. Similar representations, in which the coordinates are the real and imaginary admittance ($\mathrm{Re}\,Y$ and $\mathrm{Im}\,Y$) or capacitances ($\mathrm{Re}\,C$ and $\mathrm{Im}\,C$) are referred to as the *complex-plane admittance* and the *complex-plane capacitance plots*, respectively. Two additional ways of displaying the results are the so-called *Bode magnitude plot*, in which $\log|Z(\omega)|$ is displayed as a function of $\log\omega$ and the *Bode phase-angle plot*, in which the phase-angle, φ, is plotted versus $\log\omega$. In Figure 16.3 we can see the relations of the phase angle and the absolute value of the impedance factor to the real and imaginary components of the impedance.

When impedance spectroscopy was introduced to electrochemistry, there was some argument in the literature as to the best method of displaying EIS data. Nowadays, the software provided with most commercial instruments allows us to display the data in all the above ways and choose the representation that best suits the particular system being studied, making the above argument obsolete.

The perfect semicircle shown in Figure 16.4 is constructed by connecting the tips of the impedance vectors at different frequencies. The frequency itself is not shown, and this is one of the disadvantages of this form of presentation. This is sometimes corrected by marking the frequencies at which specific measurements were taken on

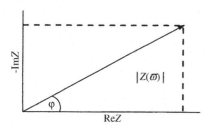

Figure 16.3 Vector representation of the impedance $Z(\omega)$ in the complex plane. (cf. Eqs. 16.3 and 16.8).

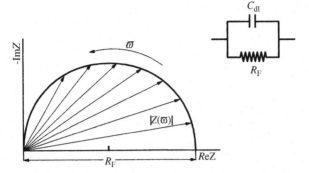

Figure 16.4 A simulated complex-plane representation of the impedance vector as a function of frequency for a simple circuit, consisting of a capacitor and resistor in parallel.

the semicircle (cf. Figure 16.5). The diameter of the semicircle is equal to the Faradaic resistance and is independent of the capacitance. As a result, plots measured for a fixed value of R_F but different values of C_{dl} cannot be distinguished in this type of presentation, even though corresponding points on the semicircle have been measured at different frequencies.

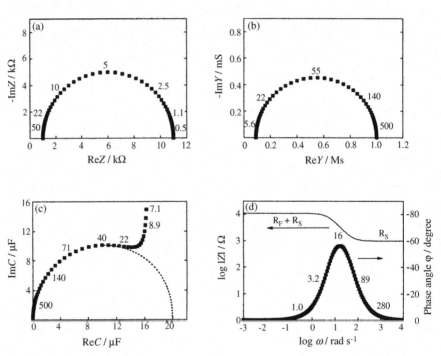

Figure 16.5 Comparison of (a) complex-plane impedance, (b) complex-plane admittance, (c) complex-plane capacitance and (d) Bode-magnitude and Bode-angle plots for the same equivalent circuit. $C_{dl} = 20\,\mu F$; $R_F = 10\,k\Omega$; $R_S = 1\,k\Omega$. Values of ω at which some of the points were calculated are shown.

For the complex-plane admittance plot the real and the imaginary parts of the admittance are defined as

$$\text{Re } Y = \frac{\text{Re } Z}{(|Z|)^2} \tag{16.13}$$

and

$$\text{Im } Y = \frac{\text{Im} Z}{(|Z|)^2} \tag{16.14}$$

Another way of presenting the data is in the complex-plane capacitance form

$$\text{Re } C = \frac{\text{Im } Z}{(|Z|)^2} \frac{1}{\omega} \tag{16.15}$$

and

$$\text{Im} C = \frac{\text{Re} Z}{(|Z|)^2} \frac{1}{\omega} \tag{16.16}$$

In Figure 16.5 the different ways of presenting the response of the same circuit are compared. The values of the components constituting the equivalent circuit were chosen as $R_F = 10 \text{ k}\Omega$, $R_S = 1 \text{ k}\Omega$ and $C_{dl} = 20 \text{ μF}$. Each way of presentation has its own advantages and disadvantages. From the impedance plot R_F and R_S can be read directly and the double-layer capacitance can be calculated, employing Eq. (16.12). The relevant time constant in this case is $\tau_c = R_F \times C_{dl} = 0.20 \text{ s}$, hence $\omega = 5 \text{ rad s}^{-1} = 0.80 \text{ Hz}$. The complex-plane admittance plot yields the values of $1/R_F$ and $1/R_S$ but it is rarely used.

On the complex-plane capacitance plot, the semicircle results from the series combination of R_S and C_{dl} while the intercept with the real axis yields the value of C_{dl}. The vertical line is due to the parallel combination of R_F and C_{dl}. While the physical meanings of complex impedance and complex admittance are clear, that of the complex capacitance is questionable. Nevertheless, presenting the data as in Figure 16.5d has some merit, because the numerical value of C_{dl} is given directly as the diameter of the semicircle.

Plotting the same data in the Bode-type representation, one notes two points:

1) Although the values of the two resistors are easily discerned, there is no region in which the circuit behaves as a pure capacitor. The slope never reaches a value of -1, and φ never even comes close to $-90°$, which one would expect for a pure capacitor.
2) The phase angle is a more sensitive test of the capacitive or resistive behavior of the system than is the plot of $\log|Z(\omega)|$ versus $\log\omega$. The detailed shape of these curves depends, of course, on the numerical values chosen for the various circuit elements. Had we used a value of $R_S = 10 \text{ Ω}$ instead of $1 \text{ k}\Omega$, the two horizontal lines in Figure 16.5d would have been much farther apart and an (almost) pure capacitive behavior would have been observed in the intermediate region.

It would seem then that the complex-plane impedance plot is the best way of presenting the data, if one is mainly interested in the value R_F and its variation with time or potential. The complex-plane capacitance plot, on the other hand, brings out more directly the value of the capacitance and its variation with the different parameters of the experiment. The Bode plots are also informative, but it could be argued that they provide a lower resolution, because they are presented on a logarithmic scale.

It should be remembered that the curves shown in Figure 16.5 are all simulated and therefore "ideal", in the sense that they follow exactly the equations derived for the assumed equivalent circuit and the numerical values of its components. In practice, the points are always scattered as a result of experimental error. Also, the frequency range over which reliable data can be collected does not necessarily correspond to the time constant of the system studied. For the case shown in Figure 16.5a the semicircle can be constructed from measurements in the range of $1 \leq \omega \leq 20$. In Figure 16.5b one would have to use data in the range of about $10 \leq \omega \leq 200$ to evaluate the numerical values of the circuit elements. From the Bode magnitude plots, R_S can be evaluated from the high end of the frequency measurements ($\omega \geq 100$), while R_F can be obtained from low frequency data ($\omega \leq 1$). The capacitance can be obtained *approximately* as $C_{dl} = 1/(\omega|Z|)$ at the inflection point (which coincides with the maximum on the Bode angle plot), but this is correct only if $\varphi = -90°$, that is, if the circuit behaved as a capacitor at this frequency. In the present example $\varphi_{max} = -58°$ and the value of C_{dl} calculated in this manner is 18.4 µF, compared to the value of 20 µF used to obtain the curves.

16.2
The Effect of Diffusion Limitation

16.2.1
The Warburg Impedance is a Constant-Phase Element

So far in this chapter we have discussed only equivalent circuits that correspond to charge transfer, namely the situation in which the faradaic resistance R_F is high and diffusion limitation is negligible. We might note in passing that EIS is inherently a *small-amplitude technique*, in which the AC component of the j/E relationship is in the linear region. This is most readily realized by maintaining the system at its open circuit potential and applying a low amplitude perturbation. It should be recalled, though, that the response to a small perturbation can be linear, even if the system is in the nonlinear region, as discussed in Section 14.12 (cf. Eq. (14.8)). The Faradaic resistance measured by EIS is the *differential* resistance, defined as

$$R_F \equiv (\partial \eta / \partial j)_\mu \qquad (16.17)$$

In the high-overpotential region, j is an exponential function of the overpotential, hence the differential Faradaic resistance also depends exponentially on potential.

$$\eta = b \log j/j_0 \tag{16.18}$$

Combining the last two equations one has, for the high-overpotential region

$$R_F = \frac{b}{j} \tag{16.19}$$

In the low overpotential region, where the current is linearly related to the over-potential

$$\frac{j}{j_0} = \frac{nF}{RT} \eta \tag{16.20}$$

hence, in this region the Faradaic resistance is given by

$$R_F = \frac{\eta}{j} = \frac{1}{j_0} \frac{RT}{nF} \tag{16.21}$$

It is very important to note the difference between the two expressions for the differential Faradaic resistance, given in Eq. (16.19) and (16.21). At high overpotential R_F is inversely proportional to the current density, but it yields no information concerning the exchange current density (and hence of the heterogeneous rate constant). For example, it is well known that j_0 for electrodeposition of Cu is about three orders of magnitude higher than that for Ni. Nevertheless, if EIS is used to measure R_F at the same current density, the same value will be observed in both cases, unless the Tafel slope is different. But the Tafel slope could change by a factor of two or three at most, while the exchange current density could vary over several orders of magnitude. In contrast, when the EIS is applied to a system at its reversible potential, or at the open circuit potential, Eq. (16.21) is applicable, and the value of j_0 can be obtained directly.

Turning now to the case in which diffusion control must be considered, the equivalent circuit takes the form shown in Figure 14.1, in which the subscript W, represents the so-called *Warburg impedance*, which accounts for diffusion limitation.

The diffusion equations have been solved for the low amplitude sine wave perturbation. It is found that the Warburg impedance is given by

$$Z_W = \frac{\sigma}{\omega^{1/2}} - j \frac{\sigma}{\omega^{1/2}} \tag{16.22}$$

where the parameter σ is defined as:

$$\sigma = \frac{RT}{(nF)^2} \frac{1}{\sqrt{2}} \left[\frac{1}{c_{b,Ox} \times D_{Ox}^{1/2}} + \frac{1}{c_{b,\,Red} \times D_{Red}^{1/2}} \right] \tag{16.23}$$

This equation can be simplified if it is assumed that the two diffusion coefficients are equal, and the bulk concentrations of the oxidized and the reduced species are also equal, yielding

$$\sigma = \frac{RT}{(n\,F)^2\,c_b} \frac{\sqrt{2}}{\times D^{1/2}} \tag{16.24}$$

If the two concentrations differ widely, Eq. (16.23) shows that the lower concentration determines the value of σ.

One should note that the real and the imaginary parts of the Warburg impedance in Eq. (16.22) depend on frequency in the same way. Therefore the phase shift generated by the Warburg impedance is independent of frequency. Plotted in the complex-plane impedance format, this leads to a straight line with a slope of unity, as shown in Figure 16.6.

Equations (16.22)–(16.24) were derived for conditions of semi-infinite linear diffusion. Thus, measurements should be conducted in quiescent solutions. In addition, if there is a thin film on the surface, through which diffusion occurs, such as the solid electrolyte interface (SEI) formed in some non-aqueous batteries, the condition of semi-infinite linear diffusion no longer applies and the part corresponding to diffusion in Figure 16.6 will be different.

We note that the Warburg impedance, which is proportional to $\omega^{-1/2}$, is in series with the faradaic resistance, R_F. At high frequencies one obtains the usual semicircle, whereas the Warburg impedance becomes predominant at low frequencies. The frequency at which the transition occurs depends on the concentrations of reactants and products, (which determine the value of σ), on the

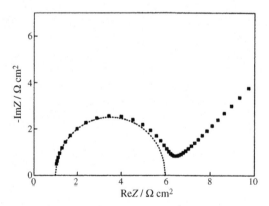

Figure 16.6 Complex-plane-impedance plot for an equivalent circuit with diffusion limitation at low frequencies. The values chosen for this simulation were: $R_S = 1\,\Omega\,cm^2$; $R_F = 5\,\Omega\,cm^2$; $C_{dl} = 20\,\mu F\,cm^{-2}$, $c_{b,Ox} = c_{b,Red} = 10\,mM$; $D_{Ox} = D_{Red} = 1 \times 10^{-5}\,cm^2\,s^{-1}$, $\sigma = 12\,\Omega\,cm^2\,s^{-1/2}$, $j_0 = 5\,mA\,cm^{-2}$, $k_{s,h} = 5 \times 10^{-3}\,cm\,s^{-1}$. Range of frequencies: $1 \times 10^{-3} \leq \omega \leq 1 \times 10^{3}$.

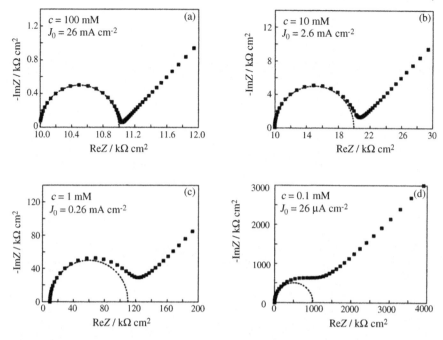

Figure 16.7 Complex-plane impedance plots showing the gradual change from charge-transfer to mass-transport control with decreasing concentration. $1 \times 10^{-3} \leq$ $\omega \leq 1 \times 10^3$ rad s^{-1}. Values used for simulation: $C_{dl} = 20\ \mu F\ cm^{-2}$, $R_S = 10\ \Omega\ cm^2$. $j_0 = 26;\ 2.6;\ 0.26;\ 0.026$ mA cm^{-2} for (a), (b), (c), and (d), respectively.

exchange current density and on the overpotential, which determine the value of the Faradaic resistance, R_F.

Four complex-plane impedance plots, calculated for the concentrations of 100, 10, 1 and 0.1 mM, respectively are shown in Figure 16.7(a–d). Note that R_F, which is proportional to $1/j_0$, also depends on concentration, therefore the scale of both ImZ and ReZ must be changed from Figure 16.7a to 16.7d. At the highest concentration shown here, a semicircle characteristic of a charge-transfer-limited process is clearly seen, with diffusion limitation becoming important only at low frequencies. As the concentration is decreased, diffusion limitation becomes gradually more important. In a 1 mM solution the initial part of the semicircle is barely seen and in a 0.1 mM solution the process is mostly diffusion controlled. The values of R_F chosen here correspond to a moderately slow reaction, having a standard heterogeneous rate constant of $k_{s,h} = 2.7 \times 10^{-3}$ cm s^{-1}.

A particular behavior may be caused by surface heterogeneity. This is tantamount to different values of R_F (and to a lesser extent also of C_{dl}). Such differences can lead to a whole range of time constants that are close to each other (since R_F tends to change gradually from site to site on a heterogeneous surface). Rather than having many semicircles, the result is often a so-called

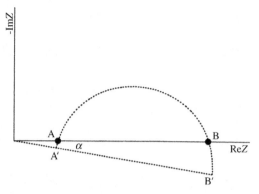

Figure 16.8 Complex-plane impedance plot with depressed semicircle. A′–B′ is the diameter of the semicircle, depressed by an angle α.

depressed semicircle, namely one with its center lying below the ReZ axis, as shown in Figure 16.8.

How should one calculate the Faradaic resistance from such a plot? Surely one cannot just take the distance between the points A and B on the ReZ axis as being equal to R_F, since this is not the diameter of the semicircle. The distance between the points A′ and B′ is equal to the diameter of the semicircle, but this line does not lie on the ReZ axis. What is the physical meaning of the angle of depression?

A result such as shown in Figure 16.8 indicates clearly that the system cannot be described correctly by a simple equivalent circuit, of the types discussed so far. Sometimes, if the depressed angle α is small (α $\leq$ 10°) the problem may perhaps be ignored, and one may obtain R_F either as the distance from A to B or from A′ to B′, which will differ in this case by a few percent. In any event, *changes* in the angle of depression or in the radius of the depressed semicircle still can be taken as an indication of variation in the properties of the interface, and may sometimes be understood by correlating to some other, independently measured, property.

16.2.2
Some Experimental Results

The plots we have shown, which are all based on simulated data, serve the purpose of illustrating the principles involved. The results of real experiments are rarely so simple and easy to interpret. This is caused by several types of factors:

1) On the one hand, the reaction may not be as simple as is assumed in the model. In other words, *the equivalent circuit assumed may not be quite equivalent.*
2) The formation of adsorbed intermediates can lead to an adsorption pseudocapacitance. The corresponding equivalent circuit will then have two different time constants that will show up as two semicircles, which could be separated or

partially overlapping, depending on the values of the corresponding resistances and capacitances.

3) The semicircle may also be distorted by experimental errors, which could arise mainly from non-uniform current distribution, caused by the geometry of the cell as a whole, or by screening of part of the working electrode with the Luggin capillary, by solution creeping into the crevice formed between the electrode and its non-conducting holder, and by changes occurring at the surface during measurement. It should be remembered in this context that the equations for EIS are based on the tacit assumption that the surface is invariant during measurement, as the frequency is scanned. This is particularly problematic when one wishes to extend the data to the very low frequency range because, as we have noted, the time taken to make each measurement is inversely proportional to the frequency.

In Figure 16.9 we show some results obtained for the anode in a direct methanol fuel cell (DMFC). Three semicircles can be distinguished, implying that there are three different processes occurring, each having its own time constant. The frequencies shown at the peaks represent highly different time constants, allowing good separation of the three processes. On the other hand, the curves are not perfect semicircles, showing added complexity in the interpretation of the experimental data.

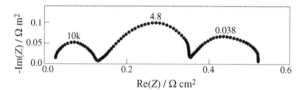

Figure 16.9 Example of complex-plane impedance results for the anode of a direct methanol fuel cell (DMFC). The numbers on the three parts of the spectrum show the frequency in Hz at the highest point that is related to the capacitance and resistance though Eq. (16.12). Reprinted with permission from J. T. Mueller and P.M. Urban, *J. Power Sources*, **75** (1998) 139.

In another study of DMFC anodes, shown in Figure 16.10, the complex-plane impedance plots were studied as a function of the current density applied. The diameters of the semicircles were found to decrease with increasing current density, as expected, but the new feature observed is an inductive branch of the curves. This can be modeled, of course, by adding an inductive element to the equivalent circuit representation, in series with the Faradaic resistance, but the physical origin of this added circuit element is still open for debate. There is a tendency to associate it with sluggish adsorption of CO, formed as an intermediate in the oxidation of methanol. However, unlike the adsorption pseudocapacitance, which is well understood (cf. Section 11.2), there is no theory for the dependence of the pseudoinductance on potential, coverage or any other measured parameter.

It should be noted that the inductive part of the complex-plane impedance plots is not unique to the response of the anode in a DMFC. In Figure 16.11 we show similar behavior for the cathode of a DMFC. Moreover, pseudoinductance at very

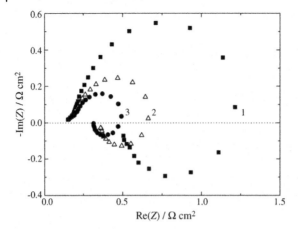

Figure 16.10 Complex-plane impedance of the anode in a DMFC as a function of applied current density. Lines 1, 2 and 3 correspond to 0.1, 0.2 and 0.3 A cm^{-2}, respectively. High values of the pseudoinductance are shown. Reprinted with permission from J. T. Mueller, P.M. Urban and W. F. Holderich, *J. Power Sources*, **84** (1999) 160.

low frequencies has already been reported in the seminal work of Epelboin *et al.* in 1975 (cf. I. Epelboin, C. Gabrielli, K. Keddam and H. Takenouti, *Electrochim. Acta*, **20**, (1975) 913).

Figures 16.9 to 16.11 are presented here only to show some of the complex behavior that one might encounter when studying an electrochemical interface by electrochemical impedance spectroscopy. The interpretation of the data shown was given by the original authors, and is outside the scope of this chapter.

In conclusion, the use of the technique of electrochemical impedance spectroscopy is straightforward only when the behavior is simple and can be modeled by an equivalent circuit that involves just a few adjustable parameters. The real picture found experimentally is rarely as simple as that shown by simulation. When complex

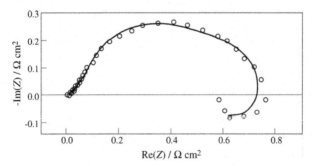

Figure 16.11 Complex-plane impedance of the cathode in a DMFC, showing pseudoinductance. Reprinted with permission from P. Piela, R. Fields and P. Zelenay, *J. Electrochem. Soc. A* 153, (2006) 1902.

shapes of the impedance spectrum are observed, one must develop a physical model and the corresponding equivalent circuit that can simulate the observed results. However, when fitting the model requires many adjustable parameters, the validity of the model may be questioned. In such cases independent experiment, such as considering the dependence of different parts of the spectrum on factors such as temperature, applied potential or current density, the concentration of the electro-active components as well as the supporting electrolyte, can be useful. However, this approach weakens the usefulness of EIS in the study of electrode kinetics: Rather than employing this technique to find the physical model, one may end up using an assumed model to explain the complex impedance spectrum.

17
The Electrochemical Quartz Crystal Microbalance

17.1
Fundamental Properties

17.1.1
Introduction

The quartz crystal microbalance is based on the piezo-electric properties of quartz, characterized by two complementary phenomena: (i) When force is applied to a quartz crystal changing its dimensions, a potential difference is generated across it. (ii) When a potential is applied across it, its dimensions are changed. When the potential applied is a periodic signal (typically a sine wave) the crystal vibrates at the frequency of the applied signal. This property is the physical basis for the quartz crystal microbalance (QCM).

When a quartz crystal (or any other solid material) vibrates, there is always a resonance frequency, which we denote f_0, at which it oscillates with minimum impedance (that is maximum admittance). The resonance frequency depends on the dimensions and on the properties of the vibrating crystal, mostly the density and the shear modulus. A quartz crystal can be made to oscillate at other frequencies, but as the distance, (on the scale of frequency), from the resonance frequency increases, the admittance decreases, until the vibration can no longer be detected. This is the basis for the analysis of the so-called (mechanical) admittance spectrum of the QCM, which is discussed below.

An important property of a vibrating quartz crystal is that its resonance frequency changes when its mass is changed, either by adding or removing weight or by interaction with the medium in which the crystal is immersed. Since frequency, and changes in it, can be measured with very high accuracy, a device based on a quartz crystal, coated on both sides with a thin layer of metal (usually but not always, gold), can serve as a mass-sensitive detector, namely a quartz crystal microbalance.

Physical Electrochemistry: Fundamentals, Techniques and Applications. Eliezer Gileadi
Copyright © 2011 WILEY-VCH Verlag GmbH & Co. KGaA, Weinheim
ISBN: 978-3-527-31970-1

17.1.2
The Fundamental Equations of the QCM

The quartz crystal microbalance was introduced in 1959 by Sauerbray, who derived the equation

$$\Delta f = -C_m \Delta m \tag{17.1}$$

where Δm is the change in mass per unit surface area and C_m is a characteristic constant given by

$$C_m = 2f_0^2 \left(\mu_q \rho_q \right)^{-1/2} \tag{17.2}$$

where $\mu_q = 2.947 \times 10^{11}$ g cm^{-1} s^{-2} is the shear modulus and $\rho_q = 2.648$ g cm^{-3} is the density of quartz.

Quartz crystals most commonly used for QCM have resonance frequencies (determined in vacuum or in H_2 or He at ambient pressure), in the range of about $(5-10)$ M Hz. For $f_0 = 6$ M Hz, (which we shall use for the sample calculation in this chapter), Eq. (17.2) yields a numerical value of

$$C_m = 8.13 \times 10^7 \text{ Hz g}^{-1} \text{ cm}^2 \tag{17.3}$$

Thus, a change of frequency of -1.0 Hz is generated by an added mass of 12.3 ng cm^{-2}. The introduction of the QCM represented a significant increase in the sensitivity of weighing materials in vacuum or in the gas phase at ambient pressure, and was soon applied as an on-line method for measuring and controlling the rate of deposition of very thin films, for example, during sputtering or physical and chemical vapor deposition.

The following simple calculation can demonstrate the high sensitivity of the QCM. A monolayer of metal deposited on the surface amounts to about 2 nmol cm^{-2}. Hence, the weight of a single atomic layer of silver, for example, is about 220 ng cm^{-2}, corresponding to a frequency shift of about 18 Hz. Even when invented in 1959, it was possible to measure a frequency change of 1 Hz, and presently available instrumentation can take us to a resolution of 0.05 Hz, corresponding to 0.6 ng cm^{-2}, allowing the detection of changes in the amount of silver deposited, corresponding to less than 0.3% of a monolayer.

Equations (17.1)–(17.3) look simple and straightforward, allowing the determination of weight with a resolution of the order of a few ng cm^{-2} [1]. There seems to be no need to calibrate such a device, because the value of C_m given in Eq. (17.3) is based only on the properties of the quartz crystal. Nevertheless, it should be remembered that there are some tacit assumptions that must be fulfilled for Eq. (17.1) to be applicable. First, the deposition of any material must be uniform all over the effective area of the crystal. Secondly, measurements must be maintained within the dynamic

---

1) Indeed, one is tempted to call it a quartz crystal nanobalance, but we shall maintain the commonly used name of QCM, in order to avoid confusion.

range of the device, which is about ± 100 kHz. This corresponds to an added weight of 1.2 mg cm^2, or to a thickness 1.2 µm of silver. Next, it is assumed that the adsorbate is rigidly attached to the surface and the surface roughness does not change during the experiment. This becomes increasingly important as the pressure of the gas increases, and could become critical when the QCM is immersed in a liquid.

Control of the temperature is also important. The crystals most commonly used are of the AT-cut type, which is defined by the angle of the crystal face used, with respect to the main crystal axis of quartz. This choice is made because it has the lowest dependence of the resonance frequency on temperature. Nevertheless, there is still some temperature dependence amounting to

$$\left(\frac{df_0}{dT}\right) = 2.4 \text{ Hz}\,^\circ\text{C}^{-1} \tag{17.4}$$

Thus, in order to maintain an accuracy of ± 0.6 ng in the determination of the added weight, one would have to maintain the temperature constant to about $\pm 0.02\,^\circ$C.

The total pressure must also be maintained constant, because the resonance frequency depends on pressure, following the equation

$$\Delta f_P = \left[1.06 \times 10^{-6} f_0\right]\Delta P \tag{17.5}$$

where the pressure is expressed in atmospheres.

17.1.3
The Effect of Viscosity

When the QCM was first introduced, it was not clear if this device could operate in condensed media, and, specifically, in aqueous solutions. This possibility was only implemented in 1980. It was shown that the interaction of vibrating QCM in a liquid leads to an added term in the change in resonance frequency, which depends primarily on the square root of the product of the viscosity and density of the fluid, namely

$$\Delta f_\eta = -C_\eta (\rho_{fl}\, \eta_{fl})^{1/2} \tag{17.6}$$

and the constant determining the effect of viscosity and density is given by

$$C_\eta = f_0^{3/2}(\pi \mu_q \rho_q)^{-1/2} \tag{17.7}$$

where ρ_{fl} and η_{fl} refer to the density and viscosity of the fluid, respectively. It is noted that both mass and viscosity represent added load, therefore they both decrease the resonance frequency. Considering that the density of liquids is usually two to three orders of magnitude higher than that of gases at ambient pressures, if is obvious why Eq. (17.6) becomes of major importance when the QCM is immersed in a liquid.

Equation (17.6) is also important when considering the effect of temperature on the shift of the resonance frequency, because the viscosity of liquids is temperature dependent, and this effect happens to be particularly high for aqueous solutions[2]. Thus, the effect of temperature on the QCM immersed in a liquid could be higher than that indicated by Eq. (17.4).

Once it was confirmed that the QCM could operate in contact with liquids, its use in electrochemistry for the study of the metal/solution interface became obvious and widespread.

17.1.4
Immersion in a Liquid

When the QCM is immersed in a liquid, the resonance frequency is decreased substantially. Based on Eq. (17.6) above, it should decrease by about 1 kHz, but in practice it is usually found to decrease by (2−3) kHz This does not imply that Eq. (17.6) is incorrect. It only shows that there must have been some tacit assumption which, alas, does not apply to real systems. In the present case, we failed to note that Eq. (17.6) applies to an ideally flat surface, while all real surfaces are rough to some extent. Now, even when a surface is highly polished, there is some roughness remaining. It is hard to describe the extent or degree of roughness, because roughness is inherently irregular. On the other hand, it is tentatively clear that a rough vibrating surface would interact with the liquid in which it is immersed much more strongly than an ideally flat surface. While it is hard to prepare an ideally smooth surface, it is easy to prepare surfaces of varying roughness. One can create increasingly rough surfaces by plating a metal at current densities close to the mass-transport-limited current density, as we shall see below. In this way one can create rough surfaces for which the resonance frequency shifts by as much as (30−40) kHz upon immersion in a dilute aqueous solution.

17.1.5
Scales of Roughness

In this context it is important to consider the various scales of roughness, and the place of the EQCM on this scale. Starting from the finest scale, we consider the double layer capacitance, adsorption of small molecules and charge transfer kinetics. All these phenomena occur within 1 nm from the surface, and are, therefore, sensitive to roughness on the atomic scale. Next comes the Gouy–Chapman diffuse double layer, which can extend up to about 10 nm in a dilute (1 mM) solution of a 1-1 electrolyte, and essentially disappears in a 1 M solution. In contrast, the Nernst diffusion layer is much thicker, in the range (5−200) μm. This is the reason why an electrode may have a high roughness factor, as far as adsorption or formation of a UPD layer is concerned, while the diffusion-limited current density can still be calculated correctly

---

2) This is caused by the fact that water has hydrogen bonds and their number decrease with increasing temperature.

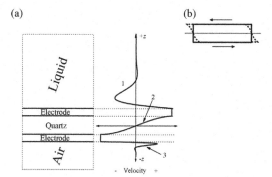

Figure 17.1 The mode of vibration of an AT-cut quartz crystal. (a) velocity vector v(z) as a function of the distance from the surface. (b) In-plane vibration of the quartz crystal. Based on M. Urbakh, V. Tsionsky, E. Gileadi and L. Daikhin in "*Piezoelectric Sensors*", Series Ed. O. S.Wolfbeis, Vol. 5, Eds C. Steinem and A Janshoff. Springer Series on Chemical Sensors and Biosensors, p. 121, (2006).

ignoring this roughness and considering the apparent (i.e., the geometrical) surface area.

The characteristic length relevant to the vibration of the EQCM is related to the depth in solution where the liquid still oscillates as a result of the oscillation of the surface of the crystal. This is given by

$$\delta = \left(\frac{\eta}{\pi \rho f_0}\right)^{1/2} \tag{17.8}$$

For a dilute aqueous solution at room temperature this yields a value of $0.23\,\mu m$, placing it between the atomic scale and the scale relevant to mass-transport-controlled processes.

The vibrations of the AT-cut quartz crystal are in the plane parallel to the surface in contact with the liquid, as shown in Figure 17.1. The velocity vector of the liquid decreases with increasing distance from the surface and is given by

$$v(z) = v_0 \exp\left(-\frac{z}{\delta}\right) \tag{17.9}$$

so it can be said that within a distance of about $1\,\mu m$ these vibrations essentially die out.

The roughness of the surface on the atomic scale is relatively easy to determine, by measurement of the maximum amount of some small molecule or atom (e.g., in the UPD region, oxide formation and adsorption of CO or of atomic hydrogen). Unfortunately, the results of such measurements are not necessarily relevant to roughness in the context of the EQCM, because of the large difference in the characteristic length for adsorption (1 nm) and for the EQCM ($0.23\,\mu m$). Figure 17.1 is not drawn to scale, but it should be noticed that the characteristic decay length in air (curve 3) is much shorter than that in liquid (curve 1).

17.2
Impedance Analysis of the EQCM

17.2.1
The Extended Equation for the Frequency Shift

An extended equation for the shift in frequency of an EQCM may be written as

$$\Delta f = \Delta f_m + \Delta f_\eta + \Delta f_P + \Delta f_T + \Delta f_R + \Delta f_{sl} \tag{17.10}$$

We have already discussed the first four terms on the right-hand side of this equation. It should be obvious that the EQCM operates as a true "microbalance" only when all terms other than Δf_m are zero. In the context of electrochemical measurements it is relatively easy to maintain constant temperature and pressure, but the term Δf_η may be difficult to determine, because the density and viscosity of a layer of the liquid adjacent to the surface may change as a result of the electrochemical process occurring at the interface, even if the viscosity and density in the bulk of the solution are maintained constant.

The effect of roughness, Δf_R is hard to assess because, as pointed out above, in practice a surface is never ideally flat, while in Eq. (17.6) Δf_η was derived for an ideally flat surfaces. Theories have been published regarding the effect of different models of roughness, but will not be discussed here. At any rate, this is of importance only when the surface roughness is known to change, or when it is changed on purpose, in order to test the theory of roughness. Otherwise the roughness could be maintained constant during the electrochemical process, so that $\Delta f_R \approx 0$.

The last term in Eq. (17.10) represents the so-called "slippage" at the interface between the surface of the EQCM and the electrolyte. This relates to yet another tacit assumption in Eq. (17.6), which was derived (as is often done in hydrodynamics) assuming the so-called *sticking-boundary condition* to solve the differential equations of hydrodynamic flow. This implies that the first layer of liquid at the interface moves at the same velocity as the surface itself, in the direction parallel to the interface. Further layers of solution move more slowly, following Eq. (17.9). Evidently, slippage leads to a smaller interaction of the vibrating surface with the liquid adjacent to it, which might be interpreted as a virtual decrease in the viscosity term, Δf_η, leading to an increase in the resonance frequency. Indeed, one is often surprised to find that adsorption of certain organic molecules leads to a positive value of Δf, implying that the apparent value of Δm is actually negative, consistent with the notion of increased slippage, caused by replacement of the hydrophilic surface of the metal by a hydrophobic surface of the organic molecules adsorbed on it.

17.2.2
Other Factors Influencing the Frequency Shift

Although Eq. (17.10) details six different factors that could affect the resonance frequency, there are additional phenomena that we might denote, somewhat naughtily, as Δf_{etc}. One of these factors is internal stress, created by absorption of hydrogen in a layer of palladium coated on one side of the QCM. It is well known that absorbed atomic hydrogen in Pd causes an expansion of the lattice. If only one side of the crystal is coated by Pd (the other side being a gold surface), the crystal will bend when hydrogen is absorbed, creating an internal stress, which shows up as a large decrease in Δf, over and above that calculated for the amount of hydrogen absorbed. If both sides are coated by Pd, the effect is much smaller and may vanish if the crystal is planar on both sides (which is not always the case). A similar effect was observed when a thin layer of copper was plated on one side of the crystal, probably caused by the stress or strain in the coated film. This may be a general phenomenon, observed when the crystal parameters of the coated metal differ from those of the gold surface of the quartz crystal.

If the surface of the sample is porous, small amounts of water may be occluded in the pores, influencing the observed frequency shift. Non-uniform current distribution during metal deposition leads to non-uniform thickness of the plated metal. These and some fine points in the design and operation of an electrochemical cell, where one side of the EQCM acts also as the working electrode, could all be lumped into the Δf_{etc} term, to interfere with the simple interpretation of the change in frequency observed as a change of mass.

17.2.3
Analysis of the Mechanical Impedance Spectrum

When the EQCM was fist introduced, the only parameter measured was the frequency of the resonance. The shift of the frequency was interpreted, often erroneously, as the change in mass calculated from Eq. (17.1). In recent years the EQCM has been studied with a frequency-response analyzer, which yields the mechanical impedance spectrum of the system. It turns out that the impedance is a complex number, so that the real and the imaginary parts can be determined separately. An equivalent way of treating the data is to plot the real part of the admittance as a function of frequency. In this way the peak represents the resonance frequency, f, while the width-at half-height, Γ, represents the imaginary part of the admittance. By combining Eq. (17.6) and 17.7 we find

$$\Delta f_{\eta} = -f_0^{3/2} \frac{(\rho_{fl} \, \eta_{fl})^{1/2}}{(\pi \, \rho_q \, \mu_q)^{1/2}} \tag{17.11}$$

where the subscripts "fl" and "q" refer to the fluid and the quartz crystal, respectively. The corresponding width of the resonance is given by the similar expression

$$\Gamma = 2 f_0^{3/2} \frac{(\rho_{fl} \, \eta_{fl})^{1/2}}{(\pi \, \rho_q \, \mu_q)^{1/2}} \tag{17.12}$$

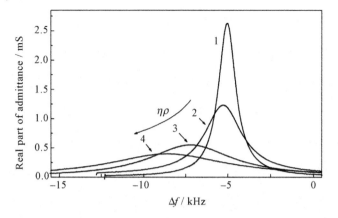

Figure 17.2 The effect of increasing viscosity of the fluid on the admittance curves. 1–dimethyl ether; 2–water; 3–40% aqueous solution of sucrose, 4–50% solution of the same. Based on data from, L. Daikhin, E. Gileadi G. Katz V. Tsionsky, M. Urbakh and D. Zagidulin, *Anal. Chem.* 74 (2002) 554.

It follows that the ratio of width-at-half-height to the shift of the resonance frequency, upon immersion of an EQCM in a fluid, is simply

$$\frac{\Gamma}{\Delta f_\eta} = -2 \qquad (17.13)$$

Note that immersing the oscillating crystal in a liquid loads it, because energy is transferred to the solution adjacent to the surface of the vibrating crystal. This leads to a lowering of the resonance frequency, just as expected for mass loading. However, the width-at-half-height increases.

In Figure 17.2 we present a series of admittance spectra, showing the effect of viscosity and density on the shape of the admittance curves. Different solutions were used, as listed in the caption to this figure. No potential was applied and no electrochemical reaction took place. Adsorption of the solvent or the solute may have taken place, but this would not account for any increase in the width, and to only a small shift in frequency, of the order of a few Hz, while the total shift of the resonance frequency shown is in the range of several kHz.

The results of a different experiment, intended to show the effect of roughness, are shown in Figure 17.3. In this case, electroplating of gold on the gold surface of the EQCM was conducted at two different current densities. Figure 17.3a shows the admittance curves measured at different times during electroplating at a low current density of 20 µA cm^{-2}. The curves are seen to be shifted to lower frequencies with increasing time of plating, indicating an increase in mass deposited, but their shape is essentially unchanged.

In Figure 17.3b electroplating of gold was conducted from the same solution, but at a much higher current density of 500 µA cm^{-2}, which is close to the

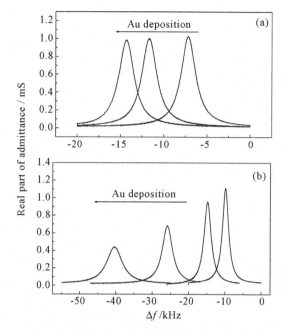

Figure 17.3 (a) Admittance spectra obtained at different times during plating of gold on the EQCM at a low current density of 20 μA cm^{-2}. (b) The same conducted at a high current density of 500 μA cm^{-2}. Based on data from V. Tsionsky, L Daikhin, M. Urbakh and E. Gileadi, *Electroanalytical Chemistry – A Series of Advances*, vol. 22; A. J. Bard and I Rubinstein, Eds, M. Dekker, N.Y., Chap. 1, Fig. 1a and 1c, (2004).

limiting current density, creating a rough surface. Here, the resonance frequency is also shifted to lower values, but the width of the resonance increases significantly, showing that this cannot be interpreted as the result of added mass by itself.

The total shift of frequency shown here can be written as

$$\Delta f = \Delta f_m + \Delta f_R \tag{17.14}$$

but we cannot separate the two terms on the right-hand side of this equation, unless we determine one of them independently.

It is important to note that determination of the width of the resonance can be critical for the use of the EQCM as a true microbalance. Relating Δf to the change in mass Δm employing Eq. (17.1) is only valid if the parameter Γ is constant in a given composition of the solution and at constant temperature and pressure. In the general case, it may be better to regard the QCM as a quartz crystal microsensor. On the other hand, measurement of both Δf and Γ can be very useful in more advance analysis of the structure of the metal/electrolyte interface, employing suitable models, which can be tested experimentally.

17.3
Use of the EQCM as a Microsensor

17.3.1
Some Applications of the EQCM

The high sensitivity of the QCM should make it an ideal tool for the study of trace amounts of impurities in the air or in liquids. This is a large field in itself and we mention it here only very briefly, to indicate some possible aspects of such applications. For a sensor to be useful it should usually be selective to the specific impurity of interest. This can be achieved by coating the surface of the gold-coated quartz crystal with a thin layer of a judiciously chosen material: A hygroscopic salt could be used to make it sensitive to humidity; a layer of palladium could make it sensitive to molecular hydrogen, a copper sulfate coating could detect ammonia in the gas, a high-surface-area platinized-platinum coating could respond selectively to methanol, and so on. In biology and medicine the use of antibodies and antigens could render the surface highly selective. The weight of a single bacterium is about three orders of magnitude below the detection limit of a QCM, but the rate of growth of colonies of bacteria could be followed.

In most such applications, the simple relationship between mass and frequency shift given by Eq. (17.1) may not apply, but the sensor could be calibrated to yield useful results.

Metal deposition might be expected to be most suitable to determine added weight, by measuring the shift in resonance frequency. The metal atoms electrodeposited become part of the crystal lattice of the substrate and can therefore be considered to be rigidly attached to the surface. Thus, slippage is not an issue. The surface is generally not ideally flat but, as long as the width of the resonance remains unchanged, it is correct to use Eq. (17.1) to determine the added weight. The sensitivity is high enough to detect a fraction of a monolayer and maintaining a current density well below the mass-transport-limited current density will usually keep the roughness unchanged.

In Figure 17.4 we show the results obtained for the deposition of copper, silver and gold. In each case the substrate is the same as the metal being deposited.

If the EQCM acts in this system as a microbalance (i.e., if Eq. (17.1) is applicable), the coordinates of Δf can be recalculated to show Δm. The slope of this line is, therefore, the equivalent weight of the metal deposited. Any systematic deviation from the calculated line can be associated with a lower current efficiency, or some other deviation of the system from the conditions under which Eq. (17.1) is applicable.

The data for silver deposition agree with the calculated line within $\pm 0.25\%$. For deposition of copper the data also follow a straight line, but the slope in this and similar experiments was found to be consistently lower than the calculated slope, by about 6–8%. This deviation from theory was not attributed to the properties of the EQCM. The electroplating of copper is commonly believed to occur in two consecutive charge-transfer steps

$$Cu^{2+}_{soln} + e^-_M \rightarrow Cu^+_{soln} \tag{17.15}$$

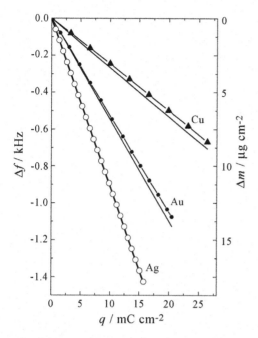

Figure 17.4 Plots of the frequency shift versus charge, during galvanostatic plating of Cu, Ag and Au, (each metal on its own metal substrate). Lines are calculated. Points represent experimental results. Based on data from *Electroanalytical Chemistry – A Series of Advances*, vol. 22; A. J. Bard and I Rubinstein, Eds, M. Dekker, N.Y., Chap. 1, Fig. 14, (2004).

followed by

$$Cu^+_{soln} + e^-_M \rightarrow Cu^0_M \tag{17.16}$$

Most of the monovalent copper formed at the solution side of the interface (at the OHP) will be reduced to metallic copper, but some may diffuse away into the solution, without being further reduced. Thus, a fraction of the electrons consumed in Eq. (17.15) may not be used to form atomic copper on the surface, and the effective charge will be less than that measured, leading to the lower slope. The same argument applies to deposition of Au, where the reactant contains Au^{3+} and the intermediate contains Au^+.

17.3.2
Plating of a Metal on a Foreign Substrate

The data shown in Figure 17.4 refer to deposition of Cu from $CuSO_4$, Ag from $AgNO_3$ and Au from $AuCl_4^-$, each on its own metal as the substrate. In the next figure we show the results of a different type of experiment, in which silver was deposited on a gold substrate, and the deposition of only a few atomic layers was recorded. A detailed description of the results shown in Figure 17.5 will not be given here. This figure is

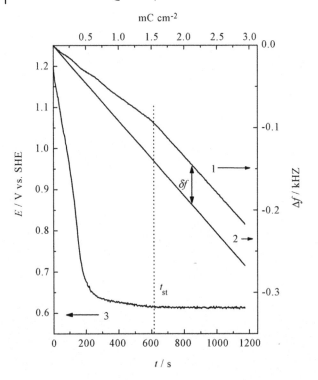

Figure 17.5 Response of the EQCM during plating of a very thin layer of silver on a gold substrate. 10 mM AgNO$_3$ + 0.5 M HClO$_4$, at $j = 2.5\ \mu$A cm^{-2}. 1 – C Experimental result; 2 – Calculated line, 3 – The dotted vertical line at t_{st} is where the potential is steady and line 1 becomes parallel to the calculated line 2. (c.f. V. Tsionsky, L Daikhin, G. Zilberman and E. Gileadi, *J. Electrochem. Soc.* **147** (2000) 567.)

presented here only to show that it is possible to follow the formation of a UPD layer by this technique, and the results for the first few layers may turn out to be complicated. For example, the potential of zero charge changes from that corresponding to Au to that corresponding to Ag as the fractional coverage by the UPD layer changes from zero to unity.

In considering Figures 17.4 and 17.5 it should be noted that a monolayer of deposited metal corresponds roughly to a charge of $(0.22 \times z \times R)$ mC cm^{-2}, where R is the roughness factor and z is the valency of the metal ion. Taking a reasonable value of $R \approx 1.5$ this shows that the amount of charge passed in Figure 17.5 corresponds to about 10 atomic layers, while in Figure 17.4 the charge corresponds to about 20–50 atomic layers for each of the three metals.

In Figure 17.5, the potential is seen to decrease rapidly, while a UPD monolayer is being formed, but interestingly, it continues to change, albeit moderately, until an amount corresponding to about five atomic layers has been deposited. This is the point at which the interface starts behaving as a pure Ag/solution interface.

18
Corrosion

18.1
Scope and Economics of Corrosion

18.1.1
Introduction

Corrosion is a common phenomenon, observed all around us. Wherever there is a metal there is bound to be, sooner or later, corrosion. This is hardly surprising, since all metals, except gold, are thermodynamically unstable with respect to their oxides in air and in water. This is manifested by the observation that metals are not found in nature in their "native" or metallic form, but rather in the form of some compound: an oxide, a sulfide, a silicate and so on[1]. The history of mankind is closely linked with the technology of reducing ores to the corresponding metals or alloys. This requires the input of energy, and the resulting product is unstable thermodynamically. Corrosion can be regarded as the natural tendency of metals to revert to a more stable state as a chemical compound of one kind or another, depending on the environment. Our technology, therefore, depends on our ability to slow down the rate of corrosion to an acceptable level.

The cost to society of corrosion and its prevention is staggering: It has been estimated to be about \$200 billion per year in the United States alone, corresponding to about \$ 700 per capita per year. Much of this amount could be saved by the proper design and choice of materials, and by the use of existing prevention methods. The problem cannot be eliminated, however, since as pointed out above, corrosion represents the natural tendency of all systems toward a state of minimum Gibbs energy.

The damage caused by corrosion is of two general kinds: esthetic and engineering. An example of the former is the development of rust spots on (so-called) stainless steel cutlery. Although rust, which is just a mixture of oxides of iron, is not harmful in any way, one would not like to eat with a rusted fork or spoon. Examples of

---

1) There are some exceptions, when base metals are found in a reducing environment, e.g., at the bottom of the oceans, but this is rather rare.

Physical Electrochemistry: Fundamentals, Techniques and Applications. Eliezer Gileadi
Copyright © 2011 WILEY-VCH Verlag GmbH & Co. KGaA, Weinheim
ISBN: 978-3-527-31970-1

engineering damage due to corrosion are countless. From car bodies to pipelines to electronic components; almost everything must be protected and, eventually, replaced because of corrosion.

In most cases corrosion can be effectively prevented by investment in the construction material, or at least slowed down to the point that the device, be it a piece of machinery or a structure, will have to be replaced for some other reason, before corrosion has become severe. Jewelry and coinage are extreme examples, but even in a chemical plant one has the choice of designing for minimum maintenance and long periods between overhauls at a high initial cost, or frequent maintenance at a lower initial investment.

There are "corrosive" environments and those which are considered benign. The combination of high humidity and high temperature favors corrosion, but above all the presence of chloride ions is detrimental to almost all metals and interferes with many methods of corrosion protection. Chloride is not the only ion that enhances corrosion, but it is the one most commonly found all around us, in seawater and even in freshwater, in the ground and in the human body. Salt spray carried by the wind from the sea is a major cause of corrosion, and it is easy to see how the importance of this factor diminishes with increasing distance inland.

As a rule, corrosion is not uniformly distributed on the exposed surface. An average rate of corrosion of 2 mpy (two thousandth of an inch per year)[2] may be concentrated in spots, leading to holes in a piece of metal (e.g., a pipeline) that is 200 mil (about 0.5 cm) thick or more. A particular type of local corrosion occurs when the worst design error, from the corrosion point of view, is made – when two different metals are in contact without isolating them from each other electrically. For example, connecting copper plates with steel rivets in effect creates a battery in which steel is the anode and copper is the cathode. Moreover, since the two metals are in intimate contact with each other, this is equivalent to having the terminals of this battery effectively shorted, leading to a high rate of discharge (i.e., corrosion). As long as the structure is totally dry, nothing will happen, but if water accumulates on the surface, corrosion of the rivets may occur, leading to critical structural damage.

In the following sections we shall discuss the basic electrochemistry of corrosion and some of the more common methods of corrosion protection.

18.1.2
The Fundamental Electrochemistry of Corrosion

The first and most fundamental step in corrosion is the oxidation of the metal to its lowest stable valence state, for example

$$Fe^0_M \rightarrow Fe^{2+} + 2e^-_M \tag{18.1}$$

This is most often followed by the formation of insoluble products, the exact nature of which depends on the metal and on the environment in which it is corroding. The

---

2) The rate of corrosion is commonly expressed in units of mpy. Note that 1 mpy = 25 μm y^{-1}.

ion formed in the initial step may be oxidized further, producing oxides or other compounds of mixed valency. In some cases (e.g., Al, Ti, Cr) the corrosion products form dense insulating layers, which prevent further corrosion. In other cases (low-carbon steel), the layer is porous, allowing the corrosion process to continue until the whole piece of metal has been consumed. When a protective layer does exist, it does not have to be very thick. About 5 nm in the case of Al and 3 nm in the case of stainless steel. Thus, as little as 10–20 molecular layers of the oxide film can provide excellent protection for long periods of time.

Anodic dissolution of a metal cannot occur by itself for any length of time, since it would lead to charging of the metal to a high negative potential[3]. The accompanying reaction in aqueous solutions is usually hydrogen evolution or oxygen reduction.

When a piece of iron is placed in 1.0 M HCl it dissolves readily, with simultaneous evolution of hydrogen. The rate of dissolution can be measured by determining the weight loss of iron, or by analysis of the solution for its ions, but one could also determine this quantity by measuring the volume of hydrogen evolved. The rates of anodic metal dissolution and of cathodic hydrogen evolution must be equal, because there can be no accumulation of electrons in the metal or the solution phase.

At what potential will this process occur? The reversible potentials for the two reactions depend on the composition of the solution in contact with the electrode surface. For the purpose of the present calculation we shall assume that after a short time the solution is saturated with molecular hydrogen and the concentration of iron is 1 µM. With these assumptions we have, for the anodic reaction:

$$E_{rev} = E^0 + \frac{2.3RT}{nF} \log \frac{[Fe^{2+}]}{[Fe^0]} = -0.617 \text{ V SHE} \tag{18.2}$$

and for the cathodic reaction, by definition $E^0 = 0.000$, hence

$$E_{rev} = E^0 - \frac{2.3RT}{F} pH = -\frac{2.3RT}{F} pH \text{ V SHE} \tag{18.3}$$

Clearly, for anodic dissolution to occur, the potential must be positive with respect to the reversible potential for the Fe^{2+}/Fe couple and negative with respect to the reversible potential for the hydrogen-evolution reaction (HER). Thus, all we could predict from this thermodynamic argument is that at pH = 0 the corrosion potential must be somewhere between −0.617 V and 0.000 V vs. SHE. The rest depends on kinetics. To proceed we need to know the exchange current densities and the Tafel slopes for the two reactions concerned. The two partial current densities are plotted in Figure 18.1 as a function of potential The potential must settle at the point where the anodic and cathodic currents are equal, which is called the corrosion or mixed potential, j_{corr}.

For the example shown in Figure 18.1, $E_{corr} = -0.52$ V SHE and $j_{corr} = 27$ mA cm^{-2}. This is quite a high current density, corresponding to fast

---

3) If one assumes a double-layer capacitance of 20 µF cm^{-2}, it would take 20 µC cm^{-2} to change the potential across the interface by 1.0 V. This amount of charge corresponds to 1×10^{-10} mol cm^{-2} or 6 ng cm^{-2} of iron.

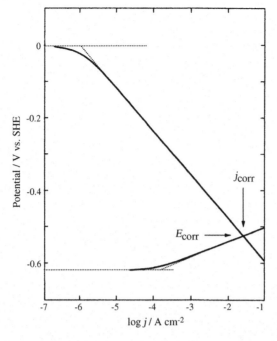

Figure 18.1 Evans diagram showing the currents for iron dissolution and hydrogen evolution at pH = 0, and the resulting values of E_{corr} and of j_{corr}. Parameters for the anodic and cathodic reactions respectively are: $j_0(an) = 1 \times 10^{-4}$ A cm^{-2}, $j_0(c) = 1 \times 10^{-6}$ A cm^{-2}; $b_{an} = 0.039$ V, $b_c = 0.118$ V, $E_{rev}(an) = -0.617$ V and $E_{rev}(c) = -0.000$ V SHE.

dissolution of the metal and vigorous hydrogen evolution, which are indeed observed when iron is dipped into a 1.0 M solution of HCl.

We can calculate the corrosion potential and the corrosion current in a straight-forward manner by writing the Tafel equation for the two partial reactions and solving for the potential at which the currents are equal:

$$ j_0(an)\exp\left[\frac{E_{corr} - E_{rev}(an)}{b_{an}}\right] = j_0(c)\exp\left[-\frac{E_{corr} - E_{rev}(c)}{b_c}\right] \qquad (18.4) $$

This type of representation, very common in corrosion studies, is referred to as an *Evans diagram*. We shall, therefore, discuss it a little further.

Consider first the effect of pH. Assuming that all the kinetic parameters are unchanged[4] and only the reversible potential for hydrogen evolution is affected, we obtain the curves shown in Figure 18.2.

As the pH is increased, the corrosion potential becomes less positive, approaching the reversible potential for the Fe^{2+}/Fe redox couple, and the corrosion current

---

4) This assumption is made here only for convenience of presentation. Usually the exchange current density for the HER. is found to be lower at intermediate pH values than in strongly acid or strong alkaline solutions.

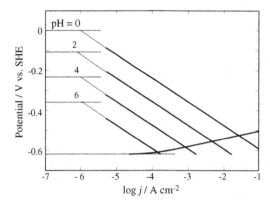

Figure 18.2 Evans diagram showing the current densities for iron dissolution and hydrogen evolution at different pH values. The kinetic parameters, which are the same as in Figure 18.1 have been arbitrarily assumed to be independent of pH.

decreases, from $j_{corr} = 27$ mA cm^{-2} at pH $= 0$ to $j_{corr} = 0.14$ mA cm^{-2} at pH $= 6$. This is hardly surprising. It simply confirms the common observation that iron corrodes faster in concentrated than in dilute acid. The interesting new insight we can gain from Figure 18.2 is that this difference is not directly related to the rate of metal dissolution in the different media. It is, in fact, determined by the different rates of hydrogen evolution, which are necessary to use up the electrons released in the process of oxidizing the metal to its ions. The process is evidently *cathode limited*, and the corrosion potential is close to the reversible potential for the *anodic process*.

Since the process is cathode limited, it is possible to slow it down by inhibiting the rate of hydrogen evolution. Many commercial corrosion inhibitors function in this manner. Considering Figure 18.2, it is easy to see that decreasing the exchange current density of hydrogen evolution by the addition of a suitable corrosion inhibitor is equivalent to increasing the pH, in terms of its effect on E_{corr} and on j_{corr}.

In practice, there is no interest in the corrosion of iron in 1 M HCl. The pH in typical environments (e.g., in the ground and in natural bodies of water) is in the range 5–9. Under these conditions the rate of hydrogen evolution may be very slow, and oxygen reduction can become the main cathodic process, controlling the rate of corrosion. We recall that the standard potential for oxygen reduction is 1.229 V, SHE, which leads to a reversible potential of 0.816 V, SHE at pH 7. Thus, corrosion occurs at a very high (negative) overpotential with respect to oxygen reduction, and the current is mass-transport limited. Figure 18.3 presents the Evans diagram for iron in neutral, aerated solutions. The exchange current densities for oxygen reduction and hydrogen evolution were taken as $j_0 = 1 \times 10^{-10}$ A cm^{-2} and $j_0 = 1 \times 10^{-8}$ A cm^{-2}, respectively. The contribution of the HER to the measured corrosion current is, therefore, quite negligible in neutral aerated solutions.

The mass-transport-limited current density for oxygen reduction is independent of the kinetic parameters for this reaction; rather, it depends on factors such as the

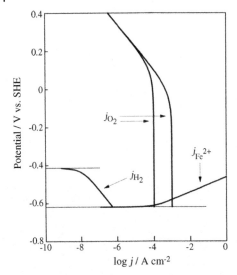

Figure 18.3 Evans diagram in neutral solution. Two values of the limiting current for oxygen reduction 10^{-4} and 10^{-3} A cm^{-2} are shown, yielding two different values for E_{corr} and j_{corr}.

concentration and the diffusion coefficient of oxygen in the medium. For example, increasing the temperature decreases the solubility of O_2 in water, but increases its diffusion coefficient. It depends also on the rate of flow of the liquid in a pipe or around a sailing ship or a structure immersed in a river.

The next question to consider in this context is the way a metal corrodes when two cathodic reactions of comparable magnitude occur in parallel on the same surface. This is shown in Figure 18.4.

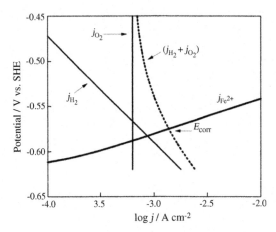

Figure 18.4 Evans diagram for the corrosion of iron in the presence of two simultaneous cathodic reactions. The dotted line represents the sum of the two cathodic currents for oxygen reduction and hydrogen evolution. pH $= 4$. $j_L = 0.63$ mA cm^{-2}. Other kinetic parameters are as in Figure 18.1.

The total cathodic current is given by

$$j_c = j_0 \exp\left[-\frac{E - E_{rev}(c)}{b_c}\right] + j_L \tag{18.5}$$

and the open circuit corrosion potential is the intersection of the line given by this equation with the line for anodic oxidation of iron, as shown in the figure.

Although hydrogen evolution and oxygen reduction are the two reactions most often involved in environmental corrosion, other electroactive materials may take part in the process, enhancing or retarding it, depending on whether they can be reduced or oxidized, respectively, in the range of potential where corrosion takes place.

18.1.3
Micropolarization Measurements

The understanding gained by considering the Evans diagrams allows us to measure the corrosion current directly. First, we must realize that the corrosion potential is in fact the open-circuit potential of a system undergoing corrosion. It represents steady state, but not equilibrium. It resembles the reversible potential in that it can be very stable. Following a small perturbation, the system will return to the open-circuit corrosion potential, just as it returns to the reversible potential. It differs from the equilibrium potential in that it does not follow the Nernst equation for any redox couple and there is both a net oxidation of one species and a net reduction of another.

Consider now the current–potential behavior of a system close to E_{corr}. Assuming that the two partial currents are in their respective linear Tafel region, one can write

$$j_c = j_0 \exp\left[-\frac{E - E_{rev}(c)}{b_c}\right] \tag{18.6}$$

The potential E near the corrosion potential can be written as

$$E = E_{corr} + \Delta E \tag{18.7}$$

Substituting in Eq. (18.6), one has

$$j_c = j_0 \exp\left[-\frac{E_{corr} - E_{rev}(c)}{b_c}\right] \times \exp\left(-\frac{\Delta E}{b_c}\right) \tag{18.8}$$

The cathodic current density at the corrosion potential is equal to the corrosion current density

$$j_{corr} = j_0 \exp\left[-\frac{E_{corr} - E_{rev}(c)}{b_c}\right] \tag{18.9}$$

Hence, Eq. (18.8) can be written in the simple form

$$j_c = j_{corr} \exp\left(-\frac{\Delta E}{b_c}\right) \tag{18.10}$$

This is very similar to the Tafel equation, written for a cathodic process as:

$$j_c = j_0 \exp\left[-\frac{\eta}{b_c}\right] \tag{18.11}$$

The corrosion current density, like the exchange current density, is an internal current, which is not observed in the external circuit. The potential difference ΔE is the difference between the applied potential and the open-circuit potential, just as η is the difference between the applied potential and the reversible potential. The big difference is that j_{corr} is equal to the anodic and cathodic currents of two *entirely different* processes, whereas j_0 represents the equal anodic and cathodic currents of the *same reaction* at the equilibrium potential.

Following the same arguments we can derive, for the anodic current density an expression equivalent to Eq. (18.10), namely

$$j_{an} = j_{corr} \exp\left(\frac{\Delta E}{b_{an}}\right) \tag{18.12}$$

The net current density observed at a potential E, close to the open-circuit corrosion potential, is hence

$$j = j_{an} - j_c = j_{corr}\left[\exp\left(\frac{\Delta E}{b_{an}}\right) - \exp\left(-\frac{\Delta E}{b_c}\right)\right] \tag{18.13}$$

For small values of $|\Delta E/b|$, we can linearize the exponents and obtain

$$\frac{j}{j_{corr}} = \Delta E\left[\frac{1}{b_{an}} + \frac{1}{b_c}\right] \tag{18.14}$$

This equation is usually written in the form:

$$R_P = \frac{1}{j_{corr}}\left(\frac{b_{an} \times b_c}{b_{an} + b_c}\right) \tag{18.15}$$

where $R_P = \Delta E/j$ is the *polarization resistance*).[5]

These equations allow us to determine the corrosion current by making current–potential measurements in the range of about $\pm 20\,\text{mV}$ around the open-circuit corrosion potential.[6]

If the cathodic reaction is mass-transport-controlled, we can derive a similar expression for the micropolarization region. For the anodic current Eq. (18.12) holds, and for the cathodic reaction one has $j_c = j_L = j_{corr}$. The total current is hence

$$j = j_{an} - j_c = j_{corr}\exp\left(\frac{\Delta E}{b_{an}}\right) - j_{corr} = j_{corr}\left[\exp\left(\frac{\Delta E}{b_{an}}\right) - 1\right] \tag{18.16}$$

---

5) The polarization resistance R_P defined here is just another name, commonly used in corrosion studies, for the Faradaic resistance R_F, which was defined in Section 5.1.2, Eq. (5.9).

6) One needs to know the values of the anodic and the cathodic Tafel slopes, to evaluate j_{corr} from Eq. (18.14) or Eq. (18.15). When these slopes are not known, a value of $b_c = 0.12\,\text{V}$ and $b_{an} = 0.04\,\text{V}$ may be used as a rough approximation.

When we linearize the exponent, this gives rise to

$$\frac{j}{j_{corr}} = \frac{\Delta E}{b_{an}} \quad \text{or} \quad j_{corr} = \frac{b_{an}}{R_P} \tag{18.17}$$

which is similar to Eqs. (18.14) and (18.15). Equation (18.17) could be derived directly from Eq. (18.14) or (18.15) by setting $b_c \to \infty$ which is the appropriate value for a mass-transport-controlled process, for which the current is independent of potential.

Experimental studies usually yield good agreement between the rates of corrosion obtained from polarization resistance measurements and those derived from weight-loss data, considering that the Tafel slopes for the anodic and the cathodic processes may not be known very accurately. It cannot be overemphasized, however, that both methods yield *the average rate of corrosion* of the sample, which may not be the most critical aspect when localized corrosion occurs. In particular, it should be noted that, at the open-circuit corrosion potential, the total anodic and cathodic *currents* must be equal, while the local *current densities* on the surface can be quite different. This could be a serious problem when most of the surface acts as the cathode and small spots (e.g., pits or crevices) act as the anodic regions. The rate of anodic dissolution inside a pit can, under these circumstances, be hundreds of times faster than the average corrosion rate obtained from micropolarization or weight-loss measurements.

18.2
Potential–pH Diagrams

18.2.1
Some Examples of Potential–pH Diagrams

A very useful method of describing the stability of metals in aqueous solutions is the potential–pH diagrams introduced by Pourbaix. These equilibrium diagrams relate the reversible potentials of reactions of interest in corrosion studies to the pH and the concentration of different ionic species in solution. We shall use a number of examples to illustrate the principles involved, starting with the most basic diagram relating to water and some of the ionic and molecular species at equilibrium with it.

To construct such diagrams, one has to identify the chemical and electrochemical reactions of interest and write the appropriate chemical equilibria and Nernst equations, respectively.[7]

In the following discussion all potentials are given against the standard hydrogen electrode (SHE), and the concentrations are written in square brackets, for clarity. The three most important equilibria for water are given below:

1) Self-ionization

$$2H_2O \to H_3O^+ + OH^- \tag{18.18}$$

---

7) Unless otherwise stated, concentrations and partial pressures, instead of activities and fugacities, respectively, are used here and in all following equations, for simplicity.

for which the equilibrium constant at 25 °C is given by

$$K_W = [H_3O^+] \times [OH^-] = 1.0 \times 10^{-14} \tag{18.19}$$

2) Hydrogen evolution, for which the Nernst equation is

$$E_{rev} = \frac{2.3RT}{2F} \log \frac{[H_3O^+]^2}{[H_2]} \tag{18.20}$$

This equation can also be written in the form

$$E_{rev} = -\frac{2.3RT}{2F} \log[H_2] - \frac{2.3RT}{F} pH \tag{18.21}$$

3) The oxygen evolution reaction, for which

$$E_{rev} = +1.229 + \frac{2.3RT}{4F} \log[O_2] - \frac{2.3RT}{F} pH \tag{18.22}$$

where $[H_2]$ and $[O_2]$ represent the partial pressures of hydrogen and oxygen.

A very simple potential–pH diagram, showing only these three equilibria, is show in Figure 18.5. The shaded area represents the region of thermodynamic stability of water. Electrolysis cannot occur inside this region. Above and below it, oxygen and hydrogen evolution are thermodynamically possible. Whether these reactions will in fact occur at a measurable rate depends on their kinetic parameters.

The region of stability of water at 25 °C is 1.229 V, independent of pH, since the reversible potentials for hydrogen and oxygen evolution change with pH in the same manner. This, incidentally, is the potential region in which the hydrogen/oxygen fuel cell can operate. Thermodynamic considerations lead us to the

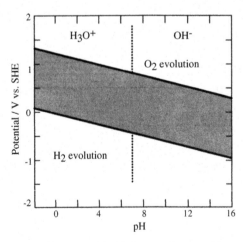

Figure 18.5 Potential–pH diagrams for water. The partial pressures of oxygen and hydrogen are taken as unity. The shaded area is the region of thermodynamic stability of water. Data from M. Pourbaix in *Atlas of Electrochemical Equilibria in Aqueous Solutions*, Pergamon Press, 1966.

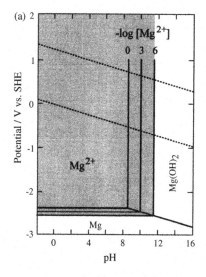

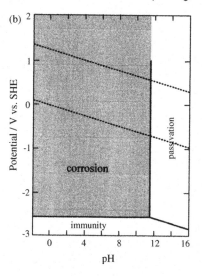

Figure 18.6 The potential–pH diagram for magnesium. (a) The detailed diagram. Lines correspond to different concentrations of Mg^{2+}, expressed as $-\log[Mg^{2+}]$. (b) Simplified form, defining regions of immunity, passivity and corrosion. Data from M. Pourbaix in *Atlas of Electrochemical Equilibria in Aqueous Solutions*, Pergamon Press, 1966.

conclusion that a water electrolyzer must operate at potentials *above* 1.229 V, whereas a hydrogen/oxygen fuel cell must operate at potentials lower than this value.[8] Typical values are $(1.6-2.0)$ V for the former and $(0.6-0.8)$ V for the latter. From the point of view of energy consumption or production, all we need to know is the cell voltage. From the point of view of corrosion, however, we shall see that the potential with respect to the reversible hydrogen electrode in the same solution is very important.

The solid lines in Figure 18.5 represent electrochemical equilibria. The dashed vertical line corresponds to the chemical self-ionization shown in Eq. (18.18), at equal concentrations of the two ions. The lines bounding the shaded area are the reversible potentials for oxygen and hydrogen evolution as functions of pH.

The effect of partial pressure of oxygen and hydrogen on the region of stability of water is rather small (cf. Eqs. (18.21) and (18.22)). For example, increasing the partial pressure of both gases from 1 atm to 10 atm will increase the potential by 44 mV, from 1.229 V to 1.273 V. This, incidentally, can be the basis for the technology of production of hydrogen or oxygen at high pressure by electrolysis of water, without the need to use a gas compressor.

Next, we consider the simple potential–pH diagram representing the behavior of magnesium in aqueous solutions, shown if Figure 18.6. For equilibrium between a solid and a soluble species, the concentration of the latter must be specified. In Figure 18.6a lines are shown for concentrations of Mg^{2+} of 1 μM, 1 mM and 1 M. In

---

8) The thermodynamic limit for a hydrogen-air fuel cell operating at ambient pressure is a little less, because the partial pressure of oxygen in air is only 0.2 atm. However, according to Eq. (18.22) the difference is only 10.3 mV.

the potential–pH diagrams these concentrations are represented as 6, 3 and 0, namely as $-\log[Mg^{2+}]$ which is similar to the definition of pH $\equiv -\log[H_3O^+]$. It is customary to simplify potential–pH diagrams by showing only the lines corresponding to a concentration of 1 µM of each soluble species. This convention is followed in all further potential–pH diagrams shown in this book. This is reasonable, in view of the fact that a moderate rate of corrosion may correspond to about 10 µA cm^{-2} or less, which cannot cause a significant accumulation of soluble corrosion products near the electrode surface, except in confined areas, such as pits or crevices. Figure 18.6b is a simplified form of Figure 18.6a.

The region of *immunity* is cathodic to the reversible potential of the metal, where it cannot be oxidized. Above it is the region of *corrosion*. Farther to the right, in alkaline media, there is a region in which the metal can be oxidized anodically, but the product is an insoluble oxide or hydroxide, in this case $Mg(OH)_2$. This is called the region of *passivity*, which is discussed below. Whether the metal will actually be passivated in this region depends on the nature of the oxide formed on its surface, in particular on its porosity and its ionic and electronic conductivity.

The potential–pH diagram for magnesium is relatively simple, because there is only one stable oxidation state of the metal ions and because magnesium is not amphoteric, namely, the oxide and hydroxide are not soluble in strong alkaline solutions.

The corrosion of aluminum represents a somewhat more complicated situation, since this metal is soluble both in acid and in alkaline media. The potential–pH diagram is shown in Figure 18.7. The most important feature in this diagram is

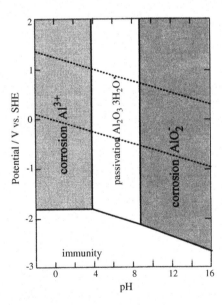

Figure 18.7 Potential–pH diagram for aluminum. The solid phase is assumed to be ($Al_2O_3 \cdot 3H_2O$). Filled areas represent regions where soluble species are stable and therefore corrosion can thermodynamically occur. Data from M. Pourbaix in *Atlas of Electrochemical Equilibria in Aqueous Solutions*, Pergamon Press, 1966.

a passivation region at intermediate pH values, with corrosion possible at both higher and lower pH. The two soluble species are Al^{3+} and AlO_2^-, and the lines representing their equilibria with the various solid phases correspond to a concentration of $1\,\mu M$, as explained above. The equilibrium between them is given by:

$$Al^{3+} + 6H_2O \rightarrow AlO_2^- + 4H_3O^+ \tag{18.23}$$

and the equilibrium constant for this reaction is given by

$$-\log \frac{\left[[AlO_2^-] \times [H_3O^+]^4\right]}{[Al^{3+}]} = pK = 20.3 \tag{18.24}$$

There is a very strong dependence on pH, and it follows from Eq. (18.24) that at pH $= 5.07$ the concentration of the two ionic species (AlO_2^- and Al^{3+}) should be equal. This should be represented by a vertical line at this pH value, like the equilibrium between H_3O^+ and OH^- in Figure 18.5. In the case of aluminum this is irrelevant, because in the range of about $4.0 \leq pH \leq 8.6$ only the solid phase is thermodynamically stable. There are two chemical equilibria, represented by the vertical lines in this figure, between the hydrated oxide $Al_2O_3 \cdot 3(H_2O)$ and the two ions, and three electrochemical equilibria between metallic aluminum, the two ions, and the oxide. We shall write here only the electrochemical equilibria and the corresponding Nernst equations.

The equilibrium with Al^{3+} is simple, and its reversible potential is independent of pH, since there are no protons or hydroxy ions involved:

$$Al^{3+} + 3e_M^- \rightarrow Al_M^0 \tag{18.25}$$

The corresponding Nernst equation is

$$E_{rev} = -1.663 + \frac{2.3RT}{3F} \log[Al^{3+}] \tag{18.26}$$

At intermediate pH values we have the equilibrium

$$Al_2O_3 + 6\,H_3O^+ + 6\,e_M^- \rightarrow 2Al_M^0 + 9\,H_2O \tag{18.27}$$

for which the reversible potential is given by

$$E_{rev} = -1.550 - \frac{2.3RT}{F} pH \tag{18.28}$$

In alkaline media the equilibrium to be considered is

$$AlO_2^- + 2\,H_2O + 3\,e_M^- \rightarrow Al_M^0 + 4(OH^-) \tag{18.29}$$

and the appropriate Nernst equation is

$$E_{rev} = -1.262 - \frac{4}{3}\frac{2.3RT}{F} pH + \frac{1}{3}\frac{2.3RT}{F} \log[AlO_2^-] \tag{18.30}$$

Aluminum represents an interesting case, which warrants further discussion. We note that the limit of the region of thermodynamic immunity lies at very negative

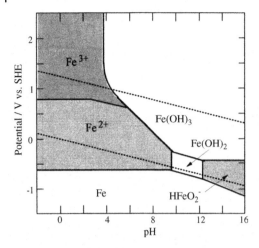

Figure 18.8 Simplified potential–pH diagram for iron. Vertical lines represent chemical equilibria. Horizontal lines correspond to electrochemical equilibria in which H_3O^+ and OH^- ions do not participate. Data from M. Pourbaix in *Atlas of Electrochemical Equilibria in Aqueous Solutions*, Pergamon Press, 1966.

potentials with respect to the lower limit of stability of water (which is the reversible potential for hydrogen evolution) at all pH values. Thus, one would expect rapid dissolution of this metal in any aqueous medium.[9]

This is indeed found in acid and alkaline solutions, but around neutral pH the oxide formed is very dense and non-conducting, and oxidation is effectively stopped after a thin layer of about 5 nm has been formed. This thin layer of oxide permits aluminum to be used as a construction material and in many other day-to-day applications. There are, of course, additional ways (e.g., anodizing and painting), to protect aluminum that is exposed to harsh environments, beyond the protection afforded by the spontaneously formed oxide film. However, the unique feature of this metal (and several others: e.g., titanium, tantalum and niobium) is that it re-passivates spontaneously when the protective layer is removed mechanically or otherwise, as long as the pH of the medium in contact with it is in the appropriate range shown in Figure 18.7.

Next we consider the Pourbaix diagram for iron, which is, of course, of paramount importance for the understanding of corrosion of ferrous alloys such as the many types of steel. This is a rather complex diagram, since two oxidation states of iron exist both as ions in solution and as solid oxides, and the metal is amphoteric to some extent. Figure 18.8 is a simplified version of the diagrams shown in the original work of Pourbaix. The two soluble species in acid solutions are Fe^{2+} and Fe^{3+}. The relevant equilibria are:

---

9) This statement is a little careless, since we cannot deduce the rate of a reaction from thermodynamic data alone. Yet when there is a very large driving force (i.e., when the system is far from equilibrium), the reaction will tend to be fast, unless some special mechanism prevents it or slows it down.

$$Fe^{+3} + e^-_M \rightarrow Fe^{2+} \qquad E^0 = +0.771 \text{ V SHE} \qquad (18.31)$$

and

$$Fe^{2+} + 2e^-_M \rightarrow Fe^0_M \qquad E^0 = -0.440 \text{ V SHE} \qquad (18.32)$$

At pH $\geq$ 12.5 pH, the anion $HFeO_2^-$ is thermodynamically stable, showing that iron can also be amphoteric to some extent.

On the basis of Figure 18.8 it can be concluded that corrosion of iron could occur for pH values of about 9 or less or for pH of about 12.5 or more. The region below pH 9 includes most natural environments with which structural materials are commonly in contact, making iron and many of its alloys vulnerable to corrosion. There are no soluble species between pH 9 and 12.5, and iron could pass directly from the immune to the passive region as the potential is increased. Also, it should be possible to passivate iron in solutions where the pH exceeds about 4, by oxidizing the surface, either chemically or electrochemically. In this case the potential can be forced to cross the active region (where Fe^{2+} is the stable species) rapidly, to reach the region of passivity, where $Fe(OH)_3$ is stable.

We shall conclude this section by making some general remarks on the advantages and limitations of potential–pH diagrams. It has already been stated that these are equilibrium diagrams. Hence, we can learn from them what *cannot happen* (e.g., a metal cannot be anodically dissolved in the region in which it is "immune", and water cannot be electrolyzed in the region of its stability, which is indicated in all such diagrams). We cannot deduce which reaction *will happen* at a measurable rate. The fact that at a certain pH and potential a metal can corrode according to its Pourbaix diagram is no proof that it actually will do so.

The regions marked as "passivation" only indicate that iron (usually an oxide or a hydroxide) is the thermodynamically stable corrosion product. Whether passivation will or will not take place depends on the nature of the oxide and on the environment in contact with it.

An important point to remember is that potential–pH diagrams are usually given for the pure elements. Now, high purity metals belong to the research laboratory, but nothing is ever constructed of a pure metal. In fact, the most important developments in the field of metallurgy have been in the area of new and improved alloys, designed to fit specific engineering requirements. The corrosion behavior of an alloy is rarely, if ever, a linear combination of the corrosion behavior of its components. Even for a given composition, the corrosion of an alloy usually depends on metallurgical factors such as the grain size and heat treatment of the material. An extreme example is the high corrosion resistance of so-called *glassy metals* or *amorphous alloys*, compared to alloys of the same composition in their usual crystalline form.

In spite of the foregoing limitations, the corrosion scientist and engineer can derive a wealth of information by consulting the relevant potential–pH diagrams. The regions of immunity, passivity, and corrosion are demarcated, and the most common corrosion products are shown. Studying the relevant diagram is an excellent way to start a new corrosion study, but it should never be the only tool used to solve the problem.

18.2.2
Passivation and Its Breakdown

Chemical passivation was discovered about 200 years ago. A piece of iron placed in concentrated nitric acid was found to be passive, while the metal dissolved readily in dilute HNO₃, with copious evolution of hydrogen. This type of behavior can be demonstrated in a very simple, yet quite spectacular, experiment. Nitric acid of various concentrations, from 1 mM to 70%, is introduced into a series of test tubes, and an aluminum wire is placed in each solution. No reaction is observed in the most dilute solutions. As the concentration is increased, however, hydrogen evolution becomes visible. At even higher concentrations, reduction of the acid takes place, in addition to hydrogen evolution. This is evidenced by the liberation of a brown gas, NO_2, which is one of the reduction products. When the concentration has reached 35%, the reaction suddenly stops. There is no gas evolution and the surface of the metal is not attacked. Accurate measurements show no weight loss when aluminum is kept in such solutions for months. Aluminum is passivated in concentrated HNO_3. A thin oxide film is formed on the surface and further attack is prevented.

Electrochemical passivation is in many ways similar to chemical passivation. As the potential of an iron sample is increased in the anodic direction, the rate of dissolution increases, reaches a maximum, and then decreases almost to zero. Further increase of the potential has little effect on the current in the passive region until passivity breaks down, whereupon the current rises rapidly with potential. The sequence of events observed on an iron electrode when its potential is swept very slowly in the positive direction is shown schematically in Figure 18.9. The peak

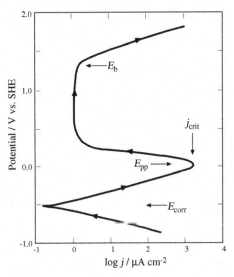

Figure 18.9 Schematic representation of the corrosion and passivation of iron in sulfuric acid. The primary passivation potential E_{PP} and the corresponding critical current density for corrosion j_{crit} are shown. Breakdown of the passive film occurs at $E > E_b$.

current density may be as high as $(1-10)$ mA cm^{-2}, whereas the current density in the passive region is of the order of $(1.0-10)$ μA cm^{-2}. The potential at which the anodic dissolution current has its maximum value, beyond which it starts to decline rapidly, is called the *primary passivation potential* E_{PP}. The corresponding current is referred to as the *critical corrosion current* j_{crit}. In the passive region, which may extend over half a volt or more, the current is nearly constant. It starts to rise again at the so-called *breakdown potential*, E_b, above which pitting corrosion occurs, along with oxygen evolution and electrochemical dissolution of the passive film. The anodic process taking place at such high potentials usually involves the transformation of the oxide to a higher oxidation state, which is often more soluble. In contrast, on so-called valve metals (e.g., Ti and Ta) the oxide continues to grow in thickness as the potential is increased. If there are no aggressive anions in solution, this anodization process can lead to very thick oxide films, up to tens of micrometers, on which oxygen evolution cannot occur.

Curves of the type shown in Figure 18.9 are obtained by sweeping the potential in the positive direction very slowly, at a rate of $(0.1-1.0)$ mV s^{-1}. Even so, steady state is not quite reached and the values of E_{PP} and E_b depend to some extent on sweep rate.

Another important quantity is the so-called *repassivation* potential, which is best explained in terms of a simple experiment. If the potential is held in the lower part of the passive region and the protective oxide is removed by scratching the surface, a large transient current is observed, but this current decays rapidly back to its value in the passive region, before the oxide has been mechanically removed. The surface is repassivated spontaneously at this potential. As the experiment is repeated at increasing anodic potentials, it is found that *repassivation* takes longer and longer, until a potential is reached beyond which the surface can no longer be repassivated. The repassivation potential is less positive than the breakdown potential. Thus, it would seem that there is a potential region where an anodic passive film cannot be formed on a bare metal surface, although an existing film is chemically and electrochemically stable. A true hysteresis of this type may indeed occur, although it has been argued that the breakdown potential and the repassivation potentials are one and the same, and the apparent difference observed between them is just a manifestation of the long induction period needed for breakdown at potentials close to E_b.

One of the unique features of a corroding metal undergoing passivation is a region of apparent negative resistance. Looking at Figure 18.9, we note that at potentials anodic to the primary passivation potential E_{pp}, the current density *decreases* with *increasing* anodic potential until it reaches the passive region. This is an unstable region in which the current keeps decreasing with time, even at constant potential, as a result of the formation and growth of the passive film.

It is interesting to consider how changes in the rate of the cathodic reaction can influence the open-circuit corrosion potential and the corresponding corrosion current. This behavior is shown schematically in Figure 18.10. Several cases can be distinguished. Line 1 crosses the anodic dissolution curve in the active region, leading to a substantial rate of corrosion. The situation is very similar to that shown in Figure 18.2: an increase in the rate of the cathodic reaction leads to a shift of

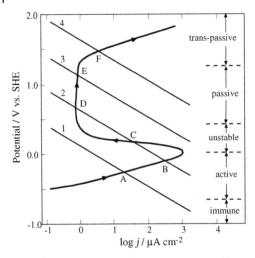

Figure 18.10 Schematic representation of the effect of changing the rate of the cathodic reaction on the corrosion potential and the corrosion current density in a system undergoing passivation.

the corrosion potential in the positive direction and an increase in the corrosion rate. Line 2 crosses the line for metal dissolution at three places. Point C is unimportant, since it is unstable, as pointed out earlier. This leaves the system with two stable corrosion potentials. Point B represents the usual situation found for an actively corroding metal. Increasing the cathodic current from line 1 to line 2 raises the corrosion rate from point A to B. But line 2 also crosses the anodic dissolution curve at point D, in the passive region. As a result, a different corrosion potential, corresponding to point D, could also be established. Note that, in the present example, the corrosion potential at point B represents a corrosion rate about 300 times that represented by point D, and, in practice, the ratio could be substantially higher.

Where will the system actually settle? That depends on the initial conditions. If the metal is initially passivated (by oxidizing it chemically or by increasing the potential in the anodic direction), it can remain passivated with the corrosion potential at D. If it is initially in the active region, it can establish its corrosion potential at B. This is referred to in the literature as unstable passivation, because passivation can be lost by transition from the passive region of potential at point D to an equally stable active corrosion potential at point B.

Line 3 represents stable passivation. The anodic and cathodic lines cross at a single point and a corrosion potential is set up at point E, well inside the passive region. However, increasing the cathodic current even more could drive the potential to point F in the trans-passive region, where corrosion and pitting could occur.

Passive films formed in aqueous solutions usually consist of an oxide or a mixture of oxides, usually in hydrated form. The oxide formed on some metals (e.g., Al, Ti, Ta, Nb) is an electronic insulator, while on other metals it has the properties of a semiconductor. Nickel, chromium, and their alloys with iron (notably the various kinds of stainless steel) can be readily passivated and, in fact, tend to be spontaneously

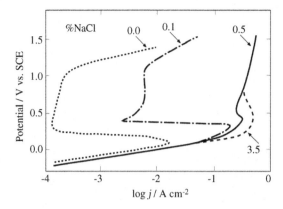

Figure 18.11 The effect of increasing the concentration of chloride ions on the passive current and on the range of potential over which passivity can be observed, for nickel in 0.5 M H_2SO_4. Data based on D.L. Piron, E.P. Koutsoukos and K. Nobe, *Corrosion*, **25**, 151, (1969).

passivated upon contact with water or moist air. It should be noted that passivation does not occur when chloride ions are introduced into the solution; and a pre-existing passive film may be destroyed.

When an experiment such as shown in Figure 18.9 is conducted in solutions of increasing concentration of NaCl, the behavior shown in Figure 18.11 is observed. Both the critical and the passivation current densities are increased and the breakdown potential becomes less positive until, at sufficiently high concentration of the Cl^- ion, passivation can no longer be observed.

If a solution containing NaCl is introduced into a cell where iron is held in the passive region, nothing happens at first, but after an induction period that depends, among others, on the concentration of Cl^- ions, the current will start to increase, and breakdown of passivity will be evidenced. Note that 3.5% NaCl is a typical concentration in sea water, so that Ni could not be passivated in see water. On the other hand, in fresh water the concentration of NaCl is of the order of $(0.003 – 0.03)$% so that passivity of Ni could be induced.

18.2.3
Localized Corrosion

18.2.3.1 Pitting Corrosion
Measurement of the average corrosion rate, per square centimeter of the sample, yields only part of the pertinent information, often a small part. Consider a piece of metal corroding in a given environment at a rate of 0.1 mm y^{-1}. This is a rather low rate, which may not worry the designer too much, if *corrosion were uniform*. Adding 1 mm to the wall thickness of a pipe, for example, would provide an additional service life of 10 years. On the other hand, if corrosion occurs on 1% of the surface, the same *average* corrosion rate would correspond to a penetration of 10 mm y^{-1} and one could not very well increase the wall thickness to 100 mm to provide a service life of 10 years.

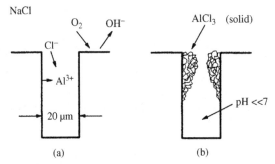

NaCl

Figure 18.12 Schematic representation of early stages in the formation of a pit. (a) The reactions taking place in and around a pit. (b) Formation of a deposit of corrosion product, partially blocking the exit of the pit.

Two important forms of localized corrosion are *pitting* and *crevice corrosion*. Although the causes of these phenomena may be quite different, the chemistry involved is similar and the following discussion is relevant to both.

Consider a pit formed in a piece of aluminum that is in contact with seawater. As we shall show, the pH of the solution inside a pit can become quite low, leading to an increased rate of corrosion, which further lowers the pH, and so on. Thus, pitting corrosion can be considered to be an autocatalytic process, with its rate increasing with time.

The main processes that take place in the pit and in its vicinity are shown schematically in Figure 18.12. At the concentration of chloride ions found in seawater, the passive layer on aluminum breaks down, and anodic dissolution of the metal can occur. This happens mostly inside the pit, where the supply of oxygen is low. On the other hand, oxygen reduction can readily take place on the surface of the metal outside the pits, where its diffusion path is short. Thus, the cathodic area is typically hundreds of times greater than the anodic area and the anodic *current density* inside the pit is hundreds of times higher.[10] Aluminum is being dissolved anodically inside the pit, forming Al^{3+} ions. To compensate for the excess positive charge, chloride ions must be transferred into the pit, as shown in Figure 18.12a. Since the current density inside the pit can be quite high, the ohmic potential drop becomes substantial. The current density at the mouth of the pit is the highest, and the rate of anodic dissolution declines with depth inside the pit. The end result is an accumulation of $AlCl_3$ near the mouth of the pit, the local concentration can exceed the solubility, leading to a precipitate of $AlCl_3$, which can partially block the pit as shown in Figure 18.12b. These are the conditions needed for the next stage-lowering the pH inside the pit.

The important point to remember in the context of pitting corrosion is that the volume of the solution inside the pit is very small. More accurately stated, the volume

---

10) Remember that at the open circuit corrosion potential, the total anodic and cathodic currents must be equal, but the current densities may be quite different.

of liquid per unit surface area is very low. For a deep cylindrical pit, this is given approximately by

$$\frac{\pi r^2 h}{2\pi rh} = \frac{r}{2} \frac{cm^3}{cm^2} \tag{18.33}$$

For a typical radius of $10\,\mu m$ this leads to a volume of about $5 \times 10^{-4}\,cm^3$ per cm^2 of surface area – about four orders of magnitude less than in a regular electrochemical cell. Inside the pits this can lead to rather unusual chemical phenomena that are not commonly encountered elsewhere.

$$Al\,Cl_3\,(solid) + 3\,H_2O \rightarrow Al(OH)_3\,(solid) + 3\,HCl \tag{18.34}$$

Partial hydrolysis, leading to the formation of species such as $Al(OH)_2^+$ and $Al(OH)^{2+}$ may also occur, but all such reactions lead to the formation of acid, making the solution inside the pit much more aggressive than outside. Similar reaction can occur during pitting corrosion of iron and its alloys. Eventually a stable pH is reached inside the pit. Its numerical value depends on the equilibrium constant in Eq. (18.34) or similar equations corresponding to other metals and alloys, depending on their composition.

Measurement of the pH inside a pit is not an easy matter, but estimates based on various calculations and on measurements in model pits lead to pH values as low as 1–2 for chromium-containing ferrous alloys and about 3.5 for aluminum-based alloys, depending on experimental conditions.

18.2.3.2 Crevice Corrosion

As the name indicates, this occurs in narrow spaces where an electrolyte can creep, usually by capillary forces, between two pieces of metal or between a metal and an insulator. This type of corrosion is usually found where two metals have been riveted together, under the head of a bolt, or under an insulating O-ring. It should be evident at this point that the chemistry and electrochemistry of crevice corrosion must be dominated by properties similar to those that dominate pitting corrosion, namely, the very small volume of the solution per unit of surface area. However, there are also two differences: first, we know exactly how a crevice is formed, while the phenomenon of pit initiation is not very well understood. The other difference entails the relative dimensions in pits and in crevices. The width and depth of a pit are of the same order of magnitude. This is certainly true in the initial stages of pit formation, but it can hold true even during more advanced stages of its propagation. In comparison, the width of a crevice is of the order of micrometers, while its depth can be of the order of millimeters. This brings to the forefront the importance of the jR_S potential drop in solution, in determining the local corrosion potential at different points in the crevice. In the Evans diagrams discussed so far (cf. Figures 18.1–18.4) this effect has been ignored. A detailed analysis, taking into account the solution resistance as a function of depth in a pit or a crevice may be rather complex, but the trends are very simple. We can view a corroding system as a battery with its terminals shorted. The current is then determined by the Gibbs energy of the reaction, which controls the potential, and by the resistance, which in this case is the sum of the solution resistance and the Faradaic resistances at the various interfaces. Clearly, some of the driving force for corrosion is dissipated by the solution resistance, and the corrosion rate is accordingly decreased.[11]

Differential aeration is also important whenever corrosion in a confined region is considered. The outer surface is accessible to oxygen and, therefore, becomes the site for the cathodic reaction. Inside a pit, the solution is rapidly depleted of oxygen, and this area becomes the site for the anodic reaction, namely, the dissolution of the metal. Since the outer surface is orders of magnitude larger than the surface inside a pit, very high *local* corrosion *rates* can take place. Differential aeration is not limited to crevices and pits, of course. One of the most common manifestations of this phenomenon is observed on partially immersed structures. Corrosion is commonly found to be most severe a short distance under the waterline, just below the oxygen-rich cathodic region, which is protected from corrosion. Farther down, the rate of corrosion is smaller because of the $j R_S$ potential drop in solution, as discussed earlier.

18.3
Corrosion Protection

The best method of corrosion protection is proper design. Unfortunately, corrosion engineers are often not members of design teams, and are left with the task of stopping the spread of corrosion, where it should not have occurred in the first place. Proper design, it must be admitted, is no minor feat. It is the best compromise among a number of factors, notably high strength, low weight, pleasing appearance and acceptable cost.

18.3.1
Bimetallic (Galvanic) Corrosion

Perhaps the worst (and most common) result of poor design is bimetallic corrosion. For example, if a copper faucet is connected to a water pipe made of low-carbon steel, a copper–iron battery is in effect formed, leading to rapid corrosion of iron, which is the more active of the two metals. Corrosion will be worst near the contact between the two metals, because the driving force farther along the tube is diminished by the potential drop across the solution resistance. This type of galvanic (bimetallic) corrosion is prevalent not only in immersed structures or in pipelines carrying an ionically conducting liquid, but also in metals exposed to humid atmospheres, where the ionic path of conductivity is established through a thin film of moisture accumulating on the surface. If the

11) Unlike in the case of batteries, the loss of driving force is a welcome effect in corrosion. It is well known, for example, that the corrosion rate in poorly conducting solutions is lower than that in highly conducting media.

use of two or more metals is unavoidable, the best way of combating bimetallic corrosion is to isolate the different metals electrically. If this is not possible, one should attempt to use metals that have very similar corrosion potentials, to ensure that the potential developed between them, upon immersion in the same solution, will be minimal.

How can we tell which will be the more active metal, or what metals will be about equal? Theory provides a very simple answer. The more positive the standard potential, the more noble the metal. On this scale, gold is the most stable and magnesium, along with most of the lanthanides, is the most active metal. Alkali metals are not considered to be of interest as construction materials, although Li/Al alloys have been used to reduce the weight, and the prevention of corrosion of such alloys poses significant challenges. Iron and zinc are more noble than aluminum and titanium on this scale, and copper is more noble than niobium. Unfortunately, this thermodynamic scale has little to do with reality, for a number of reasons. Metals such as niobium, tantalum, zirconium, titanium and aluminum form protective anodic films spontaneously when brought into contact with humid air or water. This makes them more noble than iron and even copper on a practical scale, which is what matters in engineering design. Thus, in an aluminum structure connected with steel rivets, the rivets will act as the anodes in bimetallic corrosion, although the standard potentials of aluminum and iron are -1.66 V and -0.44 V vs. SHE, respectively.

In terms of the Pourbaix potential–pH diagrams, the thermodynamic scale compares the potentials of immunity of the different metals, while the practical scale compares the potentials of passivation. But this is not enough either. The real scale depends on the environment with which the structure is in contact during service. Passivity, as we have seen, depends on pH. It also depends on the ionic composition of the electrolyte, particularly the concentration of chloride ions or other species that are detrimental to passivity. Finally, one must remember that construction materials are always alloys, never the pure metals. The tendency of a metal to be passivated spontaneously can depend dramatically on alloying elements. For example, an alloy of iron with 8% nickel and 18% chromium (known as 8–18 stainless steel) is commonly used for kitchen utensils. This alloy forms a passive layer spontaneously and should be ranked, on the practical scale of potentials, near copper. If the concentration of chromium is increased by 10%, its position with respect to copper does not change significantly. If, on the other hand, it is decreased by the same amount, the alloy will no longer function as stainless steel. Its potential in solution will be much more negative, in the vicinity of iron itself. Similarly, alloying aluminum with a few percent of tin affects its ability to form a passive film and causes its open-circuit potential to shift to much more negative values. The effect of different alloying elements on the corrosion stability is nontrivial and cannot, as a rule, be predicted from theory. The best way to determine whether two metals are going to be compatible, from the point of view of galvanic corrosion, is to measure their potentials for extended periods of time, in a medium as close as possible to that which will be encountered in service.

18.3.2
Cathodic Protection

Cathodic protection can be viewed as a form of galvanic corrosion, put to good use. In this case an active metal (most often zinc, but under certain circumstances magnesium or aluminum are also used) is employed as a *sacrificial anode*. It is attached to the steel structure being protected in one or several locations, but does not constitute part of the structure itself. The steel structure becomes the site of the cathodic reaction, and its potential is driven in the negative direction so that the rate of corrosion is slowed down to an acceptable level. The sacrificial anode is, of course, corroding at a relatively high rate and must be replaced periodically, but damage to the structure by corrosion can be minimized.

Sacrificial anodes are most commonly employed to protect the hulls of ships and boats, for offshore oil rigs and underground pipelines. They are designed to be replaced, if necessary, during routine maintenance. But these are not their only uses. Galvanized steel, which is a low-carbon steel coated with a thin layer of zinc, functions in a similar manner. The coating provides a certain degree of protection as long as it is intact. When it is partially removed, either by corrosion or abrasion, the exposed surface is still protected cathodically and does not corrode. In contrast, a Ni/Cr coating may provide much better protection as long as it is intact, but there will be severe local galvanic corrosion in the area of a scratch, since the coating will act as a large-area cathode and the exposed steel surface as a small-area anode – a very bad combination.

The theory of cathodic protection is simple and straightforward. The real engineering challenge is to design the anodes and position them so that they provide uniform current distribution on the part being protected. There are two aspects to this problem: parts of the structure that are too far away from the anode, or screened from it, may not have sufficient protection, and parts that are too close to the anode may be "overprotected". We recall that the cathodic reaction in most cases is hydrogen evolution or oxygen reduction. Both reactions lead to the formation of OH^- ions near the surface. This can weaken the bonding of paints and other nonmetallic coatings to the surface, causing delamination. Also, excessive hydrogen evolution can lead to penetration of atomic hydrogen into steel, causing embrittlement, which could lead to devastating structural damage. Thus, under certain circumstances, overprotection can be worse than no protection at all.

There are no simple equations from which the current distribution on structures having complex geometries may be obtained. A number of numerical methods can be used, however, and there is a wealth of practical experience, which can serve at least as a very good "first guess" in a calculation involving many successive iterations. The real problem lies in the fact that conditions during service life change, not always in a predictable manner. Thus, the protective coating on the hull of a ship or on a pipeline may be damaged with time, changing the effective area that needs to be protected. Rain or drought will change the conductivity of the soil in which a pipeline is buried. The supply of oxygen to the immersed parts of a boat changes dramatically when it lifts anchor to sail at full speed. The conductivity of the water changes by several orders of magnitude when a ship sails from a river into the open sea. All these factors change the current distribution. If designed for optimum protection at sea, a ship will be overprotected in fresh water and vice versa.

The best way to overcome the limitations of cathodic protection related to changes in the environment is to monitor the potential and adjust the cathodic currents accordingly. This cannot be readily done with sacrificial anodes, and the method of impressed-current cathodic protection is sometimes preferred, in spite of its higher cost.

Impressed-current cathodic protection entails the use of an external power source in combination with a stable anode. The potential of the specimen being protected is forced to negative values and its rate of corrosion is consequently reduced. The result of impressing a cathodic current on the structure is shown in Figure 18.13: For the parameters used to draw this figure we obtain $j_{corr} = 48.8\ \mu A\ cm^{-2}$ and $E_{corr} = -0.554$ V vs. SHE. Applying a cathodic current density of 72 $\mu A\ cm^{-2}$ shifts the potential to -0.58 V and the anodic current density (which represents the rate of corrosion of the cathodically protected surface) is reduced to 10 $\mu A\ cm^{-2}$, as shown by line 1.

The current flowing in the external circuit is the difference between the cathodic and the anodic currents flowing at this potential. Thus, application of a current of 72 $\mu A\ cm^{-2}$ causes a decrease in the rate of corrosion by about a factor of five. The rate of corrosion can be reduced farther by increasing the applied cathodic current, as shown by line 2 in Figure 18.13.

It is well to remember that using the impressed current cathodic protection does not alleviate the need to calculate the current distribution and locate the anodes in a way that ensures the best uniformity of current distribution on the protected structure.

Cathodic protection is usually not used by itself. A pipeline buried in the ground is painted or coated for added protection against corrosion. Ideally, such coatings

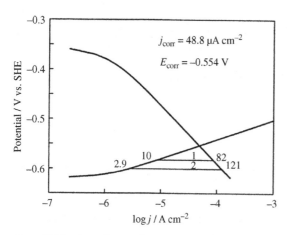

Figure 18.13 Evans diagram for iron at pH 6, showing the principle of impressed-current cathodic protection. The two horizontal lines show two levels of cathodic protection. The impressed current is the difference between the cathodic and the anodic currents shown.

should provide complete protection, but in service they never do. It is hard to tell how much of a coating is initially damaged and it is usual to determine the potential at which one wishes to operate, adjusting the impressed current accordingly. An average cathodic current density for protection may typically be $(1-2)\,\mu A\,cm^{-2}$. If only 1% of the coating is initially damaged, this corresponds to $(100-200)\,\mu A\,cm^{-2}$ on the small regions where the coating has been damaged and the underlying metal needs protection. Moreover, the potential can be monitored continuously at different locations, and the impressed current can be adjusted automatically in order to maintain the desired level of corrosion protection.

Impressed-current cathodic protection requires a little more sophistication than the use of sacrificial anodes, but it also lends itself to periodic adjustment and provides higher flexibility, particularly when structures having rather intricate shapes are considered.

18.3.3
Anodic Protection

Anodic protection makes use of the ability of iron and many of its alloys to become passive in the absence of Cl^- and other aggressive ions, as discussed in Section 18.2.2. If we extend Figure 18.9 to show the current during one cycle of the potential from the active to the transpassive region and back to the passive region, the behavior observed will be that shown schematically in Figure 18.14. Setting the potential anywhere between E_{pp}, the *primary passivation* potential, and E_{rp}, the re-passivation potential, causes the metal to be passivated. Between E_{rp} and the breakdown potential E_b, metastable pits can be formed, as indicated by the bursts of current shown in Figure 18.14.[12]

Systems behaving in the manner shown in Figure18.14 can be protected anodically. It is important to use the term "systems" rather than "metals" or "alloys" because the ability to form a stable passive film depends on both the nature of the metal or alloy and the electrolyte in contact with it, as pointed out earlier (cf. Figure 18.11). As a result, anodic protection cannot be used as universally as cathodic protection. In particular, its use is limited to situations where chloride ions are absent, a limitation that excludes it from all applications in or near the sea.

In cases where it is applicable, anodic protection has great advantages over cathodic protection, for a number of reasons. First, it requires typically only $(1-2)\,\mu A\,cm^{-2}$, about two orders of magnitude less than cathodic protection. In addition to the saving in energy, the negative side effects of cathodic protection – namely hydrogen embrittlement and delamination of nonmetallic coatings resulting from the high pH generated near the cathodic sites are eliminated.

---

12) The bursts of current result from meta-stable pits, where breakdown of the film, followed by a rapid repassivation, occurs. As the potential is increased, breakdown is more likely and repassivation is slower, until the potential E_b is reached, beyond which breakdown is the predominant process.

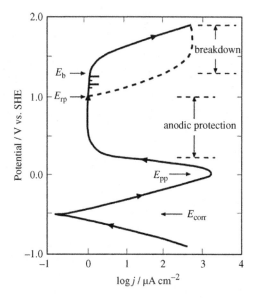

Figure 18.14 Current–potential characteristic of a system undergoing passivation. The optimum potential region for anodic protection is shown. E_b, E_{rp}, E_{pp} and E_{corr} are the breakdown potential, the re-passivation potential, the primary passivation potential, and the open circuit corrosion potential, respectively.

Since anodic passivation is performed potentiostatically, and since the currents involved are very small, uniform current distribution is easier to maintain and overprotection is not likely to occur.

Unfortunately, anodic protection is limited to certain environments, where the liquid in contact with the protected structure is well defined and known to allow passivation. There is, however, a widely used chemical form of anodic protection, entailing paints that contain strong oxidizing agents such as $K_2Cr_2O_7$ and Pb_3O_4. The corrosion protection afforded by such paints can be the result of a number of mechanisms operating in parallel. One of them is the high positive potential set up by the oxidizing agent, bringing the metal into the passive region.

18.3.4
Coatings and Inhibitors

Finally, we shall discuss very briefly coatings and inhibitors used to prevent or slow down corrosion. Coating can be considered in two groups: *active coatings* such as zinc, which acts as a sacrificial anode, even after parts of the underlying metal have been exposed to the environment, and *barrier coatings* such as paints of all sorts and protection by a more noble metal, such as nickel/chrome on steel and silver or gold on copper. Such coatings prevent corrosion by simply isolating the metal from the environment. They can be excellent as long as they are intact. However, once damaged, galvanic corrosion (in the case of more noble metals) and differential aeration (in the case of nonmetallic coatings) may lead to an increase in the rate of pitting corrosion on the exposed areas.

In this context we might mention that surface preparation is a major factor in obtaining good, adherent coatings of any type. Degreasing, chemical cleaning and, in some cases, mechanical treatment are essential steps in the preparation of the surface for coating, which often consists of several layers, for optimum protection.

Corrosion inhibitors are commonly used to prevent corrosion. There are hundreds of different inhibitors in commercial use. Some act by slowing down the cathodic reaction and others inhibit the anodic reaction. Some are ionic and some are neutral. In choosing a suitable corrosion inhibitor, it is important to know the corrosion potential with respect to the potential of zero charge, E_z. If $E_{corr} > E_z$, the excess surface charge density will be positive, and a negatively charged inhibitor may be the better choice. If E_{corr} occurs at a negative rational potential, a cationic inhibitor may be preferable. Neutral molecules can best serve as inhibitor if $E_{corr} \approx E_z$. It should be borne in mind that many of the commercial inhibitors are weak acids or bases, and their charge depends on pH. Thus, an inhibitor that acts well in a medium of low pH may be quite useless in a medium of high pH, and vice versa. When the usual aqueous medium is replaced by a non-aqueous or mixed solvent, the situation can change dramatically. The charge on the inhibitor molecule may be quite different, and E_z also depends on the solvent. The solubility of an inhibitor used in aqueous solutions may also be quite different in a non-aqueous solvent. This changes the surface coverage corresponding to a given bulk concentration. To a first approximation, the surface coverage should be similar in different solvents, when comparison is made on a normalized scale of concentration $c_b/c(sat)$, obtained by dividing the concentration in each solvent by its saturation value. Thus, a different range of concentrations of inhibitor may have to be used for each solvent.

19
Electroplating

19.1
General Observations

19.1.1
Introduction

Electrodeposition has been practiced industrially for over 150 years. For most of its history it was considered to be a low-tech empirical technology. Although many useful plating baths have been developed and additives identified for different purposes: uniformity of plating thickness, brightness, relieving of stress, increasing current efficiency, and minimizing hydrogen embrittlement, most of the progress was made by ingenious trial-and-error methods, with relatively little effort to determine the mechanism involved.

A major turning point in the above approach can be associated with the replacement of chemical vapor deposition of aluminum by electroplating of copper for wiring in ultra-large-scale integration on silicon chips, announced by IBM in 1997. This led to increased interest in electroplating in the microelectronic industry and to awareness of possible advancements through research and development, both in industry and in academia. The introduction of copper plating as an integral part of manufacturing microprocessors was no minor feat, and came after about a decade of intense research and development. Following its success, electroplating was elevated from the status of an empirical technology to that of a high technology, based on research and profound understanding of the way the different factors influence the quality of the product.

Much of what has been stated above regarding the electroplating of metals applies also to that of alloys. It was pointed out in Section 18 that metals are rarely used in the pure form. In contrast, electrodeposition leads often to the creation of coatings consisting of the pure metal, or of several layers of pure metals, one on top of the other, each serving its special purpose. Nevertheless, alloy deposition is not uncommon. Alloys of Pb–Sn for printed circuits and of Fe–Ni as soft magnets in the recording industry have been used for a very long time. More recently there is great interest in Pt–Co alloys serving as hard magnets in micro-electro-mechanical systems

Physical Electrochemistry: Fundamentals, Techniques and Applications. Eliezer Gileadi
Copyright © 2011 WILEY-VCH Verlag GmbH & Co. KGaA, Weinheim
ISBN: 978-3-527-31970-1

(MEMS). Electroplating of alloys of W and Re with Ni or Co have also gained interest in recent years where high temperature or high abrasion resistance is required.

There are several advantages to electroplating of metals and alloys, over methods of chemical or physical vapor deposition. (CVD and PVD). These include, among others, low cost, low temperature application, uniformity of thickness, or inversely, designed nonuniformity (i.e., the ability to coat only on specific areas on the surface).

Last but not least, metal deposition presents an interesting challenge to the understanding of the mechanism of charge transfer across the metal/solution interface. Alloy deposition also presents special challenges of understanding of the details of the processes taking place across the interface. There is only a limited number of cases (for example Pb–Sn) for which alloy deposition is *"normal"*, in the sense that the composition of the alloy can be predicted from the behavior of the alloying elements, each by itself. In most cases (for example Ni–Re) it is *"anomalous"* in the sense that the composition of the alloy cannot be predicted from the known thermodynamic and kinetic parameters of the alloying elements. Yet another class is *"induced codeposition"* (for example W–Ni) in which only one of the alloying elements can be electrodeposited by itself. In this specific case, W cannot be deposited alone, but it can readily be deposited as an alloy with Ni, for example.

19.1.2
The Fundamental Equations of Electroplating

The fundamental equation for the overall reaction of metal deposition is

$$M^{z+} + ze^- \rightarrow M \tag{19.1}$$

Where M is any metal with a valency z. But this equation does not specify the phase in which each species is located and, more importantly, it does not take into account the role of the solvent. Both can be accounted for by rewriting this equation as

$$\left[M(H_2O)_n\right]^{z+} + ze_M^- \rightarrow M_M^0 + n(H_2O) \tag{19.2}$$

where the number of water molecules in the hydration shell is not specified because it may be different for different metals. It is important to realize that the hydration energy of ions in solution, U_{hyd}, is very high, and can be expressed approximately by

$$U_{hyd} \approx 5 \times z^2 \text{ eV} \tag{19.3}$$

It is not implied here that all ions having a given valency possess exactly the same energy of hydration. For any given valency, smaller ions have higher energies of hydration, but the factor of z^2 does show that valency is the main factor determining this energy. Thus, for monovalent ions, the values of U_{hyd} cluster around 5 eV, for divalent ions U_{hyd} is around 20 eV and for trivalent ions it is around 45 eV.

Now, in order to deposit a metal ion from solution it is necessary to remove all its solvation shell, which requires a lot of energy. This is not a thermodynamic barrier, since much of this energy is regained in the overall process. However, the high energy of hydration is expected to play a major role in the kinetics of the deposition process, because this energy, or at least part of it, has to be supplied in order to allow the metal ion to reach the surface and interact with it. Thus, one would expect that metal deposition processes would be very slow, compared to outer-sphere charge transfer processes. This is not borne out by experiment. Indeed, the exchange current densities observed for metal deposition are often as high, or even higher, than those for outer-sphere charge transfer processes. The mechanism of charge transfer during metal deposition and dissolution, providing a mechanism consistent with this behavior is discussed in Section 19.7.

19.1.3
Practical Aspects of Metal Deposition

In this chapter the emphasis is not on the mechanistic aspects of metal deposition, but rather on its practical aspects, such as uniformity of plating, the solution chemistry, alloy deposition, and so on. Diffusion of adatoms on the surface, from their initial landing site to edges, kinks or vacancies on the surface, (where they are thermodynamically more stable, because their degree of coordination is higher) has been considered to be the rate-determining step for metal deposition in some cases. Impurities or additives adsorbed on the surface can hinder such diffusion and can control the surface morphology of the resulting deposit. These aspects of metal deposition are not discussed here.

It is interesting to consider the metal deposition process from a microscopic point of view. A rate of 20 mA cm^{-2}, corresponds to the deposition of about 50 atomic layers of metal atoms per cm^2 s^{-1}. This may be too fast for the adatoms to reach their equilibrium positions. It is indeed observed that the alloys formed during electro-deposition are not necessarily those corresponding to the phase diagram of the alloys involved at room temperature.

The interaction between the substrate and the metal being deposited can also play an important role in determining the quality of a plated product. When the crystal parameters of the two metals are different, one of two situations may be observed. The metal being plated may initially attain the crystal structure of the substrate, although this is not its most stable form. This is referred to as *epitaxial growth*. As the thickness of the deposit grows, it gradually reverts to its stable crystal structure. The stress created by the epitaxial growth can be relaxed by impurity atoms and by dislocations in the metal. Large differences between the crystal structures of the two metals do not favor epitaxial growth. In such cases a so-called *crystallization over-potential* is observed, followed by two-dimensional nucleation on the surface. The effect is similar to the formation of small crystals in a supersaturated solution or of droplets in the vapor phase[1].

Side reactions, mostly hydrogen evolution, play an important role in electroplating. As a rule, their effect is detrimental to the process because of the loss of energy and possible hydrogen embrittlement.

---

1) Note that a tenfold supersaturation is equivalent, in terms of Gibbs energy, to an overpotential of only 29.5 mV for deposition of a divalent metal.

Finally, there is the question of uniformity. Parts to be plated are rarely flat. They have grooves, edges, corners, protrusions, and so on. A good plating bath, which covers a surface uniformly irrespective of its shape, is said to have a high *throwing power*. The factors controlling the throwing power of plating baths are discussed in detail in Sections 19.2 and 19.3, where some of the practical aspects of electroplating are discussed.

19.1.4
Hydrogen Evolution as a Side Reaction

Plating from aqueous solutions is strongly influenced by the competing reaction of hydrogen evolution. Considering the standard thermodynamic potentials of a metal, it is noted that Ag ($E^0 = +0.80$ V) and Cu ($+0.34$ V SHE) have positive potentials with respect to the SHE, and therefore can be deposited even from acid solutions with 100% faradaic efficiency. Pb and Sn ($E^0 = -0.13$ V and -0.14 V SHE, respectively) can also be deposited readily, because the exchange current density for hydrogen evolution on these metals is very small, and the rate of hydrogen evolution at potentials where these metals are deposited is negligible. Proceeding to more active metals, such as Cd (-0.40 V) and Zn (-0.76 V), electroplating is still possible, but the current efficiency is usually lower. This leads, of course, to a loss of energy, but that is a minor consideration in electroplating. The worst aspect of hydrogen evolution during metal deposition is the formation of atomic hydrogen on the surface, as an intermediate. A fraction of these atoms can diffuse *into the metal*, causing hydrogen embrittlement that could cause catastrophic failure of the structure. For example, the landing gears of aircraft are made of high-strength steel, which is very sensitive to hydrogen embrittlement that can occur during electroplating by Cd, to protect it from corrosion. When the current efficiency is very low (10–20%), heavy hydrogen evolution takes place. Copious formation of hydrogen bubbles may screen parts of the surface from the solution, leading to non-uniform coating. In addition, a safety hazard may exist unless the plating shop is very well ventilated.

In the case of even more active metals, such as Al and Mg, with E^0 values of -1.66 V and -2.37 V SHE, respectively, the side reaction becomes so dominant that metal deposition from aqueous solutions can no longer take place.

It should be noted in this context that in many cases of metal and alloy deposition the bath contains some complexing agent. This can lead to better uniformity of the metal coating and improved brightness, but it also shifts the deposition potential to more negative values, leading to enhanced hydrogen evolution and a corresponding lowering of the current efficiency.

19.1.5
Plating of Noble Metals

It was stated above that Ag and Cu, with their standard potential positive with respect to the SHE, are about the easiest metals to plate. Based on this argument it would be expected that noble metals of the Pt group (Os, Ir, Pt) could also be electroplated readily. This is not the case, because these metals do not form stable hydrated ions, and most plating solutions contain a complex with a suitable ligand, driving the standard potential in the negative direction.

Choosing a suitable ligand often presents a challenge, because the complex has to fulfill two inherently opposing requirements. On the one hand, the ligands should bond to the noble metal cation strongly enough to ensure stability of the resulting complex in an aqueous solution. On the other hand, they should bond weakly enough to allow deposition of the noble metal. For the case of Pt, the two commonly employed complexes are $[PtCl_4]^{2+}$ and $[Pt(NO_2)_2(NH_3)_2]^0$. Interestingly, fresh solutions of these complexes do not perform well. In the case of the chloro-complex, aged solutions perform much better than freshly prepared ones. For the nitroso-amino complex shown above, aging is not enough. The performance of a freshly prepared solution is improved with subsequent plating operations. Recent studies indicate that the inherently conflicting requirements of the ligand have been solved in an elegant way. The freshly prepared solution is good for the stability of the bath, but is too stable for the plating operation itself. However, aging or passage of charge in the solution causes the replacement of one or more ligands by water, forming a less stable complex that allows efficient plating or the noble metal. Even so, deposition occurs at rather negative potentials where hydrogen evolution takes place, in spite of the fact that the standard potential for Pt deposition, for example, is $+1.19$ V SHE

19.2
Current Distribution in Plating

19.2.1
Uniformity of Current Distribution

A plating bath should produce a uniform thickness of plating. If metal deposition is the only reaction taking place, variations in the thickness of the deposit are an expression of the uniformity of current distribution pertaining to the specific conditions of plating. This, in turn, depends on the composition of the bath, which determines the conductivity in solution and has an influence on the kinetics of deposition. It also depends on the applied current density, the temperature and the geometry of the part being plated. Edge effects can be troublesome and the position of the counter electrode(s) with respect to the working electrode (which is the part being plated) can play a major role, sometimes overcoming the inherent limitations of the plating bath.

Nonuniformity of the current density at the edge of the substrate being plated was discussed in Section 3.2. The current distribution was discussed there for three simple geometries. In the real world, the geometry is rarely simple, and the current distribution cannot be calculated using the analytical solution of an equation. It can, however, be determined by numerical calculations. Commercial software is available to simulate different positions and shapes of the counter electrode with respect to the object being plated, eliminating the need for trial-and-error methods of obtaining the highest degree of uniformity. Incidentally, the same kind of software can be used to create any specific nonuniformity of coating thickness that may be required for specific applications.

The expected uniformity of current distribution of a given plating bath can be estimated from its conductivity and kinetics, as discussed in the next section

19.2.2
The Faradaic Resistance, R_F and the Solution Resistance, R_S

The resistance between the working electrode being plated (the cathode) and the counter electrodes (the anodes) can be represented by two resistors in series, as shown in Figure 19.1.

19.2.3
The Dimensionless Wagner Number, W_a

In considering Figure 19.1 it should be borne in mind that R_F is independent of the geometry of the cell, because the potential drop across this resistance occurs over a distance of less than 1 nm, while the solution resistance extends over the typical macro-dimensions of a plating bath, which is of the order of a few cm, namely about 10^7 nm. Thus, for the case shown in Figure 19.1a, the potential drop over R_F is only 0.10 V, while that over R_S is 1.0 V. this represents primary current distribution. Since R_S depends on the cell geometry, this will lead to nonuniformity of current

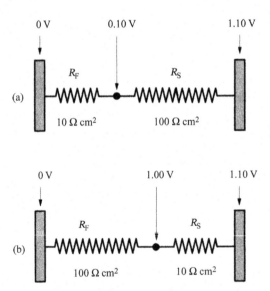

Figure 19.1 The value of R_F and R_S and the corresponding change in potential between the working and the counter electrodes at a plating rate of 10 mA cm^{-2}. (a) Primary current distribution; $W_a = 0.1$. (b) Secondary current distribution; $W_a = 10$.

distribution. In contrast, Figure 19.1b represents secondary current distribution. Most of the potential drop will occurs across the Faradaic resistance, and changes in the cell geometry have little effect on the total resistance between the anode and the cathode, leading to nearly uniform film thickness.

The Wagner number is a dimensionless parameter that helps us determine the probable level of uniformity of current distribution in electroplating. There are two limiting cases to be discussed: (i) *Primary current distribution*, where the uniformity of plating is determined exclusively by the conductivity of the solution and the geometry of the cell. (ii) *Secondary current distribution*, where the current distribution is determined by the kinetic parameters of the deposition process. As usual, reality is somewhere in between, it is neither purely primary nor purely secondary. The transition from one regime to another is characterized by the dimensionless *Wagner number*, defined as:

$$W_a = \frac{(\partial \eta / \partial j)_{c_i}}{\rho L} \tag{19.4}$$

where ρ is the specific resistivity of the solution, in units of Ω cm, and L is a characteristic length.

The choice of the characteristic length L is not always obvious. It may be the length of the electrode being plated, the distance between the working and the counter electrodes or "the dimension of the irregularity" – for example, the difference between the shortest and the longest distance between the two electrodes. The last is the best choice in most cases, since it reflects the differences in the solution resistance on different areas of the piece being plated. The exact value taken is not critical, however, since W_a should be used only as a guideline, and the actual current distribution (or variation of thickness of plating) can be found either experimentally or by obtaining a numerical solution for the specific geometry and the kinetic parameters of the metal deposition considered.

The partial derivative, taken at constant concentration (and, of course, constant temperature and pressure), is the differential Faradaic resistance, R_F, in units of Ω cm^2. The solution resistance, expressed in the same units, can be written as:

$$R_S = \rho L \tag{19.5}$$

Hence, the Wagner number can be expressed simply as

$$W_a = \frac{R_F}{R_S} \tag{19.6}$$

In the absence of mass transport limitations, the local current density at a given potential is determined by the sum of two resistances in series: the Faradaic resistance and the solution resistance. For values of $W_a \ll 1$ the solution resistance is dominant and the current distributions depends primarily on geometry. This is the realm of primary current distribution. For $W_a \gg 1$ the Faradaic resistance is predominant and secondary current distribution is observed.

It is interesting to consider the value of the Faradaic resistance at different overpotentials. Close to the reversible potential, the Faradaic resistance is given by

$$R_F = \frac{\eta}{j} = \frac{1}{j_0} \frac{RT}{nF} \tag{19.7}$$

It is independent of potential and of the applied current density, but inversely proportional to the exchange current density, because the current–potential relationship is linear in this region (cf. Section 5.2.4). In this region the Wagner number is also inversely proportional to the heterogeneous rate constant of the metal deposition reaction. Thus, fast reactions have low value of the Wagner number and tend to lead to primary current distribution. This is a rather unique situation in electrochemistry, where poor catalytic activity (i.e., low specific rate constant) is an advantage.

Considering the high-overpotential region, the relevant rate equation is

$$j = j_0 \exp\left(-\frac{\alpha_c \eta F}{RT}\right) \tag{19.8}$$

It follows that the Faradaic resistance in this region is given by

$$\frac{1}{R_F} = \left(\frac{\partial j}{\partial \eta}\right)_{c_i} = \frac{\alpha_c F}{RT} j \tag{19.9}$$

$$R_F = -\frac{1}{j} \frac{RT}{\alpha_c F} \tag{19.10}$$

It is concluded that at high overpotentials the Faradaic resistance is inversely proportional to the applied current density, but independent of the exchange current density. Assume that one compares two electrodeposition reactions having values of $j_0 = 10^{-4}$ A cm^{-2} and $j_0 = 10^{-6}$ A cm^{-2}, and deposition is conducted at a current density of $j = 10^{-2}$ A cm^{-2}. Both reactions will be in the linear Tafel region, and their Faradaic resistance will be equal, as long as their Tafel slopes are equal. This is the reason for the often-used practice of plating initially at a low current density, to increase the Wagner number and thus achieve better uniformity of current distribution, at least for a thin layer of coating in recessed areas, followed by plating at a higher current density, to keep the time needed to reach the desired thickness reasonable.

There is an important, yet mostly overlooked, point in the argument above. When the Faradaic efficiency is less that 100% (which is often the case) the value of j determining the Wagner number is the partial current density for deposition of the metal, not the applied current density. This is particularly important in the case of alloy deposition, where each metal has its own concentration, as well as its own partial and exchange current densities, as we shall see in Section 19.5.

The dependence of the Wagner number on the current density applied, for different values of the exchange current density, is shown in Figure 19.2. The region

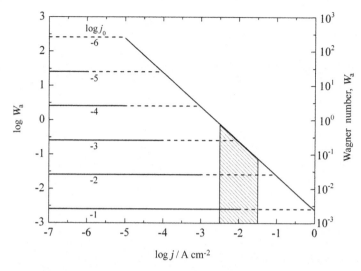

Figure 19.2 The Wagner number, calculated for different values of the exchange current density, as a function of the applied current density. $L = 2$ cm $b = 0.12$ V $\rho = 25\ \Omega$ cm, $n = 2$.

of $-\log j = 1.5{-}2.5$, (approximately to $3.0{-}30$ mA cm^{-2}) is typical for many electroplating processes and, therefore, warrants some discussion. At low current densities, where $j/j_0 \leq 0.1$, the Faradaic resistance is independent of the current density (cf. Eq. (19.7)) but is proportional to $1/j_0$. At high current densities, where $j/j_0 \geq 10$, it is independent of j_0 (cf. Eq. (19.10)), but it is proportional to $1/j$. Needless to say, the transition from one region to another is smooth, not sharp as might be implied in this figure. Consider a deposition current density of 10 mA cm^{-2}. The extrapolation of the two straight lines in Figure 19.2 cross, for $j_0 = 10^{-3}$ A cm^{-2}, at $W_a \approx 0.25$. For values of $j_0 \leq 10^{-4}$, the system will be in the region where the Wagner number is independent of j_0. At higher values of the exchange current density, where $j_0 \geq 10^{-2}$, it will be inversely proportional to j_0.

Figure 19.2 shows that at the current densities of interest in the plating industry, the Wagner number is less than unity, making it difficult to obtain good macrothrowing power. Hence uniform plating will not be attained on complex shapes of the cathode.

This leads us to discussion of the methods by which the throwing power can be increased. The most common method is to use suitable additives, as discussed in Section 19.3.3. Indeed almost all industrial plating baths contain additives of one kind or another, used to improve the uniformity of coating, as leveling and brightening agents and to relieve internal stress in the deposit.

The specific resistivity of the solution, ρ, is important. It could be decreased by adding a supporting electrolyte, but this approach is limited in scope. A value $\rho \approx 5\ \Omega$ cm measured in acid copper baths containing $CuSO_4$ and H_2SO_4 is about as low as one can go. Another approach is to add a complexing agent, as discussed for Cu deposition from a pyrophosphate salt (cf. Section 19.2.4). It should be borne in mind, however, that when a metal ion is complexed, its standard potential is shifted cathodically by a potential given by $(2.3RT/nF)\log K$, where K is the stability constant of the complex formed. As a result, hydrogen evolution could occur along with metal deposition and the current efficiency might be decreased significantly.

A decrease in current efficiency with increasing current density is observed in many cases. This can enhance the uniformity of thickness of the deposit due to a negative feedback effect. Thus, in areas on the surface where the total current density is higher, the fraction of the current consumed for metal deposition is lower. The dependence of the current efficiency on current density is governed by the kinetic parameters of the two reactions involved. Thus, in the commonly encountered situation where hydrogen evolution is activation controlled while metal deposition is partially controlled by mass transport $(0.05 \leq (j/j_L) \leq 0.7)$, the current efficiency will decrease with increasing current density.

Electrodeposition of a metal from a negatively charged complex ion can also influence the throwing power and the morphology of the deposit in other ways. Where the local current density is higher, the potential on the solution side of the interphase is more negative. This causes a decrease in the local concentration of the negative ions, which slows the reaction. In other words, a negative feedback mechanism is created, counteracting the variation of local current density caused by the primary current distribution.

19.2.4
Kinetically Limited Current Density

The case of mass transport limitation has been discussed before. This leads to a mass-transport-limited current density, independent of potential. But a limited current density could also be observed when the limitation is kinetic. This can happen when the metal ion exists in solution as a complex, although the electroactive species participating in the charge-transfer step is the free ion. An example of such a situation is the deposition of Cu from a solution containing copper pyrophosphate. A chemical step, the rate of which is independent of potential, is followed by an electrochemical step, as shown by the next two equations

$$[Cu_2P_2O_7] \xrightarrow{k(chem)} 2Cu^{2+} + P_2O_7^{4-} \tag{19.11}$$

$$2Cu^{2+} + 4e_M^- \rightarrow 2Cu_M^0 \tag{19.12}$$

A limiting current was observed for this reaction, but its value was between one and two orders of magnitude lower than that calculated for mass transport limitation, indicating that it was due to the rate of release of Cu^{2+} ions from the complex. Now the Faradaic resistance, defined as the partial derivative $\partial\eta/\partial j$ approaches infinity at the limiting current density, and so does the Wagner number, leading to secondary current distribution and, hence, to uniform thickness of the deposit.

19.3
Throwing Power

19.3.1
Macro-Throwing Power

The throwing power of a bath is a measure of its ability to produce an electroplated coating of uniform thickness on samples having complex geometries. A quantitative measure of this property can be obtained by employing the *Haring and Blum cell* shown in Figure 19.3.

In this cell, two cathodes connected electronically are positioned at *unequal* distances from two sides of an anode. The throwing power (TP) is defined as:

$$TP = \frac{K-M}{K} \times 100 \tag{19.13}$$

where K is the ratio of distances between the anode and the two cathodes (taken as 5) and M is the ratio of the coating thicknesses on the two cathodes. This is a little awkward, because ideal throwing power, defined as uniform plating thickness irrespective of geometry, corresponds to $M = 1$, yielding a value of TP = 80%, rather than 100%, which one would expect for the upper limit of such a quantity. Still, it provides a very useful quantitative scale describing one of the most important properties of plating baths. In the case of purely primary current distribution, where $W_a \ll 1$ and there is effectively no throwing power, the thickness of the deposit will simply be inversely proportional to the distance. This yields a value of $M = K$ and TP = 0, as expected.

Having defined the throwing power quantitatively, we can now proceed to discuss the physical reasons for the dependence of the throwing power on geometry and the methods available to increase the value of the TP to acceptable levels.

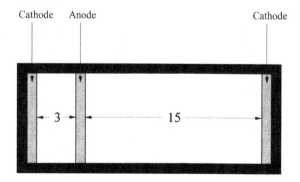

Figure 19.3 Top view of the *Haring and Blum cell* for the determination of the *macro-throwing power*.

19.3.2
Micro-Throwing Power

In the previous section we discussed *macro*-throwing power, which determines the uniformity of plating. Here, we shall discuss the *micro*-throwing power, which controls the smoothness and brightness of the deposit, and, naturally, depends on quite different factors.

Scale is very important in electrode processes. In the case of macro-throwing power the irregularities in the shape of the electrodes are on the same scale as the cell itself. The distance between the anode and the cathode may be of the order of $(1-10)$ cm and the characteristic length used to calculate the Wagner numbers would be in that range. When the appearance of the surface, particularly its brightness, is considered the scale of interest is of the order of magnitude of the wavelength of visible light. It follows from electromagnetic theory that the ratio between the light scattered from a surface and that reflected from it depends on the ratio $(h/\lambda)^2$, where h is the amplitude of the roughness and λ is the wavelength of light. If $(h/\lambda)^2$ approaches zero, one has *specular reflection*; that is, the surface reflects light like a mirror. As this ratio grows, the surface first looks dull and eventually becomes black as $(h/\lambda)^2$ exceeds unity. Thus the scale of interest, from the point of view of brightness, is of the order of magnitude of the wavelength of visible light, namely in the range $(0.4-0.8)$ µm. The ability of a plating bath to form uniform coatings on this scale of roughness is called the *micro-throwing power*.

We can see intuitively that cell geometry has little to do with micro-throwing power. If the amplitude of roughness is of the order of 1 µm or less and the distance between the anode and the cathode is a few centimeters, the variation of solution resistance at crests and valleys on the surface must be negligible. Hence, primary current distribution does not have any effect on micro-throwing power. A good way to look at it is to consider the appropriate Wagner number for this situation. Setting the characteristic length in Eq. (19.4) at a few µm, instead of a few cm, the value of W_a will increase by a factor of about 10^4, which is deep in the range of secondary current distribution.

There must be a different mechanism controlling micro- and macro-throwing power. This is *tertiary current distribution*, which depends on the rate of mass transport, itself related to the Nernst diffusion layer thickness, δ. A common observation in electroplating is that the roughness of the deposit increases with thickness. It is quite easy to produce a smooth deposit of 0.1 µm thickness, but keeping it smooth when the thickness has grown to 100 µm requires very special measures. Such an observation implies that a positive feedback mechanism is in operation, with the local current density higher at the protrusion than in recessed areas. It is easy to understand this behavior, if the plating process is assumed to be at least partially controlled by mass transport. We recall that the current density can be written as

$$j = nFD\frac{(c_b - c_s)}{\delta} \tag{4.14}$$

The Nernst diffusion layer thickness is larger in a recessed area than at a crest, hence, the local current density is smaller. As a result, recessed areas grow more slowly than crests, and the amplitude of roughness increases with time, and, hence, with thickness during plating, by a positive feedback mechanism.

It is not difficult to see how a rough surface will grow even rougher by the foregoing mechanism, but how is roughness initiated? Experiments show that even when plating is conducted on a highly polished surface, the deposit will gradually increase in roughness. We may expect that plating on an atomically flat, single-crystal surface in a highly purified solution will not produce a rough deposit, but this is of little practical interest. In practice a surface is prepared by degreasing, activation by acid (to remove oxides), but rarely by polishing. Indeed, the surface is often roughened in order to improve adhesion, but even a polished surface will not be completely flat, so that there will always be irregularities that can serve as a site for initiating the above positive feedback mechanism. In addition, real plating baths usually contain some foreign particles that could adhere to the surface during plating. These contaminants may be dust particles or solid grains of metal that fell off the anode during plating. It should be remembered here that impurity particles of sub-μm dimensions, which are often difficult to remove by filtering, are fairly large on the scale of importance here.

Another mechanism of roughness initiation may be associated with the nonuniformity of the substrate. The activation-controlled current density at sites of inclusions (such as graphite or sulfur), at grain boundaries and at different crystal faces may be different, causing uneven growth of the deposit.

19.3.3
The Use of Additives

There is great commercial incentive to produce smooth and bright deposits. Consequently, there is a vast choice of additives to improve micro-throwing power, making deposits smoother and more uniform and producing a bright metal luster. Although the properties of additives differ widely, the mechanism by which they operate is common and easy to understand.

Molecules of the additive adsorbed on the surface prevent or inhibit metal deposition. To a first approximation it can be said that the rate of metal deposition is simply proportional to the fraction of the surface that is not covered by the additive. A more detailed analysis shows that adsorption on part of the surface could also have an effect on the rate of metal deposition on the bare sites, but this refinement need not concern us here. As a rule, the molar concentration of the additive in solution is very low compared to that of the metal ion being plated. Consequently, the rate of adsorption of the additive is controlled by mass-transport limitation, while the rate of metal deposition is mostly activation controlled, with possibly some mass transport limitation involved, depending on the ratio of j/j_L, where j is the partial current density for deposition of the metal. This helps to produce a smooth surface for the same reason that a rough surface is formed in the absence of a suitable additive. On protruding parts on the surface the rate of mass transport is higher than on flat or recessed regions. Consequently, the additive is preferentially adsorbed on such regions, inhibiting the rate of metal deposition. As a result, the current density for metal deposition is higher in the recessed or flat regions, leading to leveling. Under favorable conditions this effect can actually be strong enough to reverse the trend, namely to yield a smooth deposit on an initially rough surface.

What is the fate of the additive during plating? Most of it is desorbed from the surface and released unchanged back into the solution. Some may be buried in the deposit. Alternatively, an additive may first be reduced, whereupon fragments of it are entrapped under the layers of metal being deposited. In any case the additive is consumed slowly during operation of a plating bath and must be periodically replenished. The incorporation of foreign molecules in the metal deposit affects its mechanical properties, as well as its corrosion resistance. These effects cannot generally be predicted by theory, and here the art of finding the right additive for each plating bath comes into play.

There is an optimum range of concentration over which each additive is most effective. This is also easy to understand, in terms of the mechanism just discussed. At low concentrations, the activity of each additive grows with increasing concentration, because there just is not enough material in solution to do the job: that is, coverage on the protruding areas cannot reach a sufficiently high value to induce significant leveling. In the best concentration range, coverage on protruding areas is high but in recessed areas it is relatively low, yielding the desired leveling effect. As the concentration of the additive in solution is increased further, the coverage on protruding areas reaches a limiting value and can grow no longer; the coverage on other areas keeps growing, however, until a high coverage is reached *everywhere* on the surface and leveling can no longer occur.

Finally, we might ask what determines the suitable range of concentration of an additive. It is clear that the answer is different for different additives and depends on the metal being deposited. If the adsorption isotherm for the additive on the same metal is known, a good guess would be to use a concentration of the additive that will lead to a partial coverage, in the range of $0.2 \leq \theta \leq 0.8$, because this is the range in which $\partial\theta/\partial c$ is the greatest, and one may expect to obtain the highest difference of adsorption on different regions on the surface. On the other hand, if the adsorption isotherm is not known, (which is usually the case), it is probably easier to determine the optimum range of concentration experimentally by trial and error, than to measure the isotherm and deduce the desired range of concentration from it.

The choice of a good leveling agent depends, among other things, on the excess charge density on the metal, q_M, at the potential where metal deposition is taking place. The latter depends, of course, on the potential of zero charge of the metal[2]. A positively charged additive will be preferentially adsorbed on a negatively charged

---

2) We need to be careful here in deciding "which metal". At the very beginning of plating on a foreign substrate, it will be the value of E_z for the substrate, but when the thickness of the deposit exceeds a few nm or less, it will be the value E_z for the metal being plated, because there is no longer any contact between the solution and the substrate.

surface, and vice versa. A neutral additive will be adsorbed mainly around the point of zero charge (cf. Chapter 13). The potential at which the metal is actually deposited depends on its standard potential, its concentration and the complexes it formed with ligands in the solution, as well as on the rate of deposition.

We conclude this section by noting that, although the mechanism by which different additives operate is fairly well understood, we have certainly not reached the point at which the choice of an additive can be based on its known molecular structure or even on measurement of its adsorption isotherm under equilibrium conditions. Such knowledge can be used to advantage for preliminary screening and intelligent guessing, but it cannot eliminate some degree of trial and error in identifying a good additive for a given purpose.

19.4
Plating from Nonaqueous Solutions

19.4.1
Statement of the Problem

Many metals can be plated from aqueous solutions, even though their reversible potential is cathodic with respect to the reversible hydrogen electrode (RHE) in the same solution. Hydrogen evolution can occur in such cases as a side reaction, but as long as the current efficiency is not too low, plating can be conducted on an industrial scale. One of the important reasons for this is that the exchange current density for metal deposition is usually much higher than that for hydrogen evolution, with the result that the rates of these reactions at the potential where the metal is being deposited are comparable, even if the reversible potential for metal deposition is more negative. In the presence of a suitable ligand, forming a complex with the metal ion being electroplated, the rate of metal deposition is slowed down (to achieve better throwing power) and the potential of deposition is shifted cathodically, but the high concentration of the ligand at the metal surface during plating may lower the rate of hydrogen evolution, allowing the process to occur at a reasonable current efficiency. Also, such electroplating baths usually operate in neutral or somewhat alkaline pH, and the reversible potential for hydrogen evolution is shifted in the negative direction, decreasing the overpotential for hydrogen evolution at the potential where plating is conducted. However, more active metals, such as aluminum, titanium, and magnesium, cannot be deposited from an aqueous medium at all. An attempt to do so leads to copious hydrogen evolution, but no detectable metal deposition. On the other hand, sodium and other alkali metals can be deposited on mercury from an alkaline solution, probably because of the very low exchange current density for hydrogen evolution on this metal, and because an amalgam is formed, so that the active surface is always mercury or its amalgam, not the metal being deposited.

When electroplating from an aqueous solvent is impossible, one must resort to nonaqueous systems. These present a number of technical difficulties and have been used in practice only when there was no alternative. With evolving technological development, it is anticipated that plating from nonaqueous systems may nevertheless be adopted for commercial use, and a short discussion is therefore warranted.

Since hydrogen evolution via decomposition of water is the problem, the most obvious way to proceed is to plate from a nonaqueous solution. Moreover, the solvent used should not have an active proton that could readily allow hydrogen evolution, such as an alcohol. A molten salt, such as $MgCl_2$ would seem to be a good solution to the problem. Indeed, metallic Mg is manufactured by electrolysis of anhydrous molten $MgCl_2$, and similarly Al is produced from a bath containing bauxite (Al_2O_3) dissolved in molten cryolite (Na_3AlF_6) at high temperatures since, in these baths, metal deposition is the only cathodic reaction that can take place. The quality of the deposits is usually poor, however, and the high temperature of operation (particularly in the case of Al) could damage the substrate. Thus high-temperature molten salts are suitable for manufacturing the relevant metals, but rarely used for plating them as a coating.

Refractory metals, such as Ta and Zr, can be deposited from their fluorides in a molten salt bath. In the case of Zr, for example, the bath consists of ZrF_4 or ZrF_6^{2-} in a KF/NaF/LiF mixture. The alkali fluorides are employed to increase conductivity and decrease the melting point. Even so, these baths are operated at about $800\,^\circ C$. Good deposits have been reported as long as the right valency was chosen for each metal ($+3$ for Mo and V, $+4$ for Nb and Zr and $+5$ for Ta). The bath must be operated in a pure argon atmosphere, and impurities must be strictly excluded. It should be obvious that the operation of such baths is expensive and their control is difficult. Thus, their use is limited to either research purposes or highly specialized applications, where cost is of secondary concern.

19.4.2
Methods of Plating of Aluminum

The search for a room-temperature plating bath for aluminum has been conducted for many years, in view of the excellent corrosion resistance and low toxicity of this metal. An early technological success was the so-called hydride bath, which consisted of a solution of $AlCl_3$ and $LiAlH_4$ in diethyl ether, $(C_2H_5)_2O$. A large excess of $AlCl_3$ was used ($AlCl_3/LiAlH_4 = 7:1$). The exact mechanism of Al deposition from this bath may not be known, but there can be no doubt that the negative hydride species in $LiAlH_4$ plays a crucial role, since the bath cannot be operated after it has been depleted of $LiAlH_4$, even if the concentration of $AlCl_3$ is kept constant. This technology was used on a few occasions for highly specialized purposes, mainly for the production of aluminum mirrors that are excellent reflectors of infrared radiation. It has not gained widespread commercial application because it requires the use of a highly flammable solvent and chemicals that are toxic and very sensitive to humidity and oxygen.

Aluminum can also be plated from a low-temperature molten-salt bath, employing a mixture of $AlCl_3$ and KCl. The melting point depends on composition, and the bath can be operated in the range 200–300 °C. Two anions can exist in this melt: $AlCl_4^-$ and $Al_2Cl_7^-$, their relative concentrations depending on the ratio of $AlCl_3$ to KCl in the bath. The great advantage of this system is that it dissolves chlorides of other metals, such as Ti and Mn, and allows the deposition of alloys of these metals with aluminum. The greatest disadvantage is that the bath must be operated under strictly anhydrous conditions, since $AlCl_3$ is highly hygroscopic, releasing HCl when in contact with water or moist air.

Another near-room-temperature bath for Al plating contains a metal-organic compound, $Al(C_2H_5)_3$, dissolved in toluene, with $AlCl_3$ added to provide electrolytic conductivity. This bath, operated at about 100 °C, has excellent conductivity and good throwing power. Although it has made some inroads for engineering applications, its widespread application has been limited, probably because of the need to use an expensive and dangerous metal-organic compound that ignites spontaneously upon contact with air.

A room-temperature plating bath is based on the use of Al_2Br_6 and KBr in toluene, ethyl benzene, or similar aromatic solvents. The chlorides and bromides of aluminum are covalent compounds and are highly soluble in aromatic hydrocarbons. An ionic compound such as KBr is not soluble in an aromatic hydrocarbon, but is readily dissolved in a solution of Al_2Br_6 in toluene, forming a complex ion

$$Al_2Br_6 + KBr \rightarrow K^+ + (Al_2Br_7)^- \tag{19.14}$$

but this would still lead to the formation of potassium ions, which are unstable in aromatic solvents. Thus, the real ionic species in solution must be somewhat more complex. Experiments indicated that the actual species in solution are formed in the reaction

$$3[K(Al_2Br_7)] \rightleftarrows [K_2(Al_2Br_7)]^+ + [K(Al_2Br_7)_2]^- \tag{19.15}$$

This bath has the advantage of operating at room temperature, but it is also very sensitive to moist air, forming highly corrosive and toxic HBr, and it has not been adopted by industry.

It is interesting to note that all four aluminum plating baths discussed here employ a solvent of low polarity. In fact, aluminum cannot be deposited from any polar solvent. For example, a solution of LiCl and $AlCl_3$ in acetonitrile or propylene carbonate yields a deposit of metallic lithium, but no aluminum, even though, thermodynamically, aluminum should be deposited first. The reason evidently lies in the kinetics of the process. In any polar solvent the energy of solvation of the small and highly charged Al^{3+} ion is so high that the first step in the reaction sequence – the removal of a single solvent molecule from the inner solvation shell – requires an insurmountable energy of activation. It is only when the solvent is nonpolar that this process can proceed at a significant rate.

It is not easy to find a suitable nonaqueous electrolyte for plating aluminum, titanium, and other active metals. Operating such a bath may be even more difficult. First, water, and often oxygen, must be excluded. This can be done rather easily in continuous processes, such as plating a wire or a metal sheet. In most applications, however, electroplating is typically a batch process – parts are introduced and removed from the bath routinely. Whereas the technology to perform such operations exists, it is more expensive and much less convenient than operation in an open aqueous bath. Most nonaqueous solvents are either flammable or toxic, or both. Most salts used to make up the bath are expensive, and some are quite unstable. Even a relatively inexpensive salt such as KBr can become expensive when it must be bone dry. Moreover, a current efficiency that is just a little below 100%, which may be a minor irritation in aqueous solution, could turn out to be a major problem in nonaqueous media, because the side reactions might lead to the accumulation of products that are detrimental to the operation of the bath, to say nothing of the health of the operator. Waste disposal, a problem even in aqueous plating baths, can be much more difficult and costly in nonaqueous baths.

These are some of the reasons for the failure of nonaqueous plating baths to come into general use, in spite of some clear technological advantages they can offer, in particular in making products that cannot be manufactured by other means. It is possible, however, that exacting specifications of emerging new technologies, accompanied by research and development in this field, could eventually lead to the introduction of nonaqueous plating technologies in some industrial applications.

19.5
Electroplating of Alloys

19.5.1
General Observations

There is much in common between the electroplating of a single metal and of an alloy of two or more metals. Both are inner-sphere charge transport processes, characterized by the fact that the initial states are strongly hydrated ions in solution and the final products are bare atoms that have become a part of the lattice of the metal electrode, and, in both cases, the surface of the substrate is changing as the process is proceeding. In both cases hydrogen evolution is a common side reaction, so that the current efficiency could be less that 100%, and, moreover, it could be a function of the current density applied. However, there are also significant differences that have to be taken into account when studying alloy deposition, some of which are listed below.

1) Alloy deposition, by definition, must involve at least two simultaneous electrochemical reactions occurring in parallel. In the general derivation of the equations for electrode kinetics one obtains an expression of the form

$$j = nFkc_b \left[\exp\left(\frac{\alpha_{an}F}{RT} \eta \right) - \exp\left(-\frac{\alpha_c F}{RT} \eta \right) \right] \tag{19.16}$$

where it is tacitly assumed that only a single reaction takes place and the composition and morphology of the surface do not change during the experiment. If we try to apply this equation to deposition of a binary alloy, three problems immediately emerge: First, the concentration of the two metals in solution is not necessarily the same. Indeed in most cases they have to be different, in order to achieve the desired composition of the alloy being deposited. Secondly, at any given applied (or measured) potential, the overpotentials are different for the two metals. Thirdly, the transfer coefficient (and the Tafel slopes, $b_c = 2.3RT/\alpha_c F$) of the two metals can be different. Consequently, measurement of the current–potential relationship does not allow us to draw any direct conclusions regarding the mechanism involved, although it does show in some cases the potentials where deposition of each of the two metals starts.

2) Equation (19.16) is valid only if the reaction occurs under purely activation control. When a single metal is being deposited, mass-transport limitation can be taken into account by replacing the bulk concentrations in Eq. (19.16) with the surface concentration. At steady state this is rather simple, since

$$c_s/c_b = (1-j/j_L) \tag{4.16}$$

In alloy deposition the concentrations of the two alloying elements in solution could be different: one may be deposited at a rate controlled substantially by mass transport while the other may be essentially under conditions of activation control. The number of electrons needed to reduce each of the alloying elements should also be taken into account. For example, in deposition of a Au–Ni alloy from a solution containing $AuCl_4^-$ and $NiCl_2$, at equal concentrations, the limiting current for Au would be 50% higher than that for Ni, (if the difference in the diffusion coefficient of the two ions is ignored). On the other hand, the hydrodynamic conditions, and, hence, the value of the Nernst diffusion-layer thickness are the same.

3) An attempt to calculate the extent of mass transport limitation for each of the alloying elements should be based on the partial current density for that element, never on the total current density applied. Unfortunately, this quantity cannot be derived from the current–potential relationship itself. It has to be calculated from the Faradaic efficiency (which is the measure of the rate of hydrogen evolution) and an analysis of the atomic composition of the elements in the deposit. Thus, the best one can do is to determine the *average* value of the partial current density of each alloying element. Admittedly, this limitation also applies to deposition of a single metal, unless one can be sure that hydrogen evolution does not occur at the potential where the deposition process takes place.

4) Determination of the relative concentration of the elements in the deposit, using electron dispersive spectroscopy (EDS) may not be enough, particularly when one of the alloying elements is in a high valency state, such as in WO_4^{2-}, because partial reduction to the oxide rather than the metal may be involved. It may be necessary to employ X-ray photoelectron spectroscopy (XPS), to determine the chemical state of each element in the alloy.

5) Uniformity of current distribution can always be a problem in electroplating, but in alloy plating there could be an added complication: In addition to leading to nonuniform thickness, it could also lead to nonuniform composition, since the rate of deposition of the alloying elements may depend differently on the local current density.

6) Deposition of alloys is usually conducted in the presence of one or more complexing agents. For example, in plating of a Ni–W alloy, citrate (and sometimes also ammonia) is added to the solution. Now citrate forms two different complexes with Ni, and NH_3 could form as many as five different complexes of the type $\left[Ni(NH_3)_n\right]$. In addition, the WO_4^{2-} and the Ni^{2+} ions can form two types of complexes with citrate. Adding the hydrogen evolution reaction this could add up to as many as ten cathodic reactions taking place simultaneously in parallel!

19.5.2
Some Specific Examples

Simple alloy deposition is observed for the Pb–Sn couple. The standard potentials of the two metals are $E^0 = -0.13$ V and -0.14 V vs. SHE, respectively, hence it is easy to adjust the concentrations in solution such that the reversible potentials would be the same, or close to each other. Although plating occurs at a potential negative with respect to the reversible potential for hydrogen evolution, both metals are very poor catalysts for this reaction, so that close to 100% Faradaic efficiency is readily obtained. The most important consequence of this is that the composition of the plated alloy is linearly related to the composition of the solution. This is, unfortunately, not common in alloy plating, as we show next.

Anomalous deposition can be represented by the formation of Permalloy, (a magnetic alloy used in the voice recording industry) comprising typically 80% Ni and 20% Fe. The standard potentials are $E^0 = -0.26$ V and -0.44 V vs. SHE for Ni and Fe, respectively. Hence, it might have been expected that equal concentration of the two metals in solution would yield a higher concentration of Ni in the deposited alloy, and it would be easy to reach the desired ratio of 80/20. In practice, quite the opposite is observed. It seems that the presence of Fe^{2+} ions in solution reduces the rate of deposition of Ni, so that a large excess of Ni^{2+} is needed to obtain the high Ni/Fe ratio needed in the alloy. This type of anomalous behavior is frequently encountered in alloy plating, in different variations. In some cases one can observe mutual synergism: adding one metal increases the partial current density for depositing of the other and vice versa. Sometimes mutual inhibition is observed and sometimes it is a sort of one-way relationship, as in the case of permalloy. Adding Fe^{2+} to a solution containing Ni^{2+} slows down the rate of nickel deposition, but adding Ni^{2+} to a solution of Fe^{2+} increases the rate of iron deposition.

Induced co-deposition is observed for deposition of metals that cannot be deposited at all from an aqueous solution, such as W, or can barely be deposited, with a low current efficiency and poor adherence of the deposit, such as Re. However, alloys of W with the iron-group metals can readily be formed, using, for example, a solution of $NiSO_4$ and Na_2WO_4, with citric acid added as a complexing agent. In this particular case it was shown that a Ni–W alloy is deposited from a complex containing both metals, while Ni is also deposited in parallel reactions from its complex with citrate. Very similar behavior is observed for deposition of alloys of molybdenum.

Galvanostatic versus potentiostatic measurements have a special importance in alloy deposition. This has already been discussed in Section 4.2.1 where it was pointed out that galvanostatic measurements are unique in chemical kinetics, in the sense that the total rate of the reaction (which is proportional to the current density) is held constant by the electronic circuit, and the overpotential is allowed to reach the value corresponding to the applied current density. In other words, the reaction rate is independent of the heterogeneous rate constant, the concentration, the nature of the electrocatalyst and the temperature. The effect of these variables is expressed by the overpotential developed.

Consider, for example, the deposition of an alloy of Ag with Cu from their appropriate salts, conducted galvanostatically. Having only the $AgNO_3$ in solution, the current density for deposition will be equal to the applied current density, since in this case the Faradaic efficiency is 100%. Now assume that $Cu(NO_3)_2$ is added, maintaining the concentration of $AgNO_3$ constant, and the same current density is applied. The partial current density for deposition of silver will decrease. Reversing the order, one will find that the rate of deposition of Cu will decrease upon addition of $AgNO_3$, holding the concentration of $Cu(NO_3)_2$ constant. This does not indicate that the two metals in solution inhibit the deposition of each other. It just shows that the wrong technique has been used to study whether they do or do not influence the rate of each other.

Repeating the same experiments potentiostatically, one may find that the partial current density of one metal is independent of the presence of the other in solution, because the total current density is allowed to change according to the overpotential with respect to each of the metals, and the current density measured will be the sum of the current densities of the two metals.

19.6
Electroless Deposition of Metals

19.6.1
Some Fundamental Aspects of Electroless Plating of Metals and Alloys

A superficial view would lead us to think that electrochemical and electroless plating represent essentially the same process – in both cases the metal ions are being reduced. The electrons provided for the deposition process originate in one case from an electronic device (a potentiostat or a constant-current supply) and in the other case from a reducing agent in solution, so the reduction process is the same. However, as it turns out, there are some important differences.

Electroless plating is an open-circuit process. There is only one electrode, which is the substrate being coated. In the research laboratory one often measures the open circuit potential with respect to a suitable reference electrode, but in industry this may not be considered necessary. In a sense, electroless plating is similar to corrosion. Both cases occur at open circuit, so there is no current flowing in an external circuit (indeed there is no external circuit). In corrosion, the rate of anodic oxidation of the

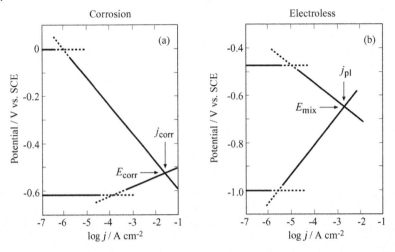

Figure 19.4 Similarity between open-circuit corrosion and electroless plating at open circuit. (a) Mixed potential in corrosion (same as in Figure 18.1). (b) Mixed potential in electroless plating. $CuSO_4$ 0.1 M, ethylenediamine tetraacetate (EDTA) 0.1 M, formaldehyde (FA) 0.05 M, pH 12.5.

metal must be equal to the rate of cathodic reduction of some species in solution (usually H_3O^+, H_2O or O_2). At steady state, a mixed potential is set up, as shown in Figures 18.1–18.4, for several possible cases.

The same logic applies to electroless plating, except that the roles are reversed. Here, the rate of anodic oxidation of a suitable reducing agent must be equal to the rate of cathodic reduction of metal ions. The two cases are shown side by side in Figure 19.4, where Figure 19.4a is just a reproduction of Figure 18.1 and Figure 19.4b shows the case of electroless plating of copper, employing formaldehyde as the reducing agent.

For the electroless deposition process corresponding to Figure 19.4b, the pH is set at 12.5 by adding NaOH, and EDTA is added as a complexing agent, to prevent precipitation of $Cu(OH)_2$. As in the case of corrosion, the mixed potential represents steady state with two reactions involved, it is not an equilibrium state. The Tafel slopes for the anodic and cathodic reaction are not in any way related to each other and the relationship $\alpha_{an} + \alpha_c = n$ *does not apply*, (although it may happen to be numerically correct). Indeed, even the number of electrons transferred may not be the same for the two reactions. The only constraint determining the position of E_{mix} is that the total anodic current must be equal to the total cathodic current. Micro-polarization at potentials close to the mixed potential can be used to determine the current density during electroless plating, following the formalism used in Section 18.1.3. Eqs. (18.6) and (18.17), can be applied, replacing j_{corr} by j_{pl}. This approach is better for the determination of the rate of electroless deposition than measuring the kinetic parameters of the two reactions (reduction of Cu and oxidation of formaldehyde) in separate solutions containing only one of these materials. The latter is theoretically correct, but it ignores the possibility that the two components in solution may interact with each other, or influence the rate of each other.

Electroless plating is applicable to the deposition of metals and of alloys. Indeed electroless plating of a Ni–P alloy, developed by Brenner in 1946, was the first example of formation of an *amorphous alloy*, often referred to as a *glassy metal*. Since then several methods have been developed for the deposition of different amorphous alloys, both by electroless and electroplating technologies, in addition to metallographic methods of preparing such alloys, which will not be discussed here,

19.6.2
Advantage and Disadvantages Compared to Electroplating

1) Eliminating the need for a counter electrode in electroless plating has a profound influence on the current distribution. Macro-throwing power is no longer an issue, because primary current distribution is caused by the nonuniformity of the electrostatic field between the object being plated and the counter electrode. Since there is no counter electrode, uniformity of plating on complex shapes can readily be achieved.

2) Micro-throwing power is a result of the interplay between the Faradaic resistance and the partial mass-transport limitation. As long as the rate of deposition of the metal is low compared to the mass-transport limited rate, the concentration at the surface is close to that in the bulk, (following the equation $c_s = c_b(1-j/j_L)$, yielding smooth and bright surfaces. In electroless deposition, the rate of deposition can be controlled by proper choice of the reducing agent and its concentration, as well as the concentration of the metal being plated. In addition, the pH and the temperature can also influence the rate of electroless plating. In any event, if electroless plating is followed by electroplating, the micro-throwing power of the bath used for electroless plating is of relatively minor importance, except in the sense that it might influence the quality of bonding to the next layer electroplated on top of it.

3) In contrast, during electroplating both macro-and micro-throwing power play a role. Moreover, since the plated layer is relatively thick (compared to that produced by electroless plating), the current density is higher, in order to increase production rate, and suitable additives are used to improve uniformity of thickness, as well as brightness and smoothness.

4) The most important feature of electroless deposition is that it can be performed on nonconducting surfaces, such as plastics, glass, semiconductors and ceramic materials. Special surface treatments are needed in order to activate the surface, and good adherent thin metallic surfaces can be produced, which can be followed by electroplating of the same or a different metal.

5) The weak point in electroless plating is that the solution is inherently unstable (since it is this thermodynamic instability that allows the bath to operate). The composition of the bath must be such that it will be stable in the bulk of the solution, and reduction of the metal will only occur on the surface that has been activated. Thus, maintaining uniformity of the product can be much more of a challenge in electroless plating than in electroplating. Nevertheless, electroplat-

ing baths are routinely used in industry, and new ones are regularly being developed to meet specific technological needs.

6) Electroless plating is usually limited to very thin layers, typically less than 1 μm. This could be an advantage, when the purpose is to create a very thin barrier layer. When a protective layer is needed (to prevent corrosion or abrasion) this is an obvious disadvantage. This, however, is not an inherent limitation of the method, and it would seem that technologies for electroless plating of thicker layers could be developed.

19.7
The Mechanism of Charge Transfer in Metal Deposition

19.7.1
Metal Deposition is an Unexpectedly Fast Reaction

The deposition of a metal is a charge transfer reaction, just like an outer-sphere charge transfer reaction, at least it would appear to be, when the reactions are written side by side, for example,

$$\left[Pb(H_2O)_n\right]^{2+}_{soln} + 2e^-_M \rightarrow Pb^0_M + n(H_2O) \tag{19.17}$$

and

$$\left[Cr(CN)_6\right]^{-3} + e^-_M \rightarrow \left[Cr(CN)_6\right]^{-4} \tag{19.18}$$

But in spite of the apparent similarity, the two reactions represent very different physical situations, and they are expected to follow different mechanisms.

It will be recalled that the hydration energy is very high, of the order of 20 eV for a divalent ion (cf. Eq. (19.3)). Assuming a hydration number of 4–6, this leads to an average energy of $(3.3-5.0)$ eV per bond, which is the range of the bonding energy in chemistry. Thus, Eq. (19.17) represents the breaking of several chemical bonds. In contrast, Eq. (19.18) shows no breaking of bonds. A minor change in the bond lengths of the ligands around the central cation and some rearrangement of water molecules around the complex ion (in the outer hydration shell) is expected, leading to a much smaller change in the Gibbs energy of the overall reaction.

Metal deposition involves the transfer of mass across the interface, while for outer-sphere charge transfer to occur only electrons have to cross the interface, while both the reactant and the product stay on the solution side of the compact Helmholtz double layer.

Based on the above differences, one would expect outer-sphere charge-transfer reactions to be many orders of magnitude faster than metal deposition. Surprisingly, such a trend is not observed experimentally. The heterogeneous rate constant for the two reactions shown in Eq. (19.17) and Eq. (19.18) are $k_h = 2.0$ and 0.9 cm s^{-1}, respectively. Even the rate of deposition of trivalent bismuth, having a very high energy of hydration, and requiring the transfer of three charges

$$\left[\text{Bi}(\text{H}_2\text{O})_n\right]^{3+}_{\text{soln}} + 3e^-_M \rightarrow \text{Bi}^0_M + n(\text{H}_2\text{O}) \tag{19.19}$$

has a rate constant that is just 17 times lower than that of the reduction of trivalent to divalent iron which is an outer sphere reaction

$$\left[\text{Fe}(\text{H}_2\text{O})_6\right]^{3+}_{\text{soln}} + e^-_M \rightarrow \left[\text{Fe}(\text{H}_2\text{O})_6\right]^{2+}_{\text{soln}} \tag{19.20}$$

19.7.2
What Carries the Charge Across the Interface During Metal Deposition?

Although metal deposition is often represented in simplified form as

$$M^{z+} + ze^- \rightarrow M \tag{19.1}$$

this representation has some profound internal contradiction (apart from the trivial correction of including explicitly water as the ligand, as shown in Eq. (19.2)). It is well known that metals do not consist of neutral atoms. They consist of positively charged metal ions "immersed" in a sea of delocalized electrons. The net result is a neutral piece of metal, of course, but the point is that at least one electron of each atom is delocalized, and constitutes the conduction band of the metal. Taking this into account, Eq. (19.1) should be rewritten (ignoring the water molecules for simplicity) as

$$M^{z+}_{\text{soln}} + ze^-_M \rightarrow M^{2+}_M + ze^-_M \tag{19.21}$$

The point of writing Eq. (19.21) in this, somewhat unusual, form is to emphasize the fact that the electrons are in the metal, both in the initial and in the final states. The only species that is moved from one phase to the other is the metal cation. Indeed the last equation should be written as

$$M^{z+}_{\text{soln}} \rightarrow M^{2+}_M \tag{19.22}$$

cancelling two identical terms on the two sides of the equation.

This is admittedly an unusual way of presenting metal deposition. At first sight it might be argued that the electrode is charge positively, but in this sense Eq. (19.22) is not different from Eq. (19.1) or (19.2), where electrons are removed from the metal to neutralize the positive ion in solution. In both cases one must bear in mind that electrolysis requires two electrodes. The electrons removed from the metal in Eq. (19.1) are replenished through the external circuit (not across the interface) from the anode, where an oxidation reaction takes place and electrons are released. The only difference between Eq. (19.1) and Eq. (19.22) is that in the former it is tacitly assumed that the cation is neutralized when it is still in the solution phase (at the OHP, to be exact) while in the latter it is assumed that the metal cation first crosses the interface and is neutralized after it has reached the metal phase.

19.7.3
Microscopic Reversibility and the Anodic Dissolution of Metals

The law of microscopic reversibility stipulates that reactions taking place in the forward and backward direction must follow the same path or, more precisely, must cross the same Gibbs energy of activation barrier. Admittedly, this is not a universal law, and there can be situations under which it does not apply, but such situations are rare, so that it can be assumed to apply, unless proven otherwise. We shall not go into the fine details of this law, except to note that it is implicitly assumed to apply whenever one writes the common relationship.

$$\frac{k_f}{k_b} = K \tag{19.23}$$

Now, we recall that the three constants in the above equation can be written as

$$k_f \propto \exp\left(-\frac{\Delta G_f^0}{RT}\right); \quad k_b \propto \exp\left(-\frac{\Delta G_b^0}{RT}\right) \quad \text{and} \quad K \propto \exp\left(-\frac{\Delta G^0}{RT}\right) \tag{19.24}$$

hence

$$\left(\Delta G_f^{0\#} - \Delta G_b^{0\#}\right) = \Delta G^0 \tag{19.25}$$

Thus, the standard Gibbs energy of a reaction is equal to the difference between the standard Gibbs energies of activation of the forward and backward reactions, which only applies when the reaction crosses the same energy barrier in both directions, namely, when the law of microscopic reversibility applies.

The purpose of introducing the concept of microscopic reversibility is to provide further evidence, confirming that charge must be transferred across the interface by metal cations, not by electrons. To prove this, consider the anodic process of metal dissolution, written commonly as

$$M_M^0 + n(H_2O) \rightarrow \left[M(H_2O)_n\right]^{z+} + ze_M^- \tag{19.26}$$

Now, oxidation in electrode kinetics is, by definition, a process in which electrons are transferred from a species in solution to the electrode. But viewing Eq. (19.26) we note that there is no such species in solution. The electrons shown on the right-hand side of this equation are in the metal and the only other species in solution (except water as the solvent) is the cation in its stable hydrated and oxidized form. Moreover, following the argument leading to Eqs. (19.21) and (19.22) above, Eq. (19.26) should be rewritten as

$$M_M^{z+} + ze_M^- + n(H_2O) \rightarrow \left[M(H_2O)_n\right]^{z+} + ze_M^- \tag{19.27}$$

and hence

$$M_M^{z+} + n(H_2O) \rightarrow \left[M(H_2O)_n\right]^{z+} \tag{19.28}$$

This shows that anodic dissolution of a metal *cannot take place* by electron transfer. On the other hand, Eq. (19.28) allows the process to occur, if charge is carried across the interface by the metal ion. This ion could start moving in the direction of the solution, under the influence of the electrical field applied, leaving behind the electron in the metal and acquiring water molecules as its effective charge increases with distance from the metal surface, as will be discussed below.

Here is where the law of microscopic reversibility comes in. If we accept that in metal dissolution the charge must be carried across the interface by the metal cations, this law immediately forces us to accept that the same applies to metal deposition!

19.7.4
Reductio Ad Absurdum

The above term in Latin, which means "reduction to the absurd", refers to a tool borrowed from logic, used already by mathematicians of the school of Euclid in ancient Greece. The logical consequence is simple, yet extremely elegant. Thus, in order to show that some statement is wrong, one assumes first that it is right, and proceeds to prove that this leads to an internal contradiction. Here, we shall use it to prove that metal deposition cannot proceed by transfer of electrons across the interface, by assuming that it can, and showing that this would lead to impossible conclusions.

We shall start with the simple case of deposition of Ag, which is written as

$$\left[Ag(H_2O)_n\right]^+ + e_M^- \rightarrow Ag_M^0 + n(H_2O) \tag{19.29}$$

It was already shown in Section 5.5 that this process could not occur in one step, because of the different time scales for the transfer of an electron and an ion across the interface. Electron transfer happens on the time scale of femtoseconds, while ion transfer occurs on the scale of nanoseconds, namely about six orders of magnitude longer. Thus, the sequence of events, based on the time each step takes, is the following

$$\left[Ag(H_2O)_n\right]^+_{soln} + e_M^- \xrightarrow{\approx 1\ fs} \left[Ag(H_2O)_n\right]^0_{soln} \xrightarrow{\approx 1\ ps} Ag^0_{soln} \tag{19.30}$$

Once the ion has been neutralized, there is no reason for the water molecules that formed the hydration shell to stay in the configuration represented as the intermediate in this sequence, and it takes about 1 ps for the molecules of water to relax back to their bulk configuration.

The following step is

$$Ag^0_{soln} \xrightarrow{\approx 0.1\ ns} Ag^0_M \tag{19.31}$$

The time taken for this step is estimated, assuming diffusion from the OHP to the surface of the metal, a distance of typically 0.6 nm, (cf. Section 5.5.2).

It is intuitively obvious that a neutral silver atom in solution should be less stable than a hydrated ion, and it turns out to be quite easy to estimate this difference, based on the following Haber cycle (where the water ligands are not shown, for simplicity)

$$(Ag^+_{soln} + e^-_M) \xrightleftharpoons{\Delta G^0 = 0} Ag^0_M$$

$$\uparrow \Delta G \qquad\qquad\qquad \downarrow \Delta G_{subl} \qquad\qquad (19.32)$$

$$Ag^0_{soln} \xleftarrow{\Delta G_1 \approx 0} Ag^0_{gas}$$

It is seen that the difference between formation of a solvated ion in solution (for which $E^0 = +0.80\,V$ vs. SHE) and a neutral silver atom in solution is close to $\Delta G_{subl}/F = -2.55\,V$, where ΔG_{subl} is the Gibbs energy for sublimation of the metal. Hence, the standard potential for the *hypothetical* reaction

$$[Ag(H_2O_n)]^+_{soln} + e^-_M \rightarrow Ag^0_{soln} \qquad\qquad (19.33)$$

is given approximately by the Gibbs energy of sublimation of the metal. This is the *absurdum*.

If silver deposition were to occur by electron transfer, it would require the creation of a highly unstable intermediate, Ag^0_{soln}, at a potential of $-2.55\,V$ vs. the reversible Ag^+/Ag electrode in the same solution. This intermediate would have to survive a very long time, compared to the time taken to form it. However, at the potential of $+0.80\,V$ there would be a very high positive overpotential of $2.55\,V$ to oxidize it. Moreover, at this potential (corresponding to $-1.75\,V$ vs. SHE), it would in effect corrode, that is, reduce water to molecular hydrogen, while it is oxidized to Ag^+_{soln} that will form the stable hydrated cation in solution.

$$2Ag^0_{soln} + 2H_2O \rightarrow 2Ag^+_{soln} + H_2 + 2(OH)^- \qquad\qquad (19.34)$$

Finally, a neutral atom in solution could diffuse towards the surface or away from it. This could lead to a longer residence time for re-oxidation of the atoms in the bulk of the solution, resulting in a reduced Faradaic efficiency, which has not been observed experimentally.

19.7.5
Migration of the Ion Across the Double Layer

Having shown that charge is transferred across the interface by the metal cations, not by the electrons, one has to propose a mechanism that would explain the observed behavior, particularly the unexpectedly high reaction rate discussed in Section 19.7.1. In the model presented below it is assumed that the ions cross the interface by migration, under the influence of the high electric field in the Helmholtz double layer, caused by application of an overpotential. This field is given by

$$\vec{E} = \frac{\eta}{\delta} \qquad\qquad (19.35)$$

where $\vec{E}$ is the electric field and δ is the thickness of the double layer, taken here as 0.6 nm. For example, an overpotential of 0.3 V yields a field of

$$\vec{E} = 5 \times 10^8 \, V \, m^{-1} \tag{19.36}$$

The velocity of a hydrated silver ion in the bulk of the solution is a product of its absolute mobility, which is $\mu_{Ag^+} = 6.4 \times 10^{-8} \, m \, s^{-1}$ (for a field of $1 \, V \, m^{-1}$) by the electrostatic field, yielding a migration velocity of

$$v_{Ag^+} = 6.4 \times 10^{-8} \, m \, s^{-1} \times 5 \times 10^8 \, V \, m^{-1} = 32 \, m \, s^{-1} \tag{19.37}$$

that would allow it to cross the interface in about 20 ps. This number is probably an underestimate, because it was calculated on the basis of the viscosity of bulk water. Even so, it is comparable to the rate of solvent rearrangement in aqueous solution. Thus, it is reasonable to assume that the hydration shell of an ion is in its lowest state of Gibbs energy, all along its path from the OHP to the surface of the metal.

19.7.6
The Mechanism of Ion Transfer

In the model presented here, ion transfer is assumed to occur in many small steps. Considering that the ion is initially hydrated and it is squeezed through a layer of water molecules adsorbed at the metal surface, the component of its movement in the direction perpendicular to the interface is expected to cause some distortion in the hydration shell around it. In this sense, each step involves some degree of solvent rearrangement. This is similar to outer-sphere charge transfer, but there are several important differences:

1) Ion transfer under the influence of the externally applied electrostatic field involves only a singe time scale – that of the movement of the hydrated ion. This is of the order of 20 ps, which is similar to the relaxation time of molecules in bulk water. Hence, there is enough time for the water molecules in the hydration shell to rearrange around the ion, as it moves in the direction perpendicular to the electrode surface.
2) Transfer of the ion can occur in numerous steps. Consequently, the change in the hydration shell in each step can be quite small, leading to a very small change in the Gibbs energy for each step. When comparing this process to outer-sphere charge transfer, it is important to note two differences:

On the one hand, the overall energy of solvent rearrangement in metal deposition is much larger than that in outer-sphere charge transfer. In the first, all the hydration shell is eventually removed; in the second, it is only rearranged to some extent. On the other hand, during ion transfer this rearrangement is assumed to occur in many small steps, corresponding to a very small change in the solvent rearrangement energy, along with small changes in the effective charge on the ion, in each step. In outer-sphere charge transfer, a full electronic charge crosses the interface, so that the entire solvent rearrangement energy must be dealt with in a single step, leading to a significant energy of activation.

3) The transfer of an ion across the interface is more complex than might have been implied so far. Assuming that the hydrated cation is initially at the OHP, at a distance of about 0.6 nm from the electrode surface, there is initially no interaction between the ion and the electrode. As the ion approaches the metal surface, some interaction between the charge on the ion and the electrons in the conduction band of the metal takes place. This leads to a decrease in the *effective charge* on the ion, which in turn causes a decrease in its hydration energy. At some point along the path of the ion towards the electrode, the lowering of the total Gibbs energy of the system caused by this effect exceeds the increase caused by solvent rearrangement associated with the distortion, and, finally, removal of the hydration shell; and the total energy of the system will start to decrease.

19.7.7
The Symmetry Factor, β

In Figure 19.5 the variation of the Gibbs energy with distance from the electrode surface is given schematically for metal deposition. Curves for different values of the overpotential are shown.

Curve 1 applies to the system at equilibrium. A value of $\bar{G}^0 = 0.5$ eV was chosen here for the electrochemical Gibbs energy of activation, and the position of the activated complex was taken to be closer to the initial state of the solvated ion (which is

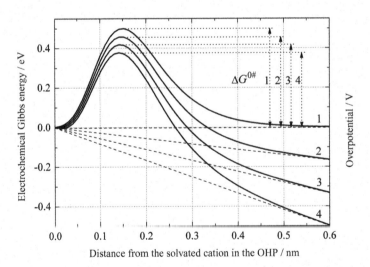

Figure 19.5 Schematic plots of the electrochemical Gibbs energy for cathodic metal deposition as a function of the distance from the solvated ion at the OHP, showing the effect of overpotential. The dotted lines represent the gradient of $\Delta\bar{G}$, which is equal to the gradient of η in this particular case. (c.f. E. Gileadi, *Israel J. Chem.* **48** (2008) 121)

at the OHP). Note that reversibility implies that $\Delta \bar{G}^0 = 0$ for transition from the initial to the final state, namely for the fully hydrated ion at the OHP to a bare ion as part of the crystal lattice of the metal. Curves 2 to 4 show the manner in which the Gibbs energy of activation is modified by applying different overpotentials.

The Gibbs energy for the overall reaction is given by

$$\Delta \bar{G}^0 = -n\eta F \tag{19.38}$$

The Gibbs energy of activation is modified by the applied overpotential so that

$$\frac{\partial \Delta \bar{G}^{0\,\#}}{\partial \eta} = \beta \frac{\partial \Delta \bar{G}^0}{\partial \eta} \tag{19.39}$$

hence, the symmetry factor β retains it usual definition

$$\beta \equiv \frac{\partial \Delta \bar{G}^{0\#}}{\partial \Delta \bar{G}^0} \tag{19.40}$$

The peak in the electrochemical Gibbs energy in Figure 19.5 was chosen to be at one quarter of the distance between the initial and final states, when the system is at equilibrium ($\eta = 0$). This leads to a numerical value of $\beta_c = 0.25$ for the cathodic process of metal deposition. The corresponding rate equation, at high cathodic overpotential and in the absence of mass-transport limitation, is given by

$$j_c = nFk_h c_b \exp\left(-\frac{\beta_c \, n \, F}{RT}\eta\right) \tag{19.41}$$

Where n is the number of unit charges transferred. Thus, the value of the cathodic transfer coefficient for $n = 2$ is given by

$$\alpha_c = n\beta_c = 0.5 \tag{19.42}$$

leading to a Tafel slope of $b_c = 2.3RT/\alpha_c F = 0.118\,\text{V}$ at room temperature, in agreement with experimental observation in some cases of deposition of divalent metals.

For the reverse reaction of metal dissolution, the corresponding values are:

$$\beta_{an} = 0.75; \quad \alpha_{an} = 1.5 \quad \text{and} \quad b_{an} = 0.039\,\text{V} \tag{19.43}$$

It should be noted that in this treatment there is no compelling reason to assign a value of $\beta_c = 0.25$ for the metal deposition process, and this should be determined by experiment. On the other hand, the requirements that

$$\beta_c + \beta_{an} = 1.0 \quad \text{and} \quad \alpha_c + \alpha_{an} = n \tag{19.44}$$

are maintained, irrespective of the value of β_c.

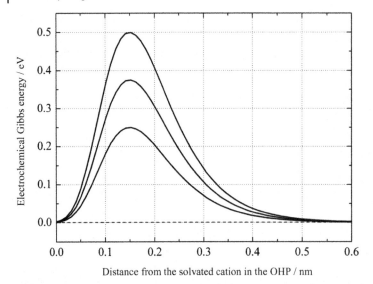

Figure 19.6 Schematic plots, showing the variation of the electrochemical Gibbs energy with distance, for systems at equilibrium, having different values of $\bar{G}^0$ and consequently of j_0. (c.f. E. Gileadi, *Israel J. Chem.* **48** (2008) 121)

19.7.8
The Exchange-Current Density, j_0

If charge transferred across the interface takes place by ionic migration generated by the electrostatic field, it is necessary to consider the influence of the metal/solution potential difference on the exchange current density. While $^{M}\Delta^{S}\phi$ is not measurable, there is no reason to assume that it would be zero at the deposition potential of any particular metal. This represents an apparent problem regarding the role of the metal/solution potential difference at equilibrium, $^{M}\Delta^{S}\phi_{equil}$, which, as we have emphasized in several places already, is not measurable.

The rate of all electrochemical processes is determined by the gradient of the electrochemical Gibbs energy, not by the gradient of electrostatic potential. Thus, the exchange current density for different metals is determined by the height of the peak of the electrochemical Gibbs energy at the reversible potential.

The three curves shown in Figure 19.6 represent schematically three electrode reactions having different values of $\Delta\bar{G}^{0\#}$, and hence different values of the heterogeneous rate constant and the corresponding exchange current densities. The curves in Figure 19.5 show the effect of overpotential on the shape of the plot of the electrochemical Gibbs energy. Two points should be emphasized.

1) When an overpotential is applied, the chemical part of the electrochemical Gibbs energy, ΔG^0, does not change, hence all the change in $\Delta\bar{G}^0$, induced by imposing an overpotential, is electrostatic. Consequently, it is correct to consider only the electrostatic field created by application of an overpotential as the driving force

for ion transfer, because it is equal to the gradient of $\Delta \bar{G}^0$, when expressed in the same units.

$$\frac{\partial \Delta \bar{G}^0}{\partial x} = nF \frac{\partial \eta}{\partial x} \tag{19.45}$$

2) There is a specific energy of activation for ion transfer across the double layer for each metal considered. Its height is changed by applying an overpotential but a definite energy of activation can be associated with ion transfer, even though it is driven by a linear electrostatic field.

The metal/solution potential difference that exists at the reversible potential $\Delta \phi_{rev}$ is one of two terms (the other being ΔG^0) that cannot be measured, but determine the numerical value of the reversible potential. The migration of ions is driven only by the electrostatic field created by the overpotential, and given by

$$\vec{E} = \eta/\delta \tag{19.36}$$

This electrostatic field is represented by the slopes of the straight lines connecting the initial and the final states in Figure 19.5. It has a value of zero for $\eta = 0$ and changes with changing overpotential. It cannot be overemphasized that the driving force is the gradient of the electrochemical Gibbs energy, $\partial \bar{G}/\partial x$ (or the corresponding gradient of the electrochemical potential $\partial \bar{\mu}/\partial x$). Equation (19.45) is a special case, in which application of an overpotential creates a situation in which the gradient of the electrochemical Gibbs energy becomes equal to the gradient of the overpotential, because the value of the chemical Gibbs energy is not affected by the overpotential, of course. This makes it possible to calculate the electrostatic field generated by applying the overpotential and consider the transfer of the metal ion as a migration. However, the true driving force is the gradient of the electrochemical Gibbs energy, which in this specific case is equal to the electrostatic field.

19.7.9
Why Are Some Electrode Reactions Fast?

It is noted above that one of the unexpected observations concerning metal deposition is that it is often a fast reaction. But what makes some chemical reactions slow, even when the decrease in the Gibbs energy of the overall reaction is negative and rather high? It is the simple fact that, in general, a chemical bond has to be broken in order to make another bond. Regarding electrode reactions, we have already noted that during outer-sphere charge transfer processes no bonds are broken – at most some are rearranged, while in metal deposition the value of $\Delta \bar{G}^0$ for removing the hydration shell is high, because the Gibbs energy of bonding of the water molecules to the central ion is comparable to that of chemical bonds. Hence, outer-sphere charge transfer processes can be expected to be fast, while metal deposition might be expected to be slow. However, there are two ways in which a chemical reaction can be fast, even if there are strong bonds broken in the process of converting the reactant to the product. The mechanism of ion transfer described here makes use of both.

One mode may be referred to as the "make-before-break" mechanism. This involves the gradual creation of the new bonds, while the old bonds are gradually weakened, in a concerted manner. A nice example of such a process (observed almost a century ago) is the spontaneous formation of adsorbed hydrogen atoms on nickel and some other metals in contact with molecular hydrogen in the gas phase, as represented by Eq. (19.46)

$$H_2 + Ni \rightarrow 2NiH \tag{19.46}$$

No energy of activation was detected, although the energy of dissociation of molecular hydrogen is quite substantial (4.52 eV).

This peculiar observation was explained by proposing that, as a molecule of H_2 approaches the surface, the length of the bond between the atoms is increased, thus increasing the Gibbs energy, while some interaction is created between each of the hydrogen atoms and the surface, leading to a decrease in the Gibbs energy. If the latter exceeds the former everywhere along the reaction coordinate, no energy of activation is observed.

In the model of ion transfer presented here, a similar interpretation can be adopted: as the ion approaches the surface, the increase in Gibbs energy associated with distortion of the solvation shell can be partially compensated for by interactions between the charge on the ion and the delocalized electrons in the conduction band of the metal, decreasing the total Gibbs energy of the system and leading to a low energy of activation.

The other mode involves reactions that advance in small steps from the initial to the final state. For example, the electrochemical reduction of oxygen is a slow process, even in the presence of the best of catalysts known today, as is well known in the field of fuel cells. Yet, this very same process takes place readily in living organisms, where it occurs in many small steps (assisted, of course, by enzymes acting as catalysts).

In the model of ion transfer discussed here, the high rates of metal deposition and dissolution are explained by combining these two effects: The ions crossing the double layer start interacting with the delocalized electrons in the metal before their hydration shell has been removed, in a "break-before-make" type mechanism. This movement occurs in many small steps, each associated with a small degree of solvent rearrangement and a minute change in the effective charge on the ion.

Summing it up, outer sphere charge transfer involves the abrupt transfer of an electron in a very short time, of the order of a femtosecond. The molecule is then rearranged to its most stable state by a much slower process taking about 10^3 fs. In metal deposition the ion carrying the charge moves much more slowly in the direction perpendicular to the electrode surface. There is no distinct point at which charge is transferred, it happens all along the way, for a period of the order of 10^6 fs. If the metal being deposited is divalent, its effective charge is reduced gradually. There is no need to consider the formation of a monovalent intermediate, although there must be a point along the path of the ion where it has a charge of $+1.0$. This, however, is not a singular point. It is just one of a great number of minute steps where the charge changes, for example from 1.05 to 1.00 and further to 0.95.

A caveat regarding the details of the movement of the ions towards the electrode is in place here. In spite of the high electrostatic field applied, (which is equal to the gradient of the electrochemical Gibbs energy, as shown in Eq. (19.36)), the ions are not moving in a straight line perpendicular to the surface. The thermal energy of the system gives rise to random motion, superimposed on the motion in the electrical field. Thus, the true physical picture would be represented by irregular motion of the ions, with a component pointing in the direction perpendicular to the surface. During metal deposition the ions could be moving in all directions, sideways, up and down, and even away from the surface. The vector of actual motion will always be the sum of two vectors: one exactly perpendicular to the surface, induced by the electrostatic field, and the other randomly distributed in all directions, as a result of the thermal energy.

20
Energy Conversion and Storage

20.1
Batteries and Fuel Cells

20.1.1
Classes of Batteries

It is appropriate to consider three classes of batteries.

20.1.1.1 Primary Batteries

Primary batteries are designed for use only once. The chemical energy is stored in the two electrodes and sometimes in the electrolyte, and is converted to electrical energy on demand. Until about the middle of the 20th century, the most important use of primary batteries was for flashlights. With the introduction of portable electronic devices, initially radios, tape recorders and cameras, and now (2010) just about everything from civilian to military applications, primary batteries are manufactured in immense quantities worldwide.

This is not necessarily an efficient way of storing energy. The total energy consumed in the manufacturing of a primary battery may be several times as much as the energy that can be retrieved from it. Yet, primary batteries offer by far the best way to store energy in small packages, and in many applications they constitute the only way to store energy[1]. The first and most commonly used battery, even today, is the Leclanché cell and, more recently, the so-called alkaline cell, which is a variant of the Leclanché cell. In both types the chemical energy is stored in the form of metallic Zn at the anode and MnO_2 at the cathode, as will be discussed below. In the second half of the 20th century primary batteries based on Li anodes and a nonaqueous solvent were introduced. Such batteries are better than the Leclanché cell in almost every aspect other than cost, so the two types dominate the world market, side by side.

---

1) It has been noted, rather sarcastically, that if all batteries would be destroyed by a magic wand, war as we know it could not be waged.

Physical Electrochemistry: Fundamentals, Techniques and Applications. Eliezer Gileadi
Copyright © 2011 WILEY-VCH Verlag GmbH & Co. KGaA, Weinheim
ISBN: 978-3-527-31970-1

20.1.1.2 Rechargeable Batteries

Rechargeable batteries, as their name indicates, are designed for multiple use. The most widely used is the lead-acid battery in our cars, and in many other applications. The chemical energy in these batteries is stored as metallic Pb in the negative electrode and as PbO_2 in the positive electrode, both immersed in a solution of about 25% sulfuric acid. Upon discharge, the anode is oxidized to $PbSO_4$ and the cathode is reduced to the very same compound. It should be noted, in passing, that during the discharge of a battery, the anode is the negative pole and the cathode is the positive pole. In a driven electrochemical cell, discussed during most of this book, it is the other way around – the anode, where oxidation takes place, is the positive terminal while the cathode, where reduction takes place, is the negative terminal.

The lead-acid battery reigned supreme among rechargeable batteries for about a century, because it is constructed from inexpensive materials and is highly reliable. When properly treated, a car battery can last for several years. Proper treatment in this context means that it is not allowed to dry out; it is never discharged completely and never overcharged. It can also deliver a large power (of 1.0–1.5 kW) for a few seconds, as needed to start a car on a very cold day.

The efficiency of storage and retrieval of energy, which is often called the *electric-to-electric* (ETE) efficiency, can be as high as 80%, depending on the way the battery is used. Lead acid batteries are very heavy (in terms of energy stored per unit weight), but this is of little consequence when a single battery is used in a car. On the other hand, it makes it essentially impossible to use such batteries as the power source for modern electrical cars, although there are some niche applications, such as golf carts and vehicles for indoors operation, such as in airport terminals or for fork lifts inside factories.

Another rechargeable battery in common use for many decades is the Ni-Cd cell, consisting of metallic Cd as the anode and trivalent nickel oxy-hydroxide as the cathode, in concentrated KOH solution. The reactions taking place during discharge of this cell are

$$NiO(OH) + H_2O + e_M^- \rightarrow Ni(OH)_2 + (OH)^- \tag{20.1}$$

$$Cd + 2(OH)^- \rightarrow Cd(OH)_2 + 2e_M^- \tag{20.2}$$

In many technological aspects this battery is superior to the lead-acid battery, but, in particular, it can be cycled about ten times more than the lead-acid battery and its weight per unit energy stored is lower. Contrary to the lead-acid battery, it is necessary to discharge it completely from time to time, in order to maintain good performance. Unfortunately, it is much more expensive.

The batteries used in modern laptops and other portable electronic devices are based on a rechargeable Li-ion battery and a nonaqueous solvent. When fully charged, the anode is some form of graphite containing atomic lithium in the ratio of approximately LiC_6, and the cathode is MnO_2 or some similar compound of the general form of $LiMn_2O_4$ or $LiFePO_4$. The electrolyte is propylene carbonate or its mixture with similar compounds, such as ethylene carbonate. During discharge Li is removed from the carbon matrix as the Li^+ ion and is intercalated into the cathode, forming compounds such as shown above.

This type of battery is much superior to any other rechargeable system. Its working potential is about 3.5 V, with prospects to increase the voltage up to 4.5 V, compared to 2.1 V for the lead-acid and 1.25 V for the Ni-Cd batteries. Its energy density is also many times higher, but so is the price. Thus, one would not consider using it to replace the lead-acid batteries in cars or the alkaline Leclanché cells for flashlights, for example, but it is used where energy density, reliability and long service life dominate over cost.

There are other methods of storing energy. These include thermal storage employing phase change or shifting of chemical equilibria, hydrogen storage as molecular H_2 or as a chemical compound (e.g. NH_3, $LiBH_4$, $LaNi_5H_{6.7}$, AlH_3, MgH_2), compressed gas, fast-turning flywheels and pumped water in hydroelectric stations). We shall not discuss the advantages and shortcomings of these methods here, except to note that they depend on the end use required and on the amount of energy being stored. For all portable electrical and electronic devices, batteries, and in some cases capacitors, are practically the only solutions.

20.1.1.3 Fuel Cells

A *fuel cell* is commonly regarded as a different type of energy conversion device, but it is just another type of battery. For example, the difference between a lead-acid battery and a hydrogen-oxygen fuel cell is in the way the energy is stored and replenished. In the battery it is all internal. The chemical energy is stored in the two electrodes and the battery is recharged, converting electrical energy to chemical energy. In the fuel cell the chemical energy is stored in containers outside the electrochemical cell. Recharging is replaced by adding the chemical energy (hydrogen and oxygen in this case) from outside containers. When the device is being used as a source of electric power, chemical energy is being converted to electrical energy in both cases.

The unique feature of batteries and fuel cells is that in them electrical energy is produced *directly* from chemical energy, bypassing the need to convert it first to heat and then construct a heat engine of one type or another, to convert the thermal energy to electrical energy. The efficiency of conversion of the latter is limited by the second law of thermodynamic expressed by the Carnot cycle[2]:

$$\eta = \frac{T_2 - T_1}{T_2} \tag{20.3}$$

In the above equation, T_2 and T_1 represent the temperatures of the hot and the cold heat reservoirs in the system. It follows that thermal energy cannot be extracted (or converted to any other form of energy) isothermally. In contrast, a fuel cell converts chemical energy directly to electrical energy isothermally, according to the simple equation

---

2) Internal combustion engines, as well as a major electrical power station, are "heat engines" in the thermodynamic sense, and their theoretical maximum efficiency is that of the Carnot cycle.

$$\Delta G = -nFE \tag{1.4}$$

This equation shows that, at the limit of reversibility, the Gibbs energy released by the system can be converted to electrical energy with 100% efficiency. A similar process, of converting chemical energy to mechanical energy takes place in the human body and most other living organisms. The energy needed to move our muscles is derived from the chemical energy of oxidation of food by the oxygen we breathe.

The ability to oxidize (burn) fuel isothermally and at high efficiency, not limited by the Carnot cycle, is very attractive. Although this principle was already demonstrated in the middle of the 19th century, serious efforts to develop fuel cells only started in 1959, when a group of scientist claimed that propane gas could be used directly to produce electric power in a fuel cell. About a decade later the first successful implementation of a H_2-O_2 fuel cell was made in the Apollo space program, aimed at sending a man to the Moon. Since then great progress has been made in the development of fuel cells. At present (2010), the favorite system being studied is a direct methanol fuel cell (DMFC). Unfortunately, it must be said that after half a century of intense research and development efforts, there is still no fuel cell for civilian applications on the market, although there may be some niche military and space applications.

20.1.2
The Theoretical Limit of Energy Per Unit Weight

Thermodynamics allows us to calculate the maximum energy density of a battery, which may be approached, but never reached, in real batteries. This type of calculation is based on the reversible potential and on the equivalent weight of the electrochemically active ingredients, ignoring the current collector, the casing, and so on. As an example, consider the Ni-Cd rechargeable battery. The reactions at the anode and the cathode are given in Eq. (20.1) and (20.2), respectively. The resulting overall reaction is hence

$$2NiO(OH) + Cd + 2H_2O \underset{\text{Charge}}{\overset{\text{Discharge}}{\rightleftharpoons}} 2Ni(OH)_2 + Cd(OH)_2 \tag{20.4}$$

The reversible cell potential in 30% KOH is 1.29 V. The electrical energy produced per mole is given by

$$nFE = 2 \times 96\,485 \times 1.29 \approx 2.5 \times 10^5 \ W\,s \approx 69\,Wh \tag{20.5}$$

The sum of the molecular weights of the reactants is 331.8, leading to a *theoretical* energy density of

$$69/0.332 = 208 \ Wh\,kg^{-1} \tag{20.6}$$

This should be compared to a practical value of about 40 Wh kg^{-1}, namely, just under 20%.

What is the purpose of this kind of calculation? In the development of a battery it is important to know how far we are from its theoretical limit, since the practical limit, which is typically half the theoretical limit, can be reached asymptotically and the effort in approaching it grows exponentially. Thus, if current technology represents only 20% of the theoretical limit, there is a very good probability that it could be developed to 50%, namely by a factor of 2.5. On the other hand, if current technology has reached 45% of the theoretical limit, it is probably close to its practical limit, and the wisdom of attempting to develop it further may be questioned.

20.1.3
How Is the Quality of a Battery Defined?

There are many parameters by which the quality of a given type of battery can be judged. Most important are the energy density, in units of $Wh\ kg^{-1}$ and the power density, in units of $W\ kg^{-1}$. But in certain applications volume is more important than weight. This applies, for example, to batteries for wrist watches and for hearing aids. For possible application for electric cars, weight is the more important parameter, but volume could also play an important role.

For primary batteries the rate of self-discharge, which determines shelf life, is critical, while for rechargeable batteries a higher rate of self-discharge may be tolerated. Safety for the user is always a major issue. Imagine the battery in your laptop catching fire (in your lap, of course, where else!), particularly if this might happen on a commercial flight. Reliability is critical for some applications, such as heart pacers or the communication system of an army unit in battle, but rather irrelevant for a battery used in a toy. Price is more important in civilian than in military and space applications, but is almost irrelevant for *in vivo* medical applications. Finally, environmental issues during manufacturing, use and disposal can no longer be ignored.

It should be remembered that primary batteries, born more than a century ago in a world obsessed by the development of new technologies, may die soon in a world obsessed by ecology and recycling. Indeed, it is almost unbelievable that in the first decade of the 21st century we still use and throw away many billions of batteries every year, while the technology to produce rechargeable batteries, which could be reused thousands of times already exists. Considered from the point of view of recycling, the transition from primary to secondary batteries is equivalent to recycling paper, aluminum or beer cans thousands of times.

20.2
Primary Batteries

20.2.1
Why Do We Need Primary Batteries?

It cannot be overemphasized that batteries (of any kind) are not sources of energy – they are devices to store energy. The energy needed to build a primary battery far exceeds the energy that can be stored in it. Moreover, primary batteries are extremely expensive, when the cost is calculated per unit of energy. For example, the cost of AA-size alkaline cells needed to store 1 kW h may be anywhere between $100 and $1000, compared to about $0.15 paid to the electric utility company for the same amount of energy. On the other hand, primary batteries carry energy in "small change" and do that more efficiently than any other device. Since the battery in your quartz-crystal watch lasts for two or three years and costs a few dollars, who cares about the cost per kW h?

20.2.2
The Leclanché and the Alkaline Batteries

The oldest commercial primary battery is the Leclanché cell, invented in 1866. In its modern version it consists of a Zn container acting as the anode (the negative electrode), where metallic Zn is oxidized to $Zn(OH)_2$. The electrolyte is NH_4Cl or $ZnCl_2$, with a small amount of water to allow electrolytic conductivity, yet leaving it in the form of a paste or gel. The open-circuit potential is about 1.5 V. The cathode material is MnO_2, mixed with carbon powder and in intimate contact with a carbon rod acting as the current collector and the positive terminal of the cell. The reactions taking place in a Leclanché cell are

$$Zn + 2H_2O \rightarrow Zn(OH)_2 + 2H^+ + 2e_M^- \qquad (20.7)$$

$$2MnO_2 + 2H^+ + 2e_M^- \rightarrow 2MnO(OH) \qquad (20.8)$$

resulting in the overall reaction

$$Zn + 2MnO_2 + 2H_2O \rightarrow Zn(OH)_2 + 2MnO(OH) \qquad (20.9)$$

Zinc is an active metal ($E^0 = -0.763$ V vs. SHE) that corrodes in aqueous solutions, giving off molecular hydrogen, according to the reaction

$$Zn + 2H_2O \rightarrow Zn(OH)_2 + H_2 \qquad (20.10)$$

This reaction, which takes place at open circuit or as a side reaction during discharge of the battery, is detrimental in two ways: it consumes one of the active materials in the cell and it produces a gas that could build up pressure and eventually rupture the cell. This is the purpose of adding HgO to the zinc anode. In contact with metallic zinc, it is reduced to metallic mercury, which amalgamates the zinc. The exchange current density for hydrogen evolution is much lower on mercury and its amalgam than on zinc, thus reducing the rate of self-discharge via the hydrogen evolution reaction. Other inhibitors have also been used, but mercury is the most effective in this respect. In early designs, the mercury content of the anode in the final product reached several percent. This amount has been gradually reduced and eventually mercury was eliminated altogether, because of its toxicity.

Leclanché cells have several disadvantages. They have a relatively short shelf life, and must be refrigerated for long-time storage. The energy density is about

$75\ \text{Wh kg}^{-1}$, which is relatively low for a primary battery, and the power density is also low. The voltage at constant load declines during discharge, which is a disadvantage for most applications, although it affords a convenient way to monitor the state of charge of the battery. On the other hand, the greatest advantage of Leclanché cells is their low price. They are therefore widely used for simple applications, where reliability and performance are not of critical importance[3].

A similar system, containing the same active materials, is the so-called *alkaline battery*. This type of battery differs from the Leclanché cell in that the electrolyte is concentrated KOH, the zinc is in the form of a high-surface-area paste and the casing is made of nickel-coated steel. It has a much higher power density, because of the higher conductivity of the electrolyte and the larger surface area of the zinc anode, and is less likely to leak. It is also more expensive. As usual, one can buy higher reliability and better performance for a higher price, and the choice depends on the end use.

20.2.3
The Li-Thionyl Chloride Battery

For about a century after the invention of the Leclanché cell (and the lead-acid rechargeable battery) the highest voltage attained in any battery was 2.1 V, and primary cells delivered a maximum voltage of about 1.5 V. This changed in the mid 1960s, when the Li-thionyl chloride ($LiSOCl_2$) battery was introduced. In this and other nonaqueous lithium batteries, a potential of about 3.0 V was achieved.

The $Li/SOCl_2$ battery consists of a lithium anode and a carbon paste cathode. The electrolyte consists of a solution of $LiAlCl_4$ in $SOCl_2$. The anode reaction is metal dissolution:

$$Li^0_M \rightarrow Li^+_{soln} + e^-_M \tag{20.11}$$

The cathode reaction is more complicated, since in this battery there is no reducible material in the cathode itself, and it is the solvent that serves as the active cathode material.

$$2\ SOCl_2 + 4\ e^-_M \rightarrow SO_2 + S + 4Cl^- \tag{20.12}$$

and the overall cell reaction is

$$4Li^0 + 2\ SOCl_2 \rightarrow SO_2 + S + 4LiCl \tag{20.13}$$

The reversible cell voltage was estimated to be 3.65 V. Assuming the above cell reaction, this leads to a theoretical energy density of $1.48 \times 10^3\ \text{Wh kg}^{-1}$. Values as high as $700\ \text{Wh kg}^{-1}$ have already been realized in commercial cells. Unlike the Leclanché cell, Li-thionyl chloride batteries are designed to be anode limited, that is, to have a stoichiometric deficiency of lithium, for obvious reasons: discarding a spent cell with a residue of metallic Li could create a serious safety hazard!

---

3) In Leclanché cells the casing is made of zinc and serves as the anode. This design increases to some extent the chances of leakage, but makes manufacturing very cheap.

Similar batteries having SO_2 or propylene carbonate (PC) as the solvent have been prepared. The former has better characteristics at low temperature. In the case of PC, the solvent does not take part in the reaction and MnO_2 is used as the cathode material. This yields a slightly lower voltage of 2.8 V, but PC is much easier to handle than either $SOCl_2$ or SO_2, both in the manufacturing process and for waste disposal.

Lithium batteries have many advantages compared to Leclanché cells and other aqueous batteries, but they are also interesting from the fundamental point of view. To begin with, the Gibbs energy of interaction of lithium with the solvent is so high that they would be expected to react violently with each other, leading to a very high rate of self-discharge at best, and to dangerous explosions at worst. However, as soon as contact between the metal and the solvent is made, a protective layer is formed, which prevents further chemical reaction. In the $Li/SOCl_2$ and Li/SO_2 cells this layer consists of LiCl, which is not soluble in the solvent. If PC is used as the solvent, the protective layer consists of Li_2CO_3. Fortunately these layers are permeable to Li^+ ions, but not to electrons. In fact, the protective layer on the surface of Li serves as an electrolyte, having a transference number of unity with respect to the positive ion. It is referred to in the literature as the solid/electrolyte interface (SEI).

A layer formed on a metal upon contact with the solution could be of three different types. If it is dense and nonconducting, it can protect the metal from corrosion, but the system cannot be used as a battery, since the metal is totally isolated from the solution. If it is electronically and ionically conducting, reduction of the solvent at the film/electrolyte interface and oxidation of the metal at the metal/film interface could proceed freely, leading to a high rate of self-discharge of the battery. It is only when the film is simultaneously an *ionic conductor* and an *electronic insulator* that the chemical pathway of spontaneous reduction of the solvent at the anode is blocked, whereas the electrochemical pathway of oxidation of the metal at the anode and reduction of the solvent (or of MnO_2) at the cathode can proceed at a sufficient rate to allow the use of the system as an anode in a battery.

In addition to the high energy density of $Li/SOCl_2$, the operating cell voltage in low-rate batteries is around 3.0 V, more than twice that of aqueous primary batteries. As a result, computer memories based on C-MOS technology can be operated using a single cell.

The rate of self-discharge of primary Li batteries is very low, about 10 times lower than that of alkaline cells, allowing storage without refrigeration for periods of 5–10 years. Another welcome feature is the stability of the voltage during discharge, which is shown in Figure 20.1a. This is an obvious advantage for the operation of any electronic device. The only drawback is that the voltage at open circuit or under load cannot be used as a measure of the state of charge of the battery, and special devices have to be developed to determine the state of charge of $Li/SOCl_2$ and other lithium-based nonaqueous batteries.

In Figure 20.2 we show the voltage of a $Li/SOCl_2$ cell during discharge at different rates. The decline in energy density with the rate of discharge is shown in the inset. These data were obtained with an AA-type cell having an energy density of about 700 Wh kg^{-1} and a calculated charge of 2.65 A h. The charge measured at a low rate of 1 mA was 2.41 Ah, which constitutes 91% of the calculated capacity, showing very

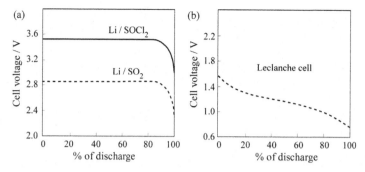

Figure 20.1 The discharge curves of (a) Li/SOCl₂ and LiSO₂ cells, (b) Leclanché cell. Note the stability of the voltage during most of the discharge process for the Li cells, compared to the decrease in voltage of the Leclanché cell. Data from *Handbook of Batteries and Fuel Cells*, Ch. 3, D. Linden, (Ed.), McGraw Hill, 1984.

efficient utilization of the active material. Some of this efficiency is lost when the discharge rate is increased, as seen in the inset in Figure 20.2. At a rate of 15 mA the charge capacity is reduced to 1.64 A h, corresponding to 62% utilization of the active material. The high energy density of this type of Li/SOCl₂ battery combined with the low rate of self-discharge and the high operating voltage makes them ideally suited to power the memory back-up in computers. In this use the battery is designed to outlast the useful lifetime of the computer itself and probably will never have to be replaced.

For low power applications, these batteries are quite safe, but high-power lithium batteries have been known to explode when accidental heating melts the lithium (m.p. 180.5 °C). This can rupture the protective SEI layer, leading to a violent reaction between the metal and the solvent, and eventually to explosion.

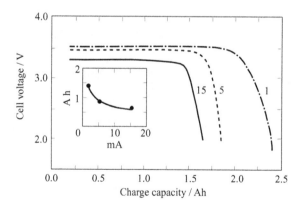

Figure 20.2 Discharge curves of a 2.65 A h AA-size Li/SOCl₂ battery designed for low-rate applications, at different rates of discharge. The numbers on the lines stand for the discharge currents in mA. The inset shows the dependence of the charge capacity on discharge rate. Data from H. Yamin, M. Pallivathikal and U. Zak, 5th International Seminar on Li Battery Technology and Applications, Florida, 1991.

The current–voltage characteristics shown in Figure 20.1 and Figure 20.2 are very important for all batteries. Ideally one would like to have a flat discharge curve, namely a potential that is almost constant throughout the discharge stage, and falling fairly sharply when the battery has been exhausted. In comparison, the voltage of a Leclanché cell declines steadily with time during discharge, as shown in Figure 20.1b.

20.2.4
The Lithium-Iodine Solid State Battery

This type of battery is used for heart pacemakers. The electrode reactions in this case are very simple, leading to the overall cell reaction:

$$2\,Li + I_2 \rightarrow 2\,LiI \tag{20.14}$$

The electrolyte is solid LiI, which is formed *in situ* during operation of the cell. The solid electrolyte interface is conducting for Li^+ ions but nonconducting for electrons. The cathode consists of a compound such as poly-2-vinylpyridine, mixed with molecular iodine and melted together until a homogeneous material has been formed. This constitutes a charge-transfer complex, allowing easy transfer of the iodine. The thickness of the solid LiI layer grows during discharge of the battery and its resistivity is quite high, but not too high, considering the very low currents (in the range of $10\,\mu A$ or less) needed to operate the pacemaker.

The cell voltage is $2.8\,V^{[4]}$, and the energy density is about $250\,Wh\,kg^{-1}$. The advantages of this type of battery are its very low rate of self-discharge and its extremely high reliability. Although it has been used in tens of thousands of cardiac patients over several decades, there has not been a single case of failure causing lethal internal injury. For the very special requirement of heart pacemakers, the foregoing advantages make this type of battery commercially viable, in spite of its very high cost[5]

20.3
Secondary Batteries

20.3.1
Self-Discharge and Cycle Life

Secondary batteries are rechargeable. The rate of self-discharge is less critical in this case, because the battery can always be recharged before use. Cycle life, which is the

4) According to tables of standard potentials, the open-circuit potential of this cell should be 3.58 V. Note, however, that standard potentials are given for aqueous solutions, whereas here the reaction occurs in the solid state where the Gibbs energy of the reaction can be quite different.

5) Most of the cost is due to the extremely high reliability demanded in this kind of application. For other uses, such as memory back-up in computers, the same battery could be manufactured at a much lower cost.

number of times the battery can be charged and discharged, is important. The efficiency of energy storage in secondary batteries is quite high, of the order of 60–90%, and a good battery can be charged and discharged a thousand times with little loss in performance. Energy density (Wh kg^{-1}), power density (W kg^{-1}) and the temperature range over which the battery can be operated are important. The energy density of a battery depends, of course, on the rate at which it is discharged, and it must be defined for a specific rate. Three factors are involved here. (i) The activation overpotential depends on the current density during discharge (i.e., on the rate of discharge). (ii) Some potential is lost as a result of the resistance of the solution and of the separator in the cell. (iii) The amount of available charge also depends on the discharge rate.

20.3.2
Battery Stacks versus Single Cells

A battery stack consists of a number of cells connected in series and packaged as a unit. This is required when the voltage of a single cell is not high enough to operate the device. It is invariably observed that the reliability in performance and the lifetime of stacks of batteries is less than that of single cells. This can be understood on the basis of very simple statistical reasoning. If the probability of failure of a single cell during the intended service life of a stack is 1%, the probability for its flawless operation is 0.99. The probability of trouble-free operation of a stack consisting of 12 cells of this type connected in series is $0.99^{12} = 0.89$. Thus, a reliability of 99% for the individual cell translates to a lower reliability of only 89% for the stack! Repeating the same calculation for a battery stack containing 100 cells in series yields a reliability of only $0.99^{100} = 0.366$. On the other hand, if the probability of failure is reduced to 0.1% for a single cell, a stack of 100 cells will have a reliability of $0.999^{100} = 0.90$.

20.3.3
Some Common Types of Secondary Batteries

20.3.3.1 The Lead-Acid Battery
The best known and most widely used secondary battery is the *lead-acid battery*, which consists of a lead anode and a lead dioxide cathode in a 25% solution of sulfuric acid. The reaction at the anode during discharge is

$$Pb + SO_4^{2-} \rightarrow PbSO_4 + 2e_M^- \qquad (20.15)$$

while at the cathode lead dioxide is being reduced, yielding the same product

$$PbO_2 + 2\,H_2SO_4 + 2\,e_M^- \rightarrow PbSO_4 + SO_4^{2-} + 2H_2O \qquad (20.16)$$

Thus the overall cell reaction is given by

$$PbO_2 + Pb + 2\,H_2SO_4 \underset{\text{charge}}{\overset{\text{discharge}}{\rightleftharpoons}} 2\,PbSO_4 + 2\,H_2O \qquad (20.17)$$

We note that sulfuric acid is being consumed during discharge, and an equivalent amount of water is formed. The resulting decrease in density of the electrolyte used to be the basis for the method used in many garages to test the state of charge of a car battery by measuring the density of the electrolyte.

The open circuit cell voltage is 2.1 V, the highest voltage for any *aqueous* battery. In fact, we would not expect a battery having such a high voltage to hold charge at all, because the voltage is high enough to electrolyze water, providing an efficient way for rapid self-discharge. Indeed, if the terminals of a lead-acid battery are connect to two platinum electrodes placed in a cell containing the same solution, copious evolution of hydrogen and oxygen is observed. But lead is not platinum. The kinetics of hydrogen evolution on lead is slow, leading to a very high overpotential for this reaction. Thus, the operation of the lead-acid battery depends, to a large extent, on the sluggishness of the hydrogen electrode reaction at the cathode. This is a case in which having a bad electrocatalyst is an asset rather than a liability! The successful operation of the lead-acid battery depends, of course, on the rate of the overall reactions shown in Eq. (20.17) in both directions.

The theoretical energy density of lead-acid batteries is only 171 Wh kg^{-1}, due to the high atomic weight of lead. The practical energy density depends on the rate of discharge but even at low rates it does not exceed about 40 Wh kg^{-1}. This represents about 23% of the theoretical value, despite massive investments in engineering, aimed at increasing the energy density of this type of battery.

As pointed out above, the rate of self-discharge of secondary batteries is not as important as it is for primary batteries. One does not expect a car battery to be fully charged after the car has been idle for several months. The quality of a secondary battery is measured, instead, in terms of its service life, which is determined by the number of times it can be charged and discharged. To understand this limitation, one must consider the structure of the electrodes in the cell and the way in which they are charged and discharged. As a rule, the active material is in the form of a high-surface-area powder, pressed onto the surface of a metallic grid, which serves as the current collector. The anode in the lead-acid battery consists of small particles of lead bonded to a lead screen[6]. During extended use, and in particular as a result of abuse, the particles may agglomerate, causing a loss of surface area. Now, we note that during discharge lead is converted into $PbSO_4$, which is an insulator. If a lead particle is coated with its oxidation product, it becomes isolated from the solution and can no longer take part in the discharge process. Agglomeration increases the size of the particles, causing a loss of active surface area, thus decreasing the effective charge capacity of the battery. The same mechanism can also cause deterioration of the cathode, since PbO_2 is electronically conducting, while the $PbSO_4$ formed on the surface of these particles during discharge is not.

---

6) The real situation is invariably more complex. The screen is not made of pure Pb, but may contain some Ca or Sb. A binding material is used to increase adhesion of the lead particles to the screen. Each manufacturer has his own secret formulations but we do no need to know these to understand how such batteries operate.

Another common mode of failure of batteries is loss of electrical contact between the active material and the current collector. There are many other ways in which batteries can fail, such as the aging of separators and accidental contact between anode and cathode. These problems are not discussed here.

The largest asset of lead-acid batteries is their low cost, compared to any other secondary battery currently available. The energy density is inherently low, but in its current application as a car battery this is tolerable. For application as the main power source of electric vehicles, an energy density of $250-500$ Wh kg^{-1} will probably be needed. This exceeds the theoretical energy density of this system. Thus it can be categorically stated that the future electric family car (assuming that there will be one) will not have a lead-acid battery as it s main source of energy. Power density is another limitation, particularly since increasing the power can decrease the energy density dramatically.

20.3.3.2 The Nickel-Cadmium Battery

The Ni-Cd battery consists of a Cd anode and a NiO(OH) cathode, in a concentrated solution of KOH.[7] The cathodic and anodic cell reactions during discharge are given in Eqs. (20.1) and (20.2), respectively. The exact composition of the fully charged cathode is not entirely clear. Some prefer to write it as NiO_x where $1.5 \leq x \leq 2$, implying that there might be some Ni^{4+} mixed with Ni^{3+}. There seems to be little doubt that the cathode can be charged beyond the level corresponding to the conversion of all the active material to trivalent nickel, but the existence of tetravalent nickel has not been confirmed. The overall reaction is

$$2NiO(OH) + Cd + 2H_2O \rightarrow \underset{\text{charge}}{\overset{\text{discharge}}{\rightleftharpoons}} 2Ni(OH)_2 + Cd(OH)_2$$

$$(20.18)$$

The cell voltage is 1.29 V and the theoretical energy density is 208 Wh kg^{-1}. Discharge curves for a Ni/Cd battery are shown in Figure 20.3. This figure shows clearly the two effects that reduce the energy obtained from a battery as a function of the rate of discharge: On the one hand, the cell voltage is decreased with increasing discharge rate. On the other hand, the fraction of the active material that can be utilized is decreased. For a 0.1C (10 h discharge) the potential is almost the same as for the reference rate of 1C, but the total charge available is about 10% higher. For a rate of 10C (6 min discharge) the output voltage is decreased to about 1.0 V and the available charge is decreased by about a factor of two.

Here we draw attention to the special way the discharge rate is defined for batteries in the literature. Thus, a discharge rate of 5C means that the cell is discharged in 1/5 h, while a rate of 0.1C indicates that it is discharged in 1/0.1 h. Full discharge is defined as the charge measured at 1C. This is somewhat arbitrary, but nevertheless quite useful as a point of reference.

7) This again is an oversimplification. The cadmium anode may contain a few percent each of iron, nickel and graphite. The NiO(OH) cathode may contain some Co and also graphite. The electrolyte is 25–30% KOH but some LiOH may be added to increase the conductivity of the electrolyte.

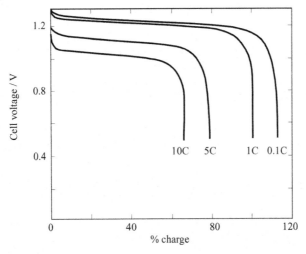

Figure 20.3 Typical discharge curves of a Ni/Cd battery at different rates. Data from B. Evjen and A.J. Catotti in *Handbook of Batteries and Fuel Cells*, Ch.17, D. Linden, editor, McGraw-Hill, 1984.

The main advantage of Ni/Cd over lead-acid batteries is the longer cycle life. At an 8 h discharge rate (0.125C), the two batteries may be nearly equal in energy density, but at a 30 min rate (2C) the Ni/Cd battery still performs well whereas the lead-acid battery can barely work, losing 80% of its capacity. The performance of high quality Ni/Cd batteries as a function of the number of cycles is shown in Figure 20.4. About 10% of the capacity is lost in the first 1000 cycles, but no further decline in capacity is observed up to about 2500 cycles. Performance deteriorates rapidly beyond 3500 cycles, and this can be considered to be the limit of the useful cycle life of such batteries. In terms of real time, Ni/Cd batteries can last 5–10 years, depending on

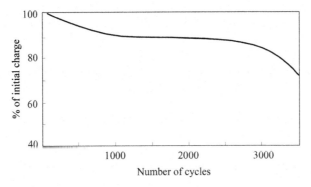

Figure 20.4 The decline in charge capacity with cycle life for a high quality sealed Ni/Cd battery. The cycle used was: 5 h charge and 3 h discharge. Discharge was stopped when the potential reached 1.0 V. Data from Wiseman in *Handbook of Batteries and Fuel Cells*, Ch. 18, D. Linden (Ed.), McGraw-Hill, 1984.

design, quality, and type of application. In comparison, lead-acid batteries typically do not last more than about 500 cycles or 5 years.

In defining the cycle life of different batteries, one should be careful to state how the test was performed. Charging back and forth from 10% to 90% may be a reasonable choice, but for a particular case that test may not be relevant. For example, a lead-acid battery in a car may be charged most of the time, with occasional heavy use (when the car is started), so that its cycles could be from 95% to 75%. In comparison, a Ni-Cd cell in a camera may be used typically until it is almost totally discharged, representing cycles of 5% to 95%.

The worst drawback of the Ni-Cd battery is its cost, but progress has been made in recent years and Ni-Cd rechargeable batteries are gradually replacing Leclanché cells and alkaline Zn-MnO$_2$ primary batteries in many simple applications, such as mechanical toys and flashlights.

20.3.3.3 The Nickel-Metal Hydride Battery (NiMH)

Storage of hydrogen in LaNi$_5$ (in the form of LaNi$_5$H$_{6.7}$) was mentioned in Section 7.3.1, but there is a long way to go between storage of dry hydrogen and building a working rechargeable battery with LaNi$_5$ as the anode and the medium for storing hydrogen. Indeed the first attempts to build such a battery with the NiO(OH) cathode were not promising. It was only the persistence and perseverance of Ovshinsky and the extensive research conducted in his company, Ovonics, that led to the development of practical NiMH rechargeable batteries that could compete successfully with Ni-Cd.

The anode of a current NiMH battery on the market has the typical AB$_5$ structure, but both A and B do not represent pure metals but rather some nonstoichiometric combination of metals of the same group as La and Ni, respectively. A typical formulation can be written as[8] [La(5.7)Ce(8.0)Pr(0.8)Nd(2.3)] [Ni(59.2)Co(12.2)Mn(6.8)Al(5.0)], where the numbers represent the relative atom% concentrations of each element. Adding up the numbers this corresponds to A(16.8)B(83.2), which could also be written as AB$_{4.95}$, very close to the LaNi$_5$ discussed above.

The unique feature of this complex formulation is that the relative concentrations of elements in the groups of A and B can be varied, in order to optimize the performance of the battery, aiming at maximum energy, highest power, longest cycle life or lowest cost. The electrolyte is 30% KOH, just as the Ni-Cd battery. The energy density is in the range$(40-110)$ Wh kg^{-1}, depending on the power requirements, which can be as high as 1900 W kg^{-1}. This battery proved to be a great success, with sales already exceeding 1 billion a year, and it is gradually replacing Ni-Cd batteries.

8) See M.A Fetcenko, S. R. Ovshinsky et al., J. Power Sources, 165 (2007) 544–555.

20.3.4
The Li-Ion Battery

The introduction of the Li-thionyl chloride primary battery represented a major advance in battery technology. The voltage of a single cell was more than twice that of the Leclanché cell and other primary batteries, energy density was increase by an even higher factor and shelf life was improved by a factor of 5 at least. It became immediately obvious that the next breakthrough would be achieved with the introduction of a rechargeable Li battery. This happened about two decades later, with the introduction of the Li-ion battery.

As mentioned in the introduction above, the Li-ion rechargeable battery comprises a graphite anode and a rather complex cathode, as discussed below. Polar organic solvents are used, such as ethylene carbonate (EC) or mixtures of several solvents, with a lithium salt serving as the electrolyte. Contrary to the safety hazards of primary lithium batteries, no metallic Li is used in the rechargeable battery. The processes inside the cell involve the transfer of Li^+ ions from the anode to the cathode during discharge and vice versa during charging. For this reason, the system was initially referred to as *the rocking chair battery* and it would seem that there was no metallic lithium involved either in the manufacturing of the battery or during its operation. Considering the processes taking place at the two electrodes, this is not true. Thus, when the battery is charged, the processes taking place can be written as

$$Li^+_{soln} + 6C + e^-_M \rightarrow LiC_6 \tag{20.19}$$

and

$$LiMn_2O_4 \rightarrow Li^+_{soln} + Mn_2O_4 + e^-_M \tag{20.20}$$

leading to the overall reaction

$$LiMn_2O_4 + 6C \underset{\text{charge}}{\overset{\text{discharge}}{\rightleftarrows}} LiC_6 + Mn_2O_4 \tag{20.21}$$

This is an odd reaction for a Li-ion battery, considering that the Li^+ ion does not even show up in the overall reaction, but it does play a role in the two partial reactions at the anode and the cathode.

One may wonder whether there is any metallic Li in this battery, either in the charged or the discharged form. Strictly speaking the answer is no, but a closer look at Eq. (20.19) shows that it does involve the reduction of a Li^+ ion to metallic lithium, except that it is not stored as such, but rather "dissolved" in the graphite. Had we replaced graphite with mercury (which would obviously be a very bad idea, both from the point of view of technology and that of the environment) the same process would lead to an amalgam of Li in Hg, which we would definitely regard as metallic Li. Considering LiC_6 as being similar to an amalgam (and hence containing in effect metallic Li dissolved in carbon) is confirmed by measurement of the open circuit potential of the Li-loaded graphite versus a piece of pure Li. This turns out to be about

$+ (0.08-0.10)$ V. Moreover, when a partially charged Li-ion battery is opened and dumped in water, copious hydrogen evolution is observed. Thus, the open circuit potential of a Li-ion battery represents the Gibbs energy involved in oxidizing a Li atom in the graphite anode and transfering the resulting Li^+ ion to the Mn_2O_4 cathode.

Actually the cathode in a Li-ion battery is generally more complex than indicated above. The process taking place during discharge could be represented by

$$Li^+ + Mn_2^{IV}O_4 \rightarrow LiMn^{III}Mn^{IV}O_4 \tag{20.22}$$

which shows that intercalation of Li^+ reduces one of the Mn atoms from the 4-valent to the 3-valent state. In some cases a fraction of the Mn atoms is replaced by another transition metal forming nonstoichiometric compounds. Changing the element added and its relative fraction in the oxide can improve the performance of the resulting Li-ion battery. It should also be noted that in the early stages of the development of Li-ion batteries, Co was used rather than Mn, but the higher price and toxicity of Co will probably lead to its replacement by Mn.

Recently, a somewhat different type of Li ion battery is being developed, based on employing $LiFePO_4$ as the cathode. In this system the intercalation of Li^+ into the structure of $FePO_4$ can be represented by reduction of the 3-valent iron atom to its 2-valent state, as represented by the equation

$$Li^+ + Fe^{III}PO_4 \rightarrow LiFe^{II}PO_4 \tag{20.23}$$

The main advantage of this system is in the lower price of Fe, compared to Co, and even to Mn, and the lower toxicity compared to Co.

Since the introduction of the Li-ion battery, the technology has developed significantly, but the basic principle remains the same. The R&D efforts were aimed at the three components of this type of battery: (i) Using different types of graphitic materials for the negative electrode, to increase the amount of Li in the graphite and the rate of its intercalation and de-intercalation; (ii) improving the positive electrode, mostly by replacing a small part of the Mn by other transition metals; and (iii) improving the stability of the solvent, to withstand the higher potential of the cell. This is a vibrant field in present day R&D, and much of it involves solid state chemistry and materials science, the details of which are in many cases proprietary, so it will not be discussed here.

In summary, it may be noted that in the first century after the development of the Leclanché cell and the lead-acid battery the field developed steadily, but slowly. In the second half of the 20th century two major breakthroughs occurred, both related to Li batteries; first the primary and then the rechargeable Li-ion battery were introduced. The open-circuit voltage rose from 2.1 V in aqueous systems to more than 4 V in non-aqueous Li batteries, and may reach an open circuit potential of 4.5 V. Considering that the whole electromotive series spans a range of about 6 V, (from Li to F_2), this technology may be approaching its limit. It should also be noted that the next challenge is to match the breakthrough in technology with a similar breakthrough in reducing the price, to allow wide use of Li-based batteries.

The above is not a comprehensive list of all the primary and rechargeable batteries available on the market. Only a representative list of some common batteries, and the chemistry involved in their operation.

20.4
Fuel Cells

20.4.1
The Energy Density of Fuel Cells

As noted in Section 20.1.1, a fuel cell is just a different type of battery, in which the chemical energy is stored in an outside container, rather than inside the battery. This can be a great advantage from the point of view of energy density since, for extended operation, the weight of the cell itself is insignificant.

20.4.1.1 The Hydrogen-Oxygen Fuel Cell
The H_2-O_2 fuel cell is theoretically the best fuel cell one could devise. The theoretical energy density is 3.66×10^3 Wh kg^{-1}, counting the weight of both hydrogen and oxygen. This should be compared to the value of 208 Wh kg^{-1} calculated for the Ni-Cd system. However, this number is misleading, because it is calculated only on the basis of the weight of the fuel and the oxidant, ignoring the weight of the fuel container. The energy density of molecular hydrogen is the highest of all elements, but in any method of storing it, the weight of the container is at least 10 times the weight of the hydrogen contained in it.

It could be argued that the energy stored in *the fuel cell proper* approaches zero, while the energy of a *fuel-cell system* approaches asymptotically the energy density of the fuels and their containers. For space applications, both hydrogen and oxygen are carried cryogenically (as the corresponding liquids) hence the energy density of the system has to include the containers of both gases. For terrestrial applications, a hydrogen-air fuel cell would be preferred, alleviating the need to store oxygen. Nevertheless, the cost of the electrode materials, the safety aspects of storing hydrogen and the low density of stored hydrogen, either as a compressed gas or cryogenically, has so far prevented this type of fuel cell from becoming widely used, although it has been employed successfully in space missions.

20.4.2
Fuel Cells Using Hydrocarbons – the Phosphoric Acid Fuel Cell (PAFC)

In the early stages of development of fuel cells, great efforts were made in studying the oxidation of all sorts of organic materials, propane gas, jet fuel and even cellulose, as prospective fuels. The (rather naïve) vision at the time was to operate a vehicle, for example, with a regular fuel tank feeding a fuel cell, in which the chemical energy would be converted directly to electrical energy at high efficiency. Although direct oxidation of hydrocarbons proved to be possible in some cases, the efficiency of converting the chemical energy to electrical energy was found to be too low to be of any practical importance.

Next, it was proposed to use common paraffin as the fuel, which is pretreated in a steam reformer, to convert it to hydrogen and CO_2, before the gas is pumped into the fuel cell. The reaction involved, taking propane as an example, is

$$C_3H_8 + 6\,H_2O \rightarrow 3\,CO_2 + 10\,H_2 \qquad (20.24)$$

The fuel cell itself contains concentrated phosphoric acid (85–100%) and it can be operated at 180–220 °C. This electrolyte rejects CO_2, therefore the gas mixture produced in the steam reformer can be pumped directly into the fuel cell, where H_2 is oxidized at the anode while CO_2 acts as an inert gas, passing through the cell without undergoing any chemical reaction. Unfortunately, the reaction does not proceed *quantitatively* according to Eq. (20.24), and some CO is produced as a side product. This is removed with a so-called *shift reactor*, in which CO is selectively oxidized to CO_2.

The boiling point of 100% phosphoric acid is above 250 °C, allowing operation of this type of fuel cell at temperatures up to 200 °C, without the need to pressurize the cell. High temperature helps the kinetics and it also diminishes the sensitivity of the electrocatalyst to poisoning.

Both electrodes contain noble metals, dispersed on high-surface-area graphite. Advances in the design of electrocatalysts made it possible to reduce the Pt loading from about 10 mg cm^{-2} in the early fuel cells used by NASA to about 0.1 mg cm^{-2} at the anode and 0.2 mg cm^{-2} at the cathodes. At such low levels of loading the cost of the noble metal may no longer be prohibitive, although it would still constitutes a significant factor in the cost of the fuel cell assembly. Moreover, if this type of fuel cell were widely used, (for example in electric cars) the cost of Pt could increase sharply. The cells could be operated at a potential of about 0.70 V, corresponding to an efficiency of $0.70/1.48 = 0.47$ [9], which is comparable to modern conventional power stations. However, the overall system efficiency is only about 40%, since the energy used to operate all the auxiliary equipment, such as pumps and various control units that should be accounted for.

The technology of building PAFC has been developed better than that of any other fuel cell systems. A few units delivering as much as 4.3 MW of electric power have been built and tested intermittently and smaller units delivering 40 kW performed well continuously for about 5 years.

An interesting possibility proposed was to operate PAFC for the combined purpose of generating electricity and heat. Placing one or more such units in an apartment building could provide electricity, hot water, heating and cooling locally for each building. Operated in this manner, the efficiency of the system has been estimated to be as high as 80%. Central power stations (both conventional and atomic) also

---

[9] For proper comparison with the efficiency of a heat engine, the efficiency of a fuel cell must be calculated on the basis of the heat content of the fuel ($\Delta H/2F = 1.48$ V), not on the basis of the Gibbs energy change $\Delta G/2F = 1.23$ V).

produce large amounts of waste heat, of course, but considering that they are usually built as far as possible from residential areas, using this thermal energy is usually not practical.

The phosphoric acid fuel cell did not make it to the market place for a number of reasons:

1) During operation of the fuel cell, water is formed at the cathode and consumed at the anode. A system for transferring water from the anode to the cathode has been designed and built, but water management on large scale units turned out to be very difficult to maintain.
2) The activity of the electrodes was found to decline with time, due to impurities introduced with the fuel and/or the air, or due to loss of active surface area as a result of agglomeration of nanoparticles of the catalyst.
3) The high initial cost of the ion-selective membrane (Nafion) and the electrode materials, which were either Pt or its alloy with another noble metal, rendered this system unsuitable for common applications, in spite of very significant advances in the technology involved, particularly in the design of the electro-catalysts both at the anode and the cathode.

20.4.3
The Direct Methanol Fuel Cell (DMFC)

It should be acknowledged that the main drive for the development of fuel cells is for future electric cars, in which the chemical energy is converted to electrical energy at a much higher efficiency than in the internal combustion engine (ICE), because, as we noted above, this energy conversion is not limited by the constraints of the Carnot cycle. Admittedly, there could be other applications, such as replacing the battery in mobile devices (laptops, fourth generation mobile phone with HD TV screens etc.), but these should be considered as side product that, on their own, would not justify the major efforts in the development of fuel cells. The vision is to develop a system where the car could be refueled in a few minutes and would have the same comfort and range afforded by present-day internal combustion engines. While the hydrogen/oxygen or hydrogen/air systems are, in principle, the best fuel cells there are, the great difficulties of transporting and storing hydrogen prevent their use. Efforts are made to build containers that could store hydrogen at a pressure of 5,000 or even 10,000 psi (350–700 atm) replacing the gas tank on conventional vehicles. But charging a tank with hydrogen at that pressure in a few minutes would generate a large amount of waste heat and would consume one quarter of the energy stored in the hydrogen, or even more. Moreover, the safety hazard of driving a vehicle with hydrogen pressurized at 10,000 psi cannot be ignored.[10]

Considering the above limitations, the best next choice would seem to be a methanol/air fuel cell, and this is indeed where the main effort in fuel cell R&D

---

10) Note that cars run on natural gas exist, but are not allowed to enter tunnels because of safety considerations.

is now (2010). Methanol can be oxidized electrochemically by a 6-electron process, yielding CO_2 and water.

$$CH_3OH + H_2O \rightarrow CO_2 + 6H^+ + 6e_M^- \tag{20.25}$$

The protons produced in this reaction are transported through the cation selective membrane to the cathode, where the reaction taking place is

$$(3/2)O_2 + 6H^+ + 6e_M^- \rightarrow 3H_2O \tag{20.26}$$

leading to the overall reaction

$$CH_3OH + (3/2)O_2 \rightarrow CO_2 + 2H_2O \tag{20.27}$$

Thus, water is consumed at the anode and produced at the cathode, and some method of transport of some of the water from the cathode to the anode must be included as part of the fuel cell system.

The equivalent weight of methanol is only 5.33. Moreover, methanol is a liquid over the whole range of temperature relevant to vehicle operation (m.p. $-97\,°C$; b.p $+64.7\,°C$). It is easy to manufacture and inexpensive. The energy density is lower than that of gasoline, (which is itself one third of that of hydrogen), but compared to hydrogen this is compensated for by the fact that the density is high $(0.792\ \text{gm cm}^{-3})$ and storage presents no problem. Considering that methanol can be manufactured from hydrogen and natural gas, or even from hydrogen and CO_2, it could also be considered as one of the ways of storing hydrogen chemically.

On the other hand, methanol is rather toxic, so why not use ethanol, which is less toxic, has similar physical properties (m.p. $-114.3\,°C$; b.p. $+78.4\,°C$; density $0.789\ \text{gm cm}^{-3}$) and is also inexpensive? It turns out that ethanol and higher alcohols cannot be oxidized all the way to CO_2, at least not at the potentials relevant to fuel cell operation. The product of electrochemical oxidation is acetic acid in the case of ethanol and the corresponding higher acid in the case of higher alcohols. It seems that breaking the C–H and the O–H bonds in methanol is easier that breaking the C–C bond in ethanol.

Finally, the choice of *direct* methanol fuel cell rather than a system operating with a steam reformer that converts methanol to hydrogen and carbon dioxide (as in the phosphoric acid fuel cell discussed above) is not practical because of the complexity of such systems, (which prevented the commercialization of PAFC.)

The development of a viable DMFC is the most vibrant field in the area of fuel cells at the present time, so we shall discuss only the main issues, not the detailed developments, which may be obsolete by the time this book is published. The three main components of any fuel cell are the anode, the cathode and the membrane

20.4.3.1 The Anode

Oxidizing methanol is much more difficult than oxidizing hydrogen, and the challenge of finding a good electrocatalyst is greater. Platinum itself is not the best choice, and it seems that an alloy of Pt–Ru is more efficient. The common wisdom at the present time is that adsorption of methanol occurs mostly on Pt sites on the surface of the alloy, and this is also the site of partial oxidation, as shown in the next two equations

$$Pt + CH_3OH_{soln} \rightarrow Pt(CH_3OH)_{ads} \tag{20.28}$$

$$Pt\text{-}CH_3OH_{ads} \rightarrow Pt\text{-}CO_{ads} + 4H^+ + 4e_M^- \tag{20.29}$$

The oxidation of CO on the Pt site is slow, but it is accelerated by oxygen atoms adsorbed on the Ru site, following the equations

$$Ru + H_2O \rightarrow Ru\text{-}OH_{ads} + H^+ + e_M^- \tag{20.30}$$

$$Pt\text{-}(CO)_{ads} + Ru\text{-}OH_{ads} \rightarrow Pt + Ru + CO_2 + H^+ + e_M^- \tag{20.31}$$

Evidently this mechanism, if it is indeed the way the system operates, requires that the Pt sites upon which methanol and later CO are adsorbed should be very close, on the atomic scale, to the Ru sites where OH is adsorbed. This requires a very intimate mixing of the two metals, so the methods of preparing the alloy catalyst may be the critical factor in its effectiveness. Alternatively, it may be assumed that there are domains of Pt sites and of Ru sites and complete oxidation of methanol to CO_2 according to reaction (20.31) only occurs at or very close to the boundaries of such domains. Should this be the case, then the role of nanotechnology in the development of better electrocatalysts becomes evident, and indeed a major effort in the development of better catalysts is being conducted in this direction.

There are two avenues of introducing methanol into the cell: either as a fairly concentrated solution in water (about 2–3 M) or as a vapor. Both methods are being studied, each having its own advantages and disadvantages.

20.4.3.2 The Polymer Electrolyte Membrane (PEM)

The solid polymer electrolyte most commonly used in fuel cells is Nafion, which is a derivative of Teflon that has been modified by replacing a certain fraction of the fluorine atoms by sulfonic acid anions $-C\text{-}SO_3^-$. This serves the purposes of an acid electrolyte, a separator and a cation-exchange membrane. Sulfonic acid is a strong acid and its anion rejects CO_2 and anions, while allowing easy passage of protons across it, which is essential for the proper operation of the cell. The chemical stability of Nafion is very high. Unfortunately, it does not prevent entirely the passage of methanol. If the fuel crosses over and reaches the positive electrode it is rapidly oxidized, consuming some of the oxygen, without generating any electrical energy. Thus, when methanol is fed into the fuel cell in an aqueous solution, its optimum concentration is determined by the need to minimize mass-transport limitation on the one hand and transfer across the Nafion membrane on the other. This is one of the reasons to use a vapor feed of methanol at the anode. It allows dynamic control of the vapor pressure by monitoring the current passed and calculating from it the actual rate of consumption of the fuel, ensuring that most of it will be consumed at the anode, preventing or at least decreasing the concentration gradient that would otherwise drive methanol across the membrane.

The worst aspect of the Nafion membrane is its high cost. Massive R&D efforts have been going on for years to develop low cost cation-selective membranes having a performance comparable to that of Nafion, at a much lower price. This is a critical issue, because the commercial success or failure of the direct methanol fuel cell hinges on our ability to develop such membranes at $(1-2)\%$ of their present cost.

20.4.3.3 The Reduction of Molecular Oxygen at the Cathode

Oxidation of the fuel is one of the problems of fuel cells. The other is the reduction of oxygen. In an acid solution, the open circuit potential for oxygen reduction should be 1.23 V vs. RHE, but in effect it is about 1.0 V. This already reduces the efficiency of energy conversion to about 80%. Moreover, the reduction of oxygen is a relatively slow process, so that further loss of efficiency will result from the activation overpotential. Mass transport limitation can also be a problem, because of the low solubility of molecular oxygen in water.

In order to be viable, a fuel cell should operate at a high current density, of the order of 1 A cm^{-2}, hence even a small Ohmic resistance between the two electrodes may cause significant loss of efficiency of energy conversion. This requires optimization of the thickness of the Nafion membrane for low resistivity combined with low permeability for methanol.

A further difficulty is that reduction of molecular oxygen could easily lead to the 2-electron reduction to hydrogen peroxide instead of the 4-electron reduction to water or OH$^-$. This is to be expected, since the reaction

$$O = O + 2e_M^- + 2H^+ \rightarrow HO\text{-}OH \tag{20.32}$$

requires the breaking of a single bond in molecular oxygen, while the reaction

$$O = O + 4e_M^- + 4H^+ \rightarrow 2H_2O \tag{20.33}$$

requires the breaking of two bonds. If the peroxide species is formed as an adsorbed intermediate that is rapidly reduced further to water, this does not pose any problem, but if it is released into the solution, it could decompose to water and oxygen in a disproportionation reaction which proceeds spontaneously in water

$$2H_2O_2 \rightarrow O_2 + 2H_2O \tag{20.34}$$

In Figure 20.5 the dependence of the rate of oxygen reduction (line 1) and hydrogen oxidation (line 2), as well as the power of a fuel cell (line 3), are shown schematically, as a function of the current density. If the application demands high energy utilization, (as might be the case in a space mission) the cell will be operated at a relatively low current density, as indicated by the vertical line at line at 0.25 A cm^{-2}. This, however, will require a higher volume and weight, to say nothing of the cost. If the application calls for maximum power output, the device will be operated at a high current density, just before mass transport limitation sets in, as indicated by the vertical line at 0.60 A cm^{-2} in Figure 20.5.

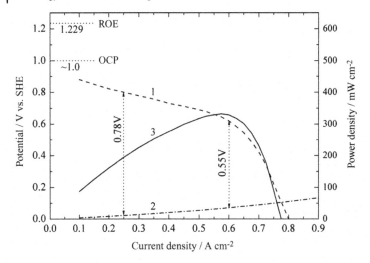

Figure 20.5 Schematic presentation of the current–potential relationship in a fuel cell. Lines 1 and 2 show the polarization curves for oxygen reduction and hydrogen oxidation, respectively. Line 3 shows the power as a function of the current density. Two possible operating points are marked, one for high energy density, the other for maximum power output.

It should be noted here that fuel cells can excel in high energy density, while batteries can provide high power density. This has led to the suggestion of employing hybrid systems, in which the fuel cells charge batteries, in electric vehicles, for example. The concept in this case is that the fuel cell will provide enough power for coasting, while the battery will provide excess power needed for passing or for climbing a hill. Another application could be for electronic devices placed in remote places, where the fuel cell charges a large capacitor. In this situation, data could be acquired using low power while a large capacitor is being charged. A larger power, needed only for a short time at relatively long intervals for transmitting the data collected, could be delivered by the capacitor.

20.4.4
High-Temperature Fuel Cells

Increasing the operating temperature of fuel cells is beneficial, in several respects. (i) The rate of the reaction at both electrodes increases with increasing temperature, decreasing the activation overpotential or effectively eliminating it. (ii) The conductivity of the ionic conductor (which is in most cases a solid) is increased, thus allowing the passage of high current densities with acceptable resistance overpotential. (iii) The rate of mass transport is also increased. (iv) The sensitivity to impurities is reduced or eliminated. For example, it was noted above that the phosphoric-acid fuel cell operating at 200 °C could tolerate up to about 2% CO in the gas stream, without significant effects of poisoning. In comparison, a solid-oxide fuel cell operating at

800–1000 °C could actually use CO as the fuel. Moreover, the heat generated in such systems can be used for co-generation of electricity and heat, increasing the overall efficiency of conversion of chemical energy to electricity and heat.

In spite of these advantages, none of these high temperature systems have gained acceptance in industry, except perhaps for some small-scale applications. Consequently they are only mentioned briefly in this chapter for their scientific interest.

20.4.4.1 The High-Temperature Solid-Oxide (HTSO) Fuel Cell

The heart of the high-temperature solid-oxide fuel cell is an yttria-stabilized zirconia, $Zr(Y)O_2$ film, which acts as a solid electrolyte, allowing high conductivity for O^{2-} ions at about 1000 °C. The fuel electrode is a porous Ni/ZrO_2 cermet (which is short for "ceramic-metal"), which serves both as the electrocatalyst and as the current collector. Actually, electrocatalysis is no longer a problem at such elevated temperatures and different fuels, including H_2, regular fossil fuels, and even CO, can be used. The air (oxygen) electrode is a layer of strontium-doped lanthanum manganite catalyst, La (Sr)MnO_3, coated with a porous doped indium oxide current collector. Cells are connected in series in a *bipolar configuration*, in which the anode of one cell is internally connected to the cathode of the next cell and only the terminal anode and cathode are connected to the external terminals of the stack. The interconnecting material, which must be a good electronic conductor, but at the same time impervious to ions, is manganese-doped lanthanum chromite, La(Mn)CrO_3.

The electrolyte is solid, but unlike the PEM, it is dry. This property eliminates many engineering problems of water management, which tend to complicate the design and operation of other types of fuel cells, considering that the critical temperature of water is 374.15 °C, so it cannot exist in the liquid phase above this temperature. Waste heat is produced at high temperature and can be used for different purposes, making the overall efficiency of electricity and heat production very high. In fact, the waste heat produced in this type of fuel cell can even be used to produce electricity in a conventional heat engine. In this mode of co-generation, fuel can be converted to electricity at an overall efficiency exceeding 60%.

Improvements in materials are being steadily made, but the more sophisticated materials developed for this purpose tend to increase the cost. Once the materials problems are overcome, the inherent simplicity of the design and operation may make HTSO fuel cells useful electrochemical energy conversion devices. Even if that goal is reached, it is unlikely that they will be applied for transportation, where the power has to be turned on and off a number of times a day. Cycling the temperature of several tightly connected but different materials, having different heat conductivities and coefficients of thermal expansion may pose an insurmountable challenge. Moreover, having to start a vehicle by first heating a part to 1000 °C, may be time consuming and wasteful of energy, particularly considering that co-generation is not an option in vehicles.

20.4.4.2 The Molten Carbonate Fuel Cell

This type of fuel cell operates at a temperature of about 600 °C, which is enough to enhance the rate of reaction at both electrodes and allow oxidation of most fuels, just as the HTSO fuel cell. The electrolyte is a mixture of alkali carbonates, chosen to reduce the melting point. The carbonate mixture rejects CO_2 and water, and the temperature is high enough to oxidize CO and most common fuels. The conductivity of the molten salt is inherently high. On the other hand, employing a very hot liquid as the electrolyte may pose serious safety hazards and it is hard to imagine that such a system (with a molten salt phase at about 600 °C) would ever be allowed in a vehicle. As in the case of the HTSO, work has been going on for several decades and a lot of very interesting scientific information has been gained, but it is highly unlikely that either of these systems could be justified from the practical point of view of energy saving, even for stationary energy storage, let alone in mobile application.

20.4.5
Why Do We Need a Fuel Cell?

It was noted in Section 20.1.1 that serious effort and generous funding for fuel cell research dates back about 50 years. In spite of that it must be admitted that fuel cell technology has not lived up to many of the early expectations. The most important stumbling block has been the development of an inexpensive electrocatalyst, suitable for work in acid solution over long periods of time. Catalysts consisting of large-surface-area graphite, coated with small amounts of platinum and bonded by Teflon, have been developed and improved over the years, to increase activity and decrease the amount of platinum needed per unit surface area, but the activity for direct oxidation of hydrocarbons is still too low for such fuel cells to be commercially viable.

The main justification for all the research and development in this field was the hope of developing electric cars using fuel cells. These were expected to consume the fuel more efficiently than internal combustion engines, thereby reducing the use of fossil fuel. Different companies have designed and built fuel-cell-driven electric cars for demonstration purposes, but none of them have made it to the civilian marketplace, although some special applications, where cost is not a major issue, exist. This is not the place to discuss all the reasons for this failure, suffice it to say that it is related to all three main components of the system: the need to use noble metals for both the anode and the cathode and the lack of a good cation-exchange membrane, comparable to Nafion in performance and lowering the cost by a factor of 50 at least. Moreover, predictions were often made based on the current price of noble metals, ignoring the fact that technical success, leading to wide use, would undoubtedly increase the price of Pt. One of the common excuses for the failure of fuel cells to become a commodity, particularly in the area of powering electric vehicles, is "the vicious circle" argument. It is claimed that fuel cells are very expensive because they are produced in small numbers. Large scale production lines could reduce the price, but the large investment needed to reduce the price is not made because there is no market, and there is no market penetration because the price is too high. This is a textbook level truism for any product, often referred to as "the economy of size", but nevertheless, risks are taken, and new products are continuously being introduced into the marketplace.

In this context, it is interesting to compare the very slow development of fuel cells to that of primary and secondary batteries. In the mid-1960s the Li-thionyl chloride primary battery was invented. This was a highly innovative item, based on a nonaqueous solvent that acts also as the cathode material, while the positive terminal is just a current collector and does not take part in the cell reaction. The operation of the anode is made possible only because a solid electrolyte interface (SEI) is formed spontaneously, and it has the unique property of being an ionic conductor for Li and an insulator for electrons (cf. Section 20.2.3). This invention led to intensive research and development of this and similar systems, using different solvents (for example propylene carbonate or SO_2). Within a few years there were primary lithium batteries on the market and within a decade it became a multibillion dollar commodity world wide.

The history of the development of the Li-ion battery is similar. Attempts to develop a rechargeable battery based on metallic lithium failed, but a totally new concept was introduced sometime in the mid-1980s (cf. Section 20.3.4). This is still a very active field of R&D, aiming at improving the performance and reducing the price, but the first commercial Li-ion batteries were introduced just a few years after it had been invented, and within a decade this too became a multibillion product world wide. Thus "the vicious circle" argument used to explain the lack of success of fuel cells applies equally well to primary and rechargeable Li batteries, but did not prevent their commercial success.

Direct methanol fuel cells are the most extensively funded (and consequently studied) fuel cells at present. Will they ever replace Li-ion batteries in applications for mobile electronic devices, such as laptops, mobile phones, video cameras and of course electric cars? Leaving prophecy to the reader, let us just make the following points:

1) A single cell in a Li-ion battery provides a voltage of about 3.6 V, and it is plausible that this value will increase to about 4.5 V within the next decade. In comparison, a DMFC operated at close to ambient temperature delivers a voltage of about 0.4 V.

2) It is often stated that the real breakthrough in DMFC development will come when catalysts that *do not contain noble metals* are found. This may be true, but remember that Li-ion batteries never required noble metal electrodes!

3) One point that is rarely if ever mentioned in regard to the future use of DMFC is the removal of excess heat. Thus, the efficiency of a Li-ion battery (in a cell phone or a laptop) is about 90%, compared to about 30% for a DMFC operating at close to body temperature. Assuming that the same electric power is provided in both cases, the amount of waste heat generated will be about three times higher in the case of the DMFC. How can this excess heat be removed from portable electronic devices? This may end up being *the thermal barrier* preventing the application of DMFCs for application in mobile electronic devices

Although R&D in the field of fuel cells may not have yet lived up to expectations, it has produced very interesting and important insights into electrochemistry, solid state science, engineering, heterogeneous catalysis and nanotechnology. It has also led to the development of some useful products related directly or indirectly to fuel cells. The so-called *metal-air* batteries use oxygen electrodes developed originally for fuel cells. The Ni-MH rechargeable battery alluded to above (cf. Section 20.3.3) makes use of hydrogen as the fuel and metal hydride as the medium for its storage at ambient pressure, as also developed in the framework of fuel cell research. A relatively low-temperature version of the HTSO fuel cell, operating at about 400 °C serves as a gas-phase detector to determine the partial pressure of oxygen. Moreover, a great deal of understanding of the theory and practice of preparing large-surface-area catalysts and operating porous electrodes has been gained in the course of research in this field. Some of this knowledge has already been put to use in improving the design and performance of existing primary and secondary batteries.

20.5
Porous Gas Diffusion Electrodes

There are two ways of increasing the catalytic activity of an electrode in a fuel cell: (i) by finding a better catalyst, which yields a higher value of the exchange current density, hence a lower overpotential at any chosen current density, and (ii) by increasing the surface area of the electrode per unit *apparent* (geometrical) surface area. Achievements in the former direction have been steady but not spectacular so far. Platinum or some of its alloys, are still the best electrocatalysts for oxygen reduction and hydrogen or methanol oxidation in acid solutions. The best electrocatalyst at the anode of DMFC seems to be a Pt–Ru alloy and most of the research effort is directed at finding the best composition and methods of preparation of such catalysts.

Development of porous electrodes having a large surface area has come a long way. The effective exchange current density has been increased very significantly, mostly by developing very high-surface area carbon and involving methods derived from nanotechnology, while the amount of noble metal needed has declined steadily, from about 10 mg cm^{-2} in the early designs to 0.1 mg cm^{-2} for the anode and 0.2 mg cm^{-2} for the cathode in state-of-the-art technology.

But catalysis is not the only reason for using porous electrodes. For a fuel cell to be economically viable, the current density should be at least 0.5 A cm^{-2}, and preferably higher. Such current densities cannot be reached at planar electrodes, even in well-stirred solutions and at substantial concentrations of the electroactive material. In H$_2$-O$_2$ fuel cells both reactants are gases that have low solubility, and vigorous stirring or pumping of the solution is not enough, and, moreover, too much of the energy produced by the fuel cell would have to be consumed for this purpose. Porous electrodes are, therefore, used also to enhance the rate of mass transport.

The way a porous electrode works can best be understood using a single pore model, shown in Figure 20.6. If one assumes that the pores are conical and are made of a material that is wetted by the solution to some extent, a meniscus will be formed, as shown in Figure 20.6a. An enlarged view of the region of contact between the solution and the gas phase inside a pore is seen in Figure 20.6b. Two factors must be taken into account: the rate of diffusion of the gas through the thin layer of the liquid to the side of the pore, which contains the electrocatalyst, and the resistance of the thin layer of the solution, which is determined by the cross-section area of the liquid phase that is in intimate contact with the pore. Near the

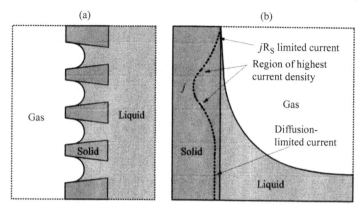

Figure 20.6 (a) The meniscus formed in conical pores of a gas-diffusion electrode. (b) The current density distribution near the three-phase boundary in one of the pores.

three-phase boundary the liquid film is very thin, allowing fast diffusion of the gas, but the solution resistance is high. Reaction at the surface of the pore in this region is therefore limited by high resistance overpotential. Far away from the edge of the meniscus, the resistance overpotential is low, but here the rate of diffusion is also low, because the layer of liquid through which the reacting gas must diffuse to reach the electrode surface is relatively thick. The reaction will be diffusion limited in this region.

Clearly, the current density must increase first and then decrease as a function of the distance from the edge of the pore, reaching a maximum somewhere in between. This is the most active area of the pore. A great saving in expensive catalyst material can be realized if this area is identified and controlled and the catalyst material is applied only to it.

It should be emphasized that the single-pore model bears only a distant resemblance to the real structure of a porous electrode, but the principles discussed above are valid. Porous electrodes are usually made by mixing small graphite particles, coated with the catalyst, with a suitable binder. The latter also determines the degree of wettability of the matrix and serves to bind the electrode material to a suitable current collector. This design is necessary to stabilize the gas/liquid interface, preventing flooding of the electrode on the one hand, and drying out on the other. Often a double-pore structure is used, with the larger particles (hence, the larger pores) on the gas phase side of the electrode; this serves to achieve and maintain better stability of the three-phase regions inside the porous structure.

It is important to understand that increasing the roughness factor of a planar electrode increases the rate of *charge transfer* but has little effect on the rate of *mass transport*. On the other hand, the use of correctly designed porous electrodes can increase the rates of both processes. Thus the use of porous electrodes will be essential whenever gaseous reactants (e.g., H_2 or O_2) are employed, even after a suitable electrocatalyst is found.

20.6
The Polarity of Batteries

The polarity of batteries may sometimes lead to confusion, which we would like to dispel here. The anode and the cathode are defined unequivocally as the electrodes where oxidation and reduction occur, respectively, but is the anode the positive or the negative terminal of a battery? This question is particularly relevant for the case of secondary batteries, where the electrode serving as the anode during discharge becomes the cathode during charging and vice versa.

Consider the lead-acid battery. The standard potentials for the two half-cells are as follows:

$$PbO_2/PbSO_4 \quad E^0 = +1.682 \, V \tag{20.35}$$

$$PbSO_4/Pb \quad E^0 = -0.359 \, V \tag{20.36}$$

The sum of these potentials yields a value of $E^0 = 2.041$ V. This is not exactly the potential of a lead-acid battery at open circuit, (but is rather close to it) because the concentration of sulfuric acid is not the standard concentration.

It is clear that the $PbO_2/PbSO_4$ electrode, which serves as the cathode during discharge, is the positive terminal of the battery and the $PbSO_4/Pb$ electrode, which acts as the anode is the negative terminal. When the battery is being charged, *the electrodes change role, but not polarity.* The $PbSO_4/Pb$ electrode is now the anode (since $PbSO_4$ is being oxidized to PbO_2), but it is still the positive terminal of the battery. The way in which the potential at each electrode is changed during charge and discharge is shown in Figure 20.7.

We must distinguish between two cases. When the battery is in the *driving mode* – that is, when the battery is the source of energy – the positive terminal is the cathode and the negative terminal is the anode. The same applies also to corrosion and to any other electrochemical process that occurs spontaneously. When the battery is *in the driven mode* – that is, when it is being charged–the positive terminal is the anode and the negative terminal is the cathode. The latter is the case in electroplating, in the electrolytic industry, and in most experiments in the research laboratory, where a current is imposed on the system and the potential is measured, or vice versa.

20.7
Super-Capacitors

20.7.1
Electrostatic Considerations

In Chapter 8 we discussed the double-layer capacitance that is always present at the metal/solution interface. The value of a capacitance, per unit surface area is given by

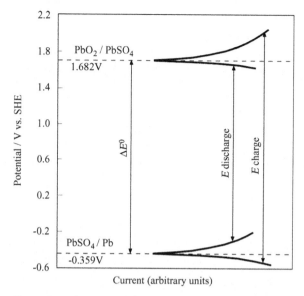

Figure 20.7 The potential of the two electrodes in a lead-acid battery during charge and discharge. Note that the polarity of the cell is not changed. (for details see P. Moran and E. Gileadi, *J. Chem. Educ.*, **66** (1989) 912.

$$C = \frac{\varepsilon_0 \varepsilon}{d} \tag{20.37}$$

where

$$\varepsilon_0 = 8.8542 \times 10^{-12} \text{ J}^{-1} \text{ C}^2 \text{ m}^{-1} \tag{20.38}$$

is the permittivity of free space and ε is the dielectric constant of the medium between the plates of the capacitor. The important feature of this capacitance for electro-chemistry is that it cannot be charged instantaneously. There is a characteristic time constant given by

$$\tau = R_S \times C_{dl} \tag{8.17}$$

Thus, when a current step is applied, the potential will approach its final value following the equation

$$E = E_\infty \left[1 - e^{-t/\tau} \right] \tag{20.39}$$

It should be noted that the double layer capacitance is not a capacitor in the usual sense, because it has only one metal plate. The other side of this capacitor is somewhere in the solution, depending on the theory assumed and on the composition of the solution. Nevertheless, it behaves as a regular capacitor, except for the fact that its capacitance is often a function of potential

Capacitors have been used in electronic devices for a century or longer. Their importance hinges on the simple equation

$$Z_C = -\frac{j}{\omega C} \tag{16.1}$$

where $j \equiv (-1)^{1/2}$ and $\omega = 2\pi f$ is the angular velocity. The impedance Z_C approaches infinity for $\omega = 0$ and decreases with increasing frequency. Thus, a capacitor can filter a DC signal, allowing an AC signal to pass, or divert high-frequency noise to the electrical ground, allowing only frequencies below a desired level to be measured.

20.7.2
The Energy Stored in a Capacitor

In this chapter we shall discuss another property of a capacitor – its ability to store energy. The charge in a capacitor is related to the voltage by

$$Q = CE \tag{20.40}$$

Thus, if the capacitance is, for example, $10\,\mu F\,cm^{-2}$, the charge needed to create a voltage difference of 1.0 V is $10\,\mu C$. The electrical energy consumed in this process, which is the energy stored in the capacitor, is equal to the increase in Gibbs energy, given by

$$\Delta G = \frac{1}{2}CE^2 = \frac{1}{2}QE \tag{20.41}$$

which, in the present example, amounts to a very small energy of $5\,\mu J\,cm^{-2}$. The energy density of a capacitor can be increased by one of two ways, either by increasing the voltage or by increasing the surface area per unit weight. The former way looks more attractive, because the energy stored is a function of the square of the voltage, but in electrochemical systems it is rather limited. In an aqueous solution one could reach about 0.8–1.2 V. Employing a nonaqueous solvent, perhaps one of those used for Li/thionyl chloride or for Li-ion batteries, could allow a voltage of 3.0–4.0 V, but the double layer capacitance, per unit surface area, is usually smaller than in aqueous solutions, so that this approach could probably lead to a factor of about ten increase in energy stored. On the other hand, increasing the *real* surface area of the electrode has been very successful, and there are porous carbon materials already available, yielding about $2.5 \times 10^3\,m^2\,g^{-1}$. Assuming a capacitance of $10\,\mu F\,cm^{-2}$ and a voltage of 1.0 V leads to an energy density of $125\,kJ\,kg^{-1} = 34.7\,Wh\,kg^{-1}$. This is, however, only a theoretical limit, (cf. Section 20.1.2), so it should be compared to a value of about 462 Wh kg for the Li-ion battery based on a $FePO_4$ cathode.

20.7.3
The Advantage of Electrochemical Super-Capacitors

The above sample calculation shows that the theoretical energy density of the current best electrochemical super-capacitor is only about 7.5% of that of the current best Li-ion battery. But the supper-capacitor has two features that no battery can match

1) Cycle life: When a battery is cycled between its charged and discharged states, chemical reactions occur at both electrodes. These reactions occur in the solid state and in most cases involve a change in the volume of the species in the transition from one state to the other. While the battery may have been designed, and even built, to have an optimal structure (particle size, crystal structure, bonding etc.) this may change just a little bit in each cycle, but after a thousand cycles the changes may accumulate to the point that the battery can no longer operate well, or that its charge capacity may decline significantly. Such aging processes are inherent to all types of rechargeable batteries, although some may last more cycles than others. In contrast, when an electrochemical super-capacitor is charged and discharged, this is merely an electrostatic process of creating an excess or deficiency of electrons on one side of the capacitor and an excess or deficiency of ions at the solution side of the interface. No chemical reaction takes place and no changes in volume or structure occur. Consequently, while battery life may extend from a few hundred to a few thousand cycles, super-capacitors can survive a thousand times as many cycles.

2) Instantaneous charge/discharge: It would be nice to think that charging and discharging an electrochemical super-capacitor is "instantaneous", but this is not quite the case. Recalling Eq. 8.17 we note that the characteristic time is a product of the capacitance and the associated solution resistance. It is hard to calculate, a priori, the solution resistance of a highly porous electrode, and there may be a range of resistances in a given capacitor. Assuming, for the sake of demonstration, an effective resistance of $0.10\ \Omega$, combined with a capacitance of 250 F this would lead to a time constant of $\tau = 25$ s. Inserting this value into Eq. (20.39) leads to a discharge from full charge to 10% in one minute. In the nomenclature used for batteries this corresponds to a discharge rate of 60C. No battery is built to survive such high rates of discharge. The characteristic time for discharge could be tailor-made to fit different applications. It may be hard to shorten it for a very large capacitor below 60 s, but it is always possible to slow it down as desired.

20.7.4
Hybrid Super-Capacitors

One of the ways to develop super-capacitors having a higher voltage is to use a chemical reaction at one of the electrodes. For example, a high-surface-area anode could be combined with a MnO_2 cathode to attain a higher voltage. This is possible, but there is no 'free lunch'. The device will be half capacitor and half battery. Moreover, since the two electrodes are effectively in series, it is the worst of the two that will dominate the behavior of the whole system. In the above example, the cycle life will be determined by that of the MnO_2 and so will the characteristic time constant for charge and discharge.

In this context it should be noted that each type of battery can be designed in different ways, to maximize either the energy density or the power density. Indeed there are systems that contain two batteries: one with high energy but low power density, charging the other that has a high power density but low energy density that in effect plays the role of a super-capacitor.

Index

Physical Electrochemistry: Fundamentals, Techniques and Applications. Eliezer Gileadi
Copyright © 2011 WILEY-VCH Verlag GmbH & Co. KGaA, Weinheim
ISBN: 978-3-527-31970-1